PHYSICAL CERAMICS

FOR ENGINEERS

To Frances, Laura, and Bruce

This book is in the
ADDISON-WESLEY SERIES IN METALLURGY AND MATERIALS
Consulting Editor
Morris Cohen

and also in the
ADDISON-WESLEY SERIES IN CERAMICS
Consulting Editor
F. H. Norton

LAWRENCE H. VAN VLACK

Department of Chemical and Metallurgical Engineering

The University of Michigan

PHYSICAL CERAMICS

FOR ENGINEERS

ADDISON-WESLEY PUBLISHING COMPANY

Reading, Massachusetts

Menlo Park, California · London · Amsterdam · Don Mills, Ontario · Sydney

Printed in the United States of America

Library of Congress Catalog Card No. 64–11884

ISBN 0-201-08068-0
DEFGHIJKLM-AL-8987654321

PREFACE

This book was written for those engineers who are encountering ceramic materials for the first time: ceramic engineering students who are taking their first course in ceramics, non-ceramic engineering students who are taking an additional course in the fast-expanding materials area (much as they might take an additional course in physical metallurgy or in polymers), and practicing engineers, without a formal background in ceramics, who find themselves involved with ceramic products.

In the past, those engineering students who had primary interest in the manufacture of ceramic products were the principal candidates for courses in ceramic materials. Recently, however, greater attention has been given to the development of new materials with specific properties, and to the necessity for modifying the properties of existing ceramic materials for new uses. Therefore, interest in ceramics as engineering materials has broadened significantly over the past few years.

Physical ceramics emerged as a subject separate from ceramic processing when it became necessary to modify the properties of ceramic materials. Thus physical ceramics gained importance gradually, starting with the early attempts to modify refractory materials or to specify glass properties, and continuing to the present attempts to control the dielectric and magnetic properties of ceramic materials. Although the subject of physical ceramics is still expanding and will continue to gain importance, the knowledge of it as an area of study is already sufficient to justify specific attention to it in certain college courses, because widely applicable basic principles have become available.

In writing this book, the author has followed a format which has proved usable in the classroom. This pattern includes, first, attention to the internal structure of ceramic materials, because this knowledge is necessary before the origin and variation in properties can be understood. Second, consideration is given to the various properties commonly encountered and required of ceramic materials. Finally, the commonly used procedures for processing ceramic products are summarized. This last subject is not, strictly speaking, in the physical ceramics area; however, unless a ceramic product can be made in a satisfactory manner, it will have little or no use to the design engineer, even though its properties may have potentialities.

Part 1, which is concerned with the nature of ceramic materials, pays first attention to the atomic interactions within ceramic compounds. The principles involved are either those which are encountered in introductory science courses or those which can be derived directly from such courses. Chapters 3 and 4 consider the packing arrangement of atoms in both crystalline and noncrystalline phases. More attention is given to binary and ternary compounds and less to the more complicated silicates than in most ceramics courses. This procedure permits exposition of required principles using relatively simple compounds as illustrations. Furthermore we find in these compounds many of the ceramic products currently of interest for electrical, magnetic, and mechanical applications. Surfaces and interfaces are considered to be of sufficient importance to be given a separate chapter, as are the equilibrium relationships between phases and in ceramic microstructures.

In Part 2—Properties and Behavior of Ceramic Materials—separate attention is given to the mechanical, thermal, electrical, magnetic, and optical properties, and their nuclear applications. In each case attempts are made to relate these properties specifically to the structures of the materials and to the nature of the interatomic attractions.

The third and final part of the book—Processing of Ceramic Products—is divided into two chapters, the first dealing with the forming of ceramic products, and the second with thermal treatments used on ceramic products. These topics follow naturally after the previous topics on the nature of ceramic products and their properties. However, this does present a departure from the typical ceramics text, in which initial attention is given to processing, probably for the reason that authors feel that the product must be made before it can be tested.

Any author is aware that he can never write a textbook without the help and tolerance of all his associates. These include his own family, his fellow faculty members, the students in his classes, his professional counterparts in other schools, and finally—but certainly of utmost importance—his secretarial help. I should like to use this opportunity to express my appreciation to each person for his contributions. I should also like to give thanks in advance to those readers of this text who will, I trust, provide me with corrections and suggested modifications.

Ann Arbor, Michigan L.H.V.V.
April 1964

CONTENTS

Part 1

THE NATURE OF CERAMIC MATERIALS

Since an understanding of ceramic materials may most easily be gained if we have a knowledge of their structure and composition, initially this book will pay attention to the *nature* of ceramic materials (Part 1). A knowledge of their structural and chemical nature will enable us to understand the origin and usefulness of their *properties* (Part 2), and provide us with a basis for producing ceramic materials and controlling their properties (Part 3).

Hence Part 1 of this text will have as its subject the internal structure of ceramic materials, a structure which will be examined on several levels. In Chapter 2, concerned with the atomic level, attention will be given to *interatomic bonding*. In Chapter 3 we will examine *crystalline ceramic compounds*. And in Chapter 4, instead of discussing only those crystalline phases which have long-range atomic order, we shall deal with *glasses and amorphous phases*, which have only short-range order. *Surfaces and interfaces* have structures that are sufficiently different from those of bulk phases to warrant separate treatment, which will be given in Chapter 5. In Chapter 6, attention will be directed toward the *equilibria and reactions* among different structures. Finally, in Chapter 7, we will consider the geometric relationships among grains and phases that are found in *ceramic microstructures*.

Chapter 1

INTRODUCTION

1–1 Ceramic materials

An engineer employs ceramic materials in many production and design applications. Examples of the more common types of ceramic materials include glasses, high-frequency dielectric materials, some of the newer magnetic materials, abrasives, cermets, refractories, heavy clay products, vitreous enamels, and whitewares.

To examine the components of these materials, we look at the elements of the periodic table, which may be divided into two general categories: metallic and nonmetallic (Fig. 1–1). Metallic atoms easily lose their valence electrons and become positive ions. Nonmetallic atoms, on the other hand, hold their valence electrons more firmly than metallic ones and, in fact, will accept or share additional electrons. In common terminology, ceramic materials are considered to be substances which are usually composed of *compounds of metallic and nonmetallic elements.* Thus MgO, $BaTiO_3$, SiO_2, and SiC are relatively simple ceramic compounds. More complicated ceramic compounds include clays, magnetic spinels, mullite, and amorphous glasses, to name but a few.

When we look at the metal–nonmetal spectrum (Fig. 1–2), we observe that some compounds, such as Al_2O_3, are easily distinguishable as ceramic materials. We also note that many materials contain only metallic elements, and others only nonmetallic elements. There is no sharp division between the two extremes, a fact which permits us to anticipate properties of many ceramic compounds on the basis of the metallic tendencies of the elements contained in the compound. Nonmetallic elements are frequently classified as organics. With such a spectrum it is to be expected that there will be materials which are intermediate. Further, there will be some materials containing intermediate elements which may combine with either metallic or nonmetallic elements to form compounds. Silicon, for example, behaves like a metal atom in quartz (SiO_2), like a borderline metal atom in SiC, and like a borderline nonmetal atom in $MoSi_2$. The first two are commonly categorized as ceramic materials; the latter material is commonly left to the metallurgist as an intermetallic compound. Other amphoteric elements are indicated in the overlap zone of Fig. 1–1.

← Metals — — — Nonmetals →

IA	IIA	IIIB	IVB	VB	VIB	VIIB	VIII			IB	IIB	IIIA	IVA	VA	VIA	VIIA	0
1 H 1.00797																1 H 1.0079	2 He 4.0026
3 Li 6.939	4 Be 9.012											5 B 10.811	6 C 12.011	7 N 14.007	8 O 15.9994	9 F 18.998	10 Ne 20.183
11 Na 22.990	12 Mg 24.312											13 Al 26.98	14 Si 28.086	15 P 30.97	16 S 32.064	17 Cl 35.453	18 Ar 39.95
19 K 39.102	20 Ca 40.08	21 Sc 44.96	22 Ti 47.90	23 V 50.94	24 Cr 52.00	25 Mn 54.94	26 Fe 55.85	27 Co 58.93	28 Ni 58.71	29 Cu 63.54	30 Zn 65.37	31 Ga 69.72	32 Ge 72.59	33 As 74.92	34 Se 78.96	35 Br 79.91	36 Kr 83.80
37 Rb 85.47	38 Sr 87.62	39 Y 88.91	40 Zr 91.22	41 Nb 92.91	42 Mo 95.94	43 Tc 99	44 Ru 101.07	45 Rh 102.91	46 Pd 106.4	47 Ag 107.87	48 Cd 112.40	49 In 114.82	50 Sn 118.69	51 Sb 121.75	52 Te 127.60	53 I 126.90	54 Xe 131.30
55 Cs 132.90	56 Ba 137.34	57–71 La series*	72 Hf 178.49	73 Ta 180.95	74 W 183.85	75 Re 186.2	76 Os 190.2	77 Ir 192.2	78 Pt 195.1	79 Au 196.97	80 Hg 200.59	81 Tl 204.37	82 Pb 207.19	83 Bi 208.98	84 Po 210	85 At 210	86 Rn 222
87 Fr 223	88 Ra 226	89– Ac series†															

58 Ce 140.12	59 Pr 140.91	60 Nd 144.24	61 Pm 147	62 Sm 150.35	63 Eu 151.96	64 Gd 157.25	65 Tb 158.92	66 Dy 162.50	67 Ho 164.93	68 Er 167.26	69 Tm 168.93	70 Yb 173.04	71 Lu 174.97
90 Th 232.04	91 Pa 231	92 U 238.03	93 Np 237	94 Pu 239	95 Am 241	96 Cm 242	97 Bk 249	98 Cf 252	99 Es 254	100 Fm 253	101 Md	102 No	103 Lw

FIG. 1–1. Table of elements (simplified) showing metallic and nonmetallic elements. Ceramic materials usually contain phases that are compounds of metals and nonmetals. The metallic elements readily lose their valence electrons, while the more stable nonmetallic elements can accept or share additional electrons.

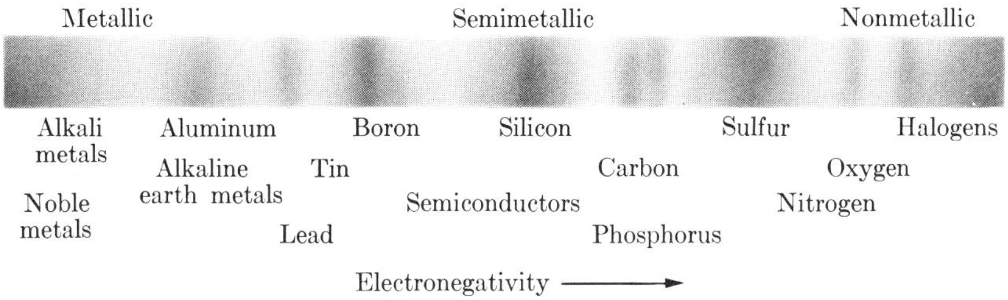

FIG. 1-2. The metal-nonmetal spectrum.

1-2 Ceramics versus metals

Although there is no sharp boundary between metals and ceramics or between organics and ceramics, it is convenient to make comparisons between typical ceramic and typical metallic materials, because the average engineer is quite familiar with metallic materials. Historically, ceramic materials predate metals. Archaeologists have found glass and brick to be among the earliest of man-made materials. However, metals have been better understood than ceramics because metals are somewhat simpler in structure. Thus metals can serve as a guide to the structure and behavior of ceramic materials, although we must realize that there are limits to generalizations. We should not make specific comparisons of the merits of ceramics and of metals without giving attention to the service requirements of the engineering design.

(1) As a class, metallic materials are better electrical and thermal conductors than are ceramic materials, a fact which is related to the freedom of the valence electrons within the metal. Conversely, ceramic materials have better dielectric properties than do metals.

(2) As a class, ceramic materials are more stable than metals are, in both chemical and thermal environments. This generality may be explained on the basis of the lower free energy of ceramic compounds compared with that of metallic structures. (Attention will be given to the energy relationships later on in Part 1 of this book, where we shall note that there are exceptions to this last generalization; however, most exceptions can be accounted for.)

(3) Metals, as a class, have comparable tensile and compressive strengths, whereas ceramic materials are usually appreciably stronger in compression than in tension. The contrast in this respect between ceramics and metals is a consequence of differences in structure, and of the effect of structure on plastic deformation by shear.

As one would expect, there is no sharp division between metallic and ceramic materials; hence no attempt will be made to classify all inorganic materials into one or the other category. For example, TiC is defined as a ceramic compound containing metallic titanium and nonmetallic carbon. With this in mind, are Fe_3C and SiC intermetallic or ceramic compounds? Let us note that the general class of ceramic materials merges into the class of metallic materials.

1-3 Ceramics versus organics

It is unfortunate that many engineers have only an empirical knowledge of organic materials. Yet the introductory concepts of organic structures are relatively simple, and involve the sharing of electrons by adjacent nonmetallic atoms. If this sharing involves the repetitious arrangement of two or more neighboring atoms, a continuous structure, called a polymer, may be formed, and many useful engineering materials result. Because ceramic compounds may contain nonmetallic or semimetallic elements which share electrons, polymerization may also be found in some (but not all) ceramic materials. Among the ceramic materials, the glasses are the most highly polymerized (see Sections 2-6 and 4-2).

As with ceramics and metals, there is a gradation between ceramic and organic polymers. A silicone, for example, possesses such an intermediate status (Section 4-7). Furthermore, in making the following general comparisons between organic and ceramic materials, we ought to expect that there will be exceptions; however, there is usually a structural explanation for these exceptions.

(1) Both ceramic and organic materials crystallize less easily than metals do. The similarity between ceramics and organics in this respect is due to their greater structural complexity, as compared with metals. Note also that ionic ceramic materials, for example $CaSiO_3$, crystallize more readily than those of more complicated structures, such as the framework silicates. Various factors affecting crystallization will be discussed in subsequent chapters.

(2) Because organic materials ordinarily contain numerous atoms which share electrons with only one (or two) neighbors, their temperature stability is usually less than that of ceramic materials. This difference may be attributed to the fact that organic materials have strong primary and weaker secondary bonds (Section 2-4).

(3) As a class, ceramic materials exhibit more rigidity and less plasticity than do organic materials, a contrast which can be partially explained on the basis of bonding, and which will be discussed more extensively in the next chapter.

1-4 Background principles

It is assumed that the reader has an elementary knowledge of certain principles and concepts which are important in the study of ceramic materials. These concepts include: (1) the periodic table and its general arrangement (Fig. 1-1); (2) the kinetic theory of matter, particularly the fact that the thermal vibrations of atoms within a material increase with increasing temperature; (3) the concept of coulombic attraction and repulsion, i.e., the fact that unlike charges attract and like charges repel; (4) the energy-stability relationship, i.e., the lower the energy state the more stable the state of the material; and (5) such properties of materials as strength, ductility, thermal expansion, electrical conductivity, etc.

Each of the above items will be discussed in this book. Attempts will be made to correlate them with the nature and behavior of those ceramic materials which are of interest to engineers.

References for Further Reading

1–1. ALFREY, T., "Molecular Structure and Mechanical Behavior of High Polymers," *The Science of Engineering Materials*, J. E. Goldman, editor. New York: John Wiley & Sons, 1957, pp. 469–487. This short review provides a good basis for understanding the source of polymer properties.

1–2. BILLMEYER, F. W., *Textbook of Polymer Chemistry*. New York: Interscience Publishers, 1957. Introduction to polymerization structures and reactions.

1–3. CHALMERS, B., *Physical Metallurgy*. New York: John Wiley & Sons, 1959. Advanced undergraduate textbook of physical metallurgy, with emphasis on the science of metals.

1–4. GUY, A. G., *Elements of Physical Metallurgy*. Reading, Mass.: Addison-Wesley, 1959. Advanced undergraduate textbook of physical metallurgy, with emphasis on the origin of properties.

1–5. JASTRZEBSKI, Z., *Nature and Properties of Engineering Materials*. New York: John Wiley & Sons, 1959. Introductory textbook, with emphasis on the properties of various materials, both metallic and nonmetallic.

1–6. KOENIG, J. H., and E. J. SMOKE, "Characteristic Properties of Modern Ceramics," *Modern Materials*, Vol. I, H. H. Hausner, editor. New York: Academic Press, 1958, pp. 243–271. Introductory ceramics for engineers.

1–7. ROGERS, BRUCE A., *The Nature of Metals*. Ames, Iowa: Iowa State University Press; and Cleveland: American Society for Metals, 1951. Introductory metallurgy, much of which applies equally well to ceramics. Background material for physical ceramics.

1–8. SCHMIDT, A. X., and C. A. MARLIES, *High-Polymer Theory and Practice*. New York: McGraw-Hill, 1948, pp. 190–199. Comparisons between the structures of silicates and polymers.

1–9. TOBOLSKY, A. V., *Properties and Structure of Polymers*. New York: John Wiley & Sons, 1960. Advanced-level presentation of viscoelastic properties of polymers.

1–10. VAN VLACK, L. H., *Elements of Materials Science*, second edition. Reading, Mass.: Addison-Wesley, 1964. Introductory textbook on materials. The contents of this book are assumed as background for physical ceramics.

Chapter 2

ATOMIC INTERACTIONS IN CERAMIC MATERIALS

2–1 Introduction

The atomic bonding within ceramic solids is a result of interatomic reactions. In this connection there are three important considerations: first, the elementary concepts of energy levels of electrons associated with individual atoms, and of atoms in polyatomic solids. Second, interatomic spacings. Third, classification of bond types according to the freedom and distribution of the outer valence electrons when the atoms interact.

2–2 Electronic structure of atoms

Introductory science courses emphasize the fact that the electrons which surround an atomic nucleus do not all possess the same energies; therefore, we divide the electrons into shells, or groups with different energy levels. The first or lowest quantum shell contains a maximum of two electrons; the second shell contains a maximum of 8; the third, 18; and the fourth, 32. Thus, the maximum number of electrons in a given shell is $2n^2$, where n is the principal quantum number of the shell.

Although the concept of the quantum shell is a convenient one, and will be used frequently in succeeding sections, it is oversimplified, because it implies that all electrons within a shell are equivalent. A more complete treatment is necessary if the properties of the materials are to be understood. However, this concept can be utilized without rigorous explanation. We can approach the subject by way of the Pauli exclusion principle, which states that no more than two interacting electrons in an atom can have the same orbital quantum numbers, and even these two electrons are not identical, because they possess inverted magnetic behaviors, or "spins." Since the electrons in a quantum shell do not all possess the same energy, as mentioned above, it is desirable to subdivide the shells into subshells. Here again, subshells with the lowest energy are filled before those with higher energy; and, as with the main quantum shells, the upper energy levels within a given subshell may require more energy than the lowest energy level in a succeeding subshell.

Electronic notation. Experimental verification of the electronic groupings and subgroupings was initially obtained through spectrometric studies, where it was concluded that a quantum of energy is required to move an electron from a lower energy level to a next higher energy level. Conversely, a quantum of energy (called a photon) is released when an electron drops into a lower level. The energy E possessed by the photon may be calculated directly from the photon wavelength λ:

$$E = hc/\lambda = h\nu, \tag{2–1}$$

where h is Planck's constant and c is the velocity of light. The wave frequency is c/λ or ν.

Spectrographic notations. The sharpest spectral lines are produced by the electrons which drop to the lowest energy level within a given quantum shell. Early spectrographic experimenters used the notation s to describe the electrons that are at the lowest energy levels within each shell. A notation such as $1s^2$ indicates that two electrons (of opposite magnetic spins) are located in the low energy position of the *first* quantum shell (i.e., the K-shell). Similarly, $2s^2$ indicates that two electrons are located in the lowest energy position of the *second* quantum shell (L-shell). Two is, as we have said, the maximum number of electrons that can exist in the s-subshell.

Each succeeding shell has two or more subshells which are designated p, d, and f. The maximum numbers of electrons in these subshells are 6, 10, and 14, respectively. Neon, which has its L-shell (or second quantum shell) completely filled, possesses the electron notation of $1s^22s^22p^6$, which indicates that two electrons are in the first shell and eight in the second, with two in its lower subshell and six in the next subshell. The extension of the notation scheme indicated above may be obtained from Table 2–1.

EXAMPLE 2–1. Iron has 26 protons. Two electrons go into the $4s$-subshell before the $3d$-subshell is filled. Show the electronic notation for a single iron atom, and for ferrous and ferric iron ions.

Answer:

$$\mathrm{Fe} = 1s^22s^22p^63s^23p^63d^64s^2; \tag{2–2}$$

$$\mathrm{Fe}^{2+} = 1s^22s^22p^63s^23p^63d^6; \tag{2–3}$$

$$\mathrm{Fe}^{3+} = 1s^22s^22p^63s^23p^63d^5. \tag{2–4}$$

From the above, the sequential progression of levels becomes apparent. Again it should be emphasized that there can be an overlap of energy levels, that is, certain quantum groups may start to be filled before the preceding groupings are completed. This means that there are groups of transition elements, such as the scandium-to-nickel series of Table 2–1, in which the outer or valence subshell is partially filled before a full quota of 10 electrons is present in the $3d$-subshell. The same situation occurs in the $4d$-subshell of the yttrium-to-palladium series. Furthermore, it is apparent that the whole series of rare-

TABLE 2-1

ELECTRON QUANTUM NUMBERS*

Element		K ($n = 1$)	L ($n = 2$)		M ($n = 3$)			N ($n = 4$)				O ($n = 5$)				P ($n = 6$)			Q ($n = 7$)
Symbol	Number	$1s$	$2s$	$2p$	$3s$	$3p$	$3d$	$4s$	$4p$	$4d$	$4f$	$5s$	$5p$	$5d$	$5f$	$6s$	$6p$	$6d$	$7s$
H	1	1																	
He	2	2																	
Li	3	2	1																
Be	4	2	2																
B	5	2	2	1															
C	6	2	2	2															
N	7	2	2	3															
O	8	2	2	4															
F	9	2	2	5															
Ne	10	2	2	6															
Na	11	2	2	6	1														
Mg	12	2	2	6	2														
Al	13	2	2	6	2	1													
Si	14	2	2	6	2	2													
P	15	2	2	6	2	3													
S	16	2	2	6	2	4													
Cl	17	2	2	6	2	5													
Ar	18	2	2	6	2	6													
K	19	2	2	6	2	6		1											
Ca	20	2	2	6	2	6		2											
Sc	21	2	2	6	2	6	1	2											

Ti	22	2	2	6	2	6	2	2					
V	23	2	2	6	2	6	3	2					
Cr	24	2	2	6	2	6	5	1					
Mn	25	2	2	6	2	6	5	2					
Fe	26	2	2	6	2	6	6	2					
Co	27	2	2	6	2	6	7	2					
Ni	28	2	2	6	2	6	8	2					
Cu	29	2	2	6	2	6	10	1					
Zn	30	2	2	6	2	6	10	2					
Ga	31	2	2	6	2	6	10	2	1				
Ge	32	2	2	6	2	6	10	2	2				
As	33	2	2	6	2	6	10	2	3				
Se	34	2	2	6	2	6	10	2	4				
Br	35	2	2	6	2	6	10	2	5				
Kr	36	2	2	6	2	6	10	2	6				
Rb	37	2	2	6	2	6	10	2	6		1		
Sr	38	2	2	6	2	6	10	2	6		2		
Y	39	2	2	6	2	6	10	2	6	1	2		
Zr	40	2	2	6	2	6	10	2	6	2	2		
Nb	41	2	2	6	2	6	10	2	6	4	1		
Mo	42	2	2	6	2	6	10	2	6	5	1		
Tc	43	2	2	6	2	6	10	2	6	6	1		
Ru	44	2	2	6	2	6	10	2	6	7	1		
Rh	45	2	2	6	2	6	10	2	6	8	1		
Pd	46	2	2	6	2	6	10	2	6	10			

(continued)

* Adapted from Hume-Rothery, W., and G. V. Raynor, *The Structure of Metals and Alloys*, 3rd edition, London: Institute of Metals (1954), pp. 14–16.

TABLE 2–1 (*Continued*)

ELECTRON QUANTUM NUMBERS*

Element		$K\ (n = 1)$	$L\ (n = 2)$		$M\ (n = 3)$			$N\ (n = 4)$				$O\ (n = 5)$				$P\ (n = 6)$			$Q\ (n = 7)$
Symbol	Number	1s	2s	2p	3s	3p	3d	4s	4p	4d	4f	5s	5p	5d	5f	6s	6p	6d	7s
Ag	47	2	2	6	2	6	10	2	6	10		1							
Cd	48	2	2	6	2	6	10	2	6	10		2							
In	49	2	2	6	2	6	10	2	6	10		2	1						
Sn	50	2	2	6	2	6	10	2	6	10		2	2						
Sb	51	2	2	6	2	6	10	2	6	10		2	3						
Te	52	2	2	6	2	6	10	2	6	10		2	4						
I	53	2	2	6	2	6	10	2	6	10		2	5						
Xe	54	2	2	6	2	6	10	2	6	10		2	6						
Cs	55	2	2	6	2	6	10	2	6	10		2	6			1			
Ba	56	2	2	6	2	6	10	2	6	10		2	6			2			
La	57	2	2	6	2	6	10	2	6	10		2	6	1		2			
Ce	58	2	2	6	2	6	10	2	6	10	2	2	6			2			
Pr	59	2	2	6	2	6	10	2	6	10	3	2	6			2			
Nd	60	2	2	6	2	6	10	2	6	10	4	2	6			2			
Pm	61	2	2	6	2	6	10	2	6	10	5	2	6			2			
Sm	62	2	2	6	2	6	10	2	6	10	6	2	6			2			
Eu	63	2	2	6	2	6	10	2	6	10	7	2	6			2			
Gd	64	2	2	6	2	6	10	2	6	10	7	2	6	1		2			
Tb	65	2	2	6	2	6	10	2	6	10	8	2	6	1		2			
Dy	66	2	2	6	2	6	10	2	6	10	10	2	6			2			
Ho	67	2	2	6	2	6	10	2	6	10	11	2	6			2			
Er	68	2	2	6	2	6	10	2	6	10	12	2	6			2			
Tm	69	2	2	6	2	6	10	2	6	10	13	2	6			2			
Yb	70	2	2	6	2	6	10	2	6	10	14	2	6			2			
Lu	71	2	2	6	2	6	10	2	6	10	14	2	6	1		2			

Hf	72	2	2	6	2	6	10	2	6	10	14	2	6	2		2			
Ta	73	2	2	6	2	6	10	2	6	10	14	2	6	3		2			
W	74	2	2	6	2	6	10	2	6	10	14	2	6	4		2			
Re	75	2	2	6	2	6	10	2	6	10	14	2	6	5		2			
Os	76	2	2	6	2	6	10	2	6	10	14	2	6	6		2			
Ir	77	2	2	6	2	6	10	2	6	10	14	2	6	7		2			
Pt	78	2	2	6	2	6	10	2	6	10	14	2	6	8		2			
Au	79	2	2	6	2	6	10	2	6	10	14	2	6	10		1			
Hg	80	2	2	6	2	6	10	2	6	10	14	2	6	10		2			
Tl	81	2	2	6	2	6	10	2	6	10	14	2	6	10		2	1		
Pb	82	2	2	6	2	6	10	2	6	10	14	2	6	10		2	2		
Bi	83	2	2	6	2	6	10	2	6	10	14	2	6	10		2	3		
Po	84	2	2	6	2	6	10	2	6	10	14	2	6	10		2	4		
At	85	2	2	6	2	6	10	2	6	10	14	2	6	10		2	5		
Rn	86	2	2	6	2	6	10	2	6	10	14	2	6	10		2	6		
Fr	87	2	2	6	2	6	10	2	6	10	14	2	6	10		2	6		1
Ra	88	2	2	6	2	6	10	2	6	10	14	2	6	10		2	6		2
Ac	89	2	2	6	2	6	10	2	6	10	14	2	6	10		2	6	1	2
Th	90	2	2	6	2	6	10	2	6	10	14	2	6	10		2	6	2	2
Pa	91	2	2	6	2	6	10	2	6	10	14	2	6	10	2	2	6	1	2
U	92	2	2	6	2	6	10	2	6	10	14	2	6	10	3	2	6	1	2
Np	93	2	2	6	2	6	10	2	6	10	14	2	6	10	5	2	6		2
Pu	94	2	2	6	2	6	10	2	6	10	14	2	6	10	6	2	6		2
Am	95	2	2	6	2	6	10	2	6	10	14	2	6	10	7	2	6		2
Cm	96	2	2	6	2	6	10	2	6	10	14	2	6	10	7	2	6	1	2

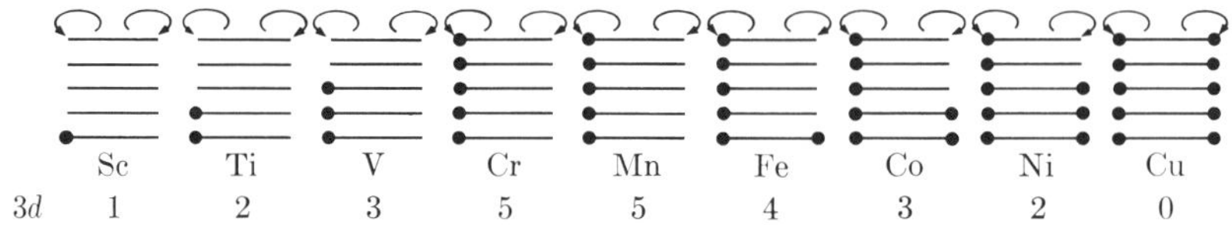

FIG. 2-1. Bohr magnetons. As additional 3*d*-electrons surround the atom, they are aligned to give it a magnetic moment. After all the levels are filled with one alignment, additional electrons must have the opposite spin, and therefore the magnetic moment (Bohr magnetons) is decreased. (There is only one 4*s*-electron in chromium and copper. Cf. Table 2-1.)

earth elements occurs as a result of successive addition of electrons to the 4*f*-subshell in spite of the fact that there are eight or more electrons in the fifth or *O*-shell. Finally, the addition of 5*f*-electrons starts a second rare-earth series which includes uranium and neighboring elements.

Bohr magneton. The partial filling of the 3*d*-subshells is of interest because it affects the magnetic characteristics of atoms, as shown in Fig. 2-1. You may note that the one to ten electrons in the 3*d*-level tend to align themselves with parallel spins. Thus, as the number of electrons increases, there is an increased unbalance in the direction of spins, reaching a maximum with manganese, which has five more electrons spinning in one direction than in the other. This provides the atom with a magnetic moment which enables it to serve as a small magnet. The unbalance in magnetic spins indicates the number of Bohr magnetons. Thus manganese has five Bohr magnetons, and iron has four when the atoms possess all of their 3*d*-electrons.

EXAMPLE 2-2. Utilizing the procedure of Fig. 2-1, indicate the number of Bohr magnetons in ferric iron.

Answer: See Fig. 2-2.

FIG. 2-2. Ferric iron. Trivalent iron has lost both of its 4*s*-electrons and one 3*d*-electron (×). Therefore it has five Bohr magnetons.

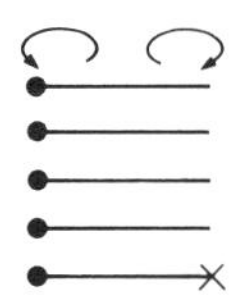

Energy bands. Although the previous discussion has centered around single atoms, the general rule that only two interacting electrons may have the same energy level applies to polyatomic materials as well. This means that when electrons from two or more atoms do interact, it is necessary to form new energy levels which are discrete, although only infinitesimally different (Fig. 2-3). Thus each energy level becomes expanded into an energy band, and each band contains as many discrete levels as there are interrelated atoms in the material.

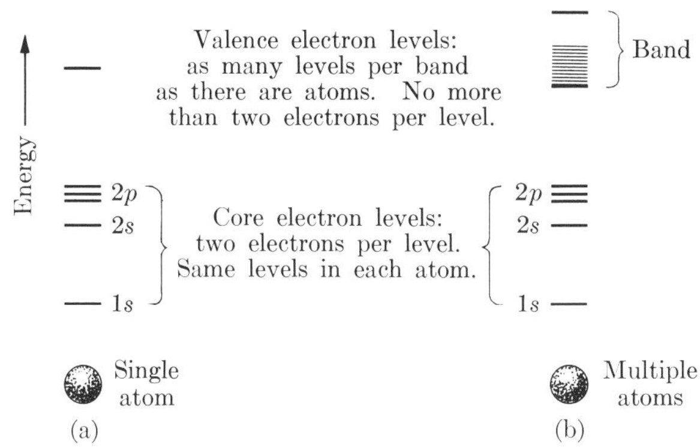

FIG. 2-3. Energy levels and energy bands (sodium): (a) individual atoms; (b) polyatomic materials. Each energy band has as many discrete levels as there are interacting atoms.

Because alkali metals have a valence of one, their energy bands are only half full (Fig. 2-4), and each level of the lower half of the band contains two electrons. Because the band is not filled, electrons can receive additional energy for conduction. Alkaline earth metals (e.g., beryllium, magnesium, and calcium) have two valence electrons per atom. This number would be sufficient to fill the first energy band with two electrons in each level; however, there is an overlapping, because the lowest levels of the second band lie below the highest levels of the first band. Some electrons "overflow" into the levels of the second band (Fig. 2-5). Again, it is possible to readily energize electrons into higher levels for conduction.

The adjacent energy bands in materials do not always overlap, so that an energy gap may be present (Fig. 2-6a). This situation does not affect the electrical conductivity of the material which maintains vacant levels within the band. For example, aluminum, with energy gaps between the second and

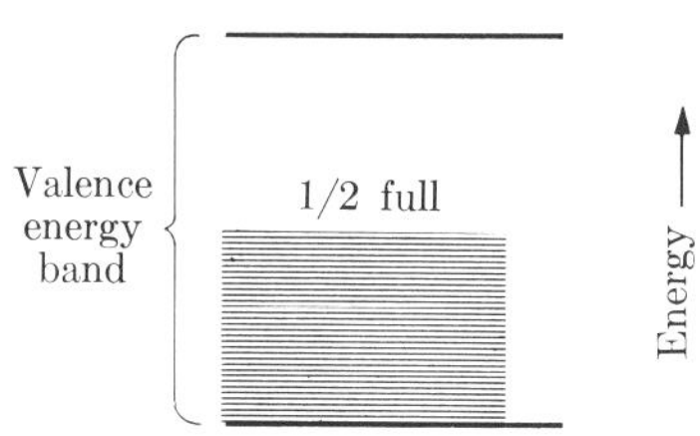

FIG. 2-4. Energy band of sodium. There are as many energy levels in the band as there are atoms. Since each energy level may contain only two electrons (opposite spins), and since sodium has only one valence electron per atom, the valence energy band is only half filled.

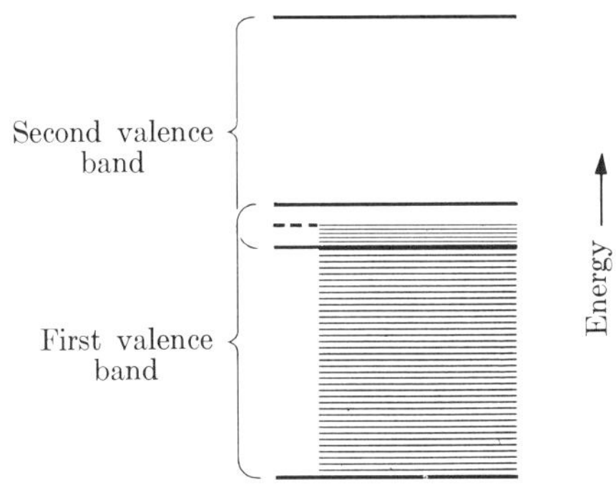

FIG. 2–5. Energy bands of magnesium. Nominally, all the valence electrons could go into the first valence band because magnesium is divalent. However, there is an overlap; therefore less energy is required for electrons to fill some levels in the second band than to reach the top levels in the first band. The positions of lowest energy are filled first.

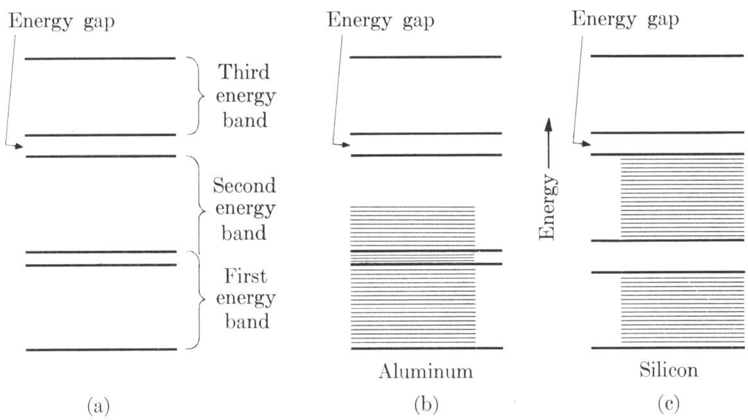

FIG. 2–6. Energy gaps. In silicon and in almost all ceramic materials, there are forbidden energy levels between energy bands. These gaps do not affect the conduction in metals, because there are extra conducting levels in the top of the unfilled band. However, if the band is filled, energy must be provided to raise the electron into the next band for conduction.

third bands (Fig. 2–6b), has excellent conductivity because its outer electrons within the second band can be energized by small electric fields. However, silicon and almost all ceramic materials have a gap above the filled energy bands (Fig. 2–6c). The only way to move electrons across this gap to higher levels is to supply enough energy to enable them to jump the gap. This requires an energy of several electron volts, so that there is only a very small probability that an electric field, or thermal energy, can energize the electron enough to cross the gap. Thus Fig. 2–6(c) represents the electron-energy relationship in insulating materials.

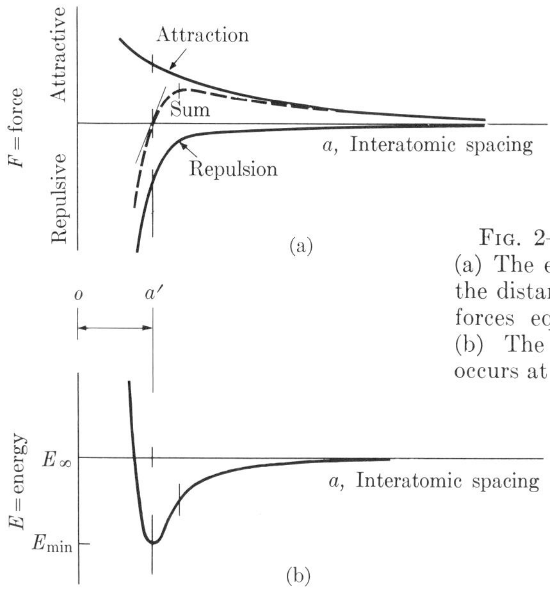

FIG. 2–7. Interatomic distances. (a) The equilibrium spacing o–a' is the distance at which the attractive forces equal the repulsive forces. (b) The lowest potential energy occurs at distance o–a'.

2–3 Interatomic spacings

All scientific evidence indicates that the lowest energy condition is the most stable. This factor is responsible for the natural tendency of atoms to assume positions in which there is a minimum of free energy. Such minimum-energy positionings are the subject of this section.

The groupings of atoms in close coordination are due to the attractive forces between atoms; there are also interatomic repulsive forces. The *equilibrium spacing* between atoms is that distance at which the attractive and repulsive forces are balanced (Fig. 2–7a). Furthermore, because these net attractive and repulsive forces must be overcome, energy is required to increase or decrease the distance between the atoms. The increase in energy is equal to the integrated area along the "sum" curve of Fig. 2–7(a), where the letters are defined for the following relationship:

$$E - E_{\min} = \int_{a'}^{r} F\,dr. \tag{2–5}$$

An integration of Eq.(2–5) is shown in Fig. 2-7(b). The "energy trough" of this figure is typical of all interatomic bondings. It is common to use the energy at infinite spacing as the datum energy, because two or more atoms have zero interaction at that distance.

An examination of Fig. 2–7(a) will reveal that attractive forces are generally effective over a longer range than repulsive forces. Furthermore, at close range the repulsive forces become exceptionally high and exceed the attractive forces. If this were not true, two atoms would "fuse" into one and the distance between

TABLE 2-2
SELECTED* IONIC RADII†
(Based on Goldschmidt)

Ion	Radius, A	Ion	Radius, A
Al^{3+}	0.57	Mn^{2+}	0.91
Ba^{2+}	1.43	Na^{+}	0.98
Be^{2+}	0.34	Ni^{2+}	0.78
Ca^{2+}	1.06	O^{2-}	1.32
Cl^{-}	1.81	$(OH)^{-}$	1.32
Cr^{3+}	0.64	P^{5+}	0.2-0.4
Cs^{+}	1.65	S^{2-}	1.74
F^{-}	1.33	Si^{4+}	0.39
Fe^{2+}	0.83	Sr^{2+}	1.27
Fe^{3+}	0.67	Ti^{4+}	0.64
K^{+}	1.33	Zn^{2+}	0.83
Li^{+}	0.78	Zr^{4+}	0.87
Mg^{2+}	0.78		

* Additional values presented in Appendix D.

† Approximate values for CN = 6. The radii are increased about 3% for CN = 8, and decreased about 9% for CN = 4. Also the ions become nonspherical when they possess anisotropic surroundings, such as O^{2-} in $BaTiO_3$.

them would decrease to zero. Although such atomic fusion can be realized for lightweight elements under the special conditions of a hydrogen bomb, such a reaction cannot occur in the solid engineering materials.

Atomic and ionic radii. Although atoms and ions are not "hard spheres," it is convenient to use the concept of atomic and ionic radii as a measure of the effective sphere of influence. In a single-component metal, e.g., pure aluminum, all atoms are identical. As a result, the atomic radius is considered to be half the interatomic distance between nearest neighbors. Because this interatomic distance involves two different elements in compounds, it is impossible to measure the radii of the two different atoms or ions independently. However, a radius of 1.33 A has been assigned to the F^- ion (by Goldschmidt, on theoretical bases). With such an assignment, it is possible to approximate radii of other ions. The resulting radii are given in Table 2-2. These figures must be considered as approximations only; however, they are useful because they permit coordination calculations (Section 3-2) which help to determine structural relationships.

The size of an atom varies slightly with the number of neighbors (coordination number). When there are fewer neighbors there is greater attraction between adjacent atoms, and a smaller interatomic distance. The valence of an ion has a great influence on the ionic radius. An example is iron, which has a radius of 0.83 A when it is ferrous (Fe^{2+}) and 0.67 A when it is ferric (Fe^{3+}).

These radii are in contrast to the radius of 1.24 A for iron in the metallic form, where the valence may be considered zero. In effect, as electrons are removed from an atom, the remaining electrons are drawn more closely to the atom's nucleus, thus reducing the repulsive forces between the electrons of adjacent atoms (Section 2-5), and decreasing the interatomic distances. Conversely, the atomic radius is increased when the atom accepts electrons to become an anion. A consequence of the above changes in radii is that anions are larger than all but the heavier of the monovalent and divalent cations.

The radius of the hydroxyl ($(OH)^-$) ion is of special interest because it has the same dimension as O^{2-}. We can anticipate this similarity, since the $(OH)^-$ ion has the same number of electrons and therefore the same electronic repulsion as an O^{2-} ion, the only difference being the presence of a small proton which may be aligned in a position which is farthest removed from the rest of the cations.

2-4 Types of bonds

The outermost or valence electrons are responsible for most of the energy relationships which produce atomic bonding in molecules, liquids, or solids. Four general categories of bonds are encountered: (1) metallic, (2) ionic, (3) covalent, and (4) van der Waals (Fig. 2-8). These bonds possess different degrees of directionality, with the metallic bond equally effective in all three coordinate directions. This is also true for the coulombic attractions of the ionic bond; however, an ionically bonded material has coulombic repulsive forces between like atoms which are second neighbors. The covalent bond is very specifically oriented and is most effective between the two atoms of a pair. The van der Waals bonds, although weak, are polar and therefore directional.

The term "primary bond" is commonly applied to the metallic, ionic, and covalent types. These bonds have energies approximating the magnitude of 100 kcal/mole. In contrast to primary bonds, the secondary bonds (van der Waals) generally have energies of less than 10 kcal/mole.

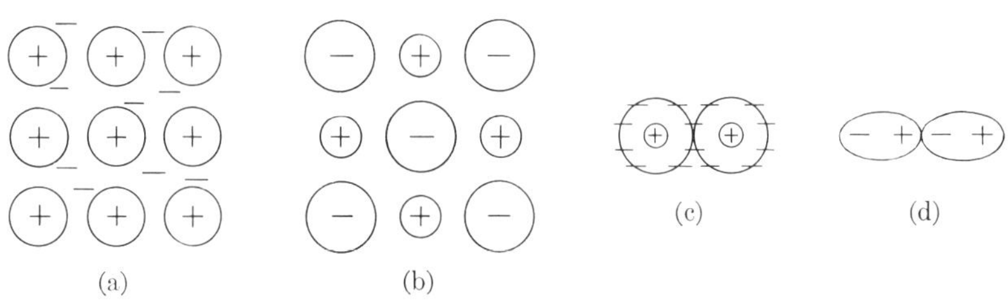

FIG. 2-8. Bond categories. (a) Metallic. Positive "ions" are held together by "free" electrons. (b) Ionic. Positive and negative ions are held together by coulombic attraction. (c) Covalent. Electrons are shared with electron pairs between two positive "ions." (d) Van der Waals. The centers of positive and negative charges are not coincident. Therefore an apparent coulombic attraction exists.

It should be emphasized at this point that there is almost always more than one type of bond in a ceramic solid. As an example, the Ca—O bonding within gypsum must be different from the S—O bonding. These heterogeneities in bonding produce materials with pronounced anisotropic properties.

We shall not consider metallicly bonded solids in detail, other than to note that (1) the free electrons, and (2) the general absence of bonding directionality have significant effects on the properties of metals. The other three bond categories are more pertinent to ceramic materials, and will be considered in appropriate detail in the following sections.

2–5 Ionic solids

Materials which derive most of their coherency from coulombic forces of attraction between charged atoms are classified as ionic solids; prototypes are the alkali halides. Admittedly, these particular solids find only very specialized use as engineering materials; however, they illustrate the nature of the ionic bond, which is found in many ceramic materials, and also they are nearly identical to more widely used ceramics, such as MgO.

The net coulombic attraction within an ionic solid may be represented by the upper curve of Fig. 2–7(a). As the positive and negative ions are brought closer together, energy is released. Not until the electron shells around the ions begin to interact are the net coulombic forces of attraction balanced by those of repulsion [lower curve of Fig. 2–7(a)]. These forces may be calculated for simple ionic solids, and furthermore may be verified experimentally. When we include the coulombic energies, the electron repulsion energies, and the other minor energies, the correspondence between calculation and experiment is close enough for substantial credence to be given to our present concepts concerning ionic solids.

Coulombic energy. The potential energy E, corresponding to two point charges, is

$$E_C = \frac{(Z_1 e)(Z_2 e)}{a_{1-2}}. \tag{2–6a}$$

In this equation, a_{1-2} is the distance between the two points; Z_1 and Z_2 are the number of charges at each point, and may be either positive or negative. The electronic charge e is 1.6×10^{-19} coul, or 4.8×10^{-10} esu. Unlike charges give a negative value to E, indicating that energy is given off as the distance is decreased. The above equation is simplified when two unlike, monovalent ions are considered:

$$E_C = -e^2/a. \tag{2–6b}$$

The energy of two positive (or two negative) monovalent ions increases as these are brought closer together:

$$E_C = +e^2/a. \tag{2–6c}$$

TABLE 2–3

MADELUNG CONSTANTS*

Structure	Composition	Reference figure	Madelung constant
Halite	NaCl	3–9b	1.747
	CsCl	3–9a	1.763
Sphalerite	ZnS	3–9c	1.638
Wurtzite	ZnS	3–9d	1.641
Fluorite	CaF_2	3–10	5.04
Rutile	TiO_2	3–12a	4.82
Cuprite	Cu_2O	3–12b	4.12
Perovskite	$CaTiO_3$	3–13	12.38

* Adapted from Dick, B. G., "Madelung Constants," *Encyclopedia of Science and Technology* **8,** New York: McGraw-Hill (1960), pp. 14–15.

Madelung constant. The potential energy of an ion within an ionic solid is not the same as the potential energy of a single pair of ions, because there are many neighbors, like as well as unlike. The Madelung constant is the ratio of the potential energy of an ion within a three-dimensional solid to the energy between a single pair of ions. These ratios are given in Table 2–3, in which it is indicated that the removal of a sodium ion from the interior of a NaCl crystal requires 1.747 times as much energy as it does to separate it from a single Cl^- ion. This is one of the reasons that diatomic NaCl molecules are found less often than NaCl crystals. Using the equation

$$E_C = -1.747e^2/a, \tag{2–7a}$$

we may calculate the net coulombic energy for NaCl to be 206 kcal/gm-mole.

EXAMPLE 2–3. Calculate the Madelung constant for a hypothetical linear array of alternating positive and negative monovalent ions (Fig. 2–9a).

Answer:

$$\begin{aligned} E_C &= -2e^2/a + 2e^2/2a - 2e^2/3a + 2e^2/4a - \cdots \\ &= -2\frac{e^2}{a}\left(1 - \tfrac{1}{2} + \tfrac{1}{3} - \tfrac{1}{4} + \cdots\right) \\ &= -2\frac{e^2}{a}\ln 2 \\ &= -1.385\, e^2/a \\ &= -Ae^2/a; \\ A &= 1.385. \end{aligned} \tag{2–7b}$$

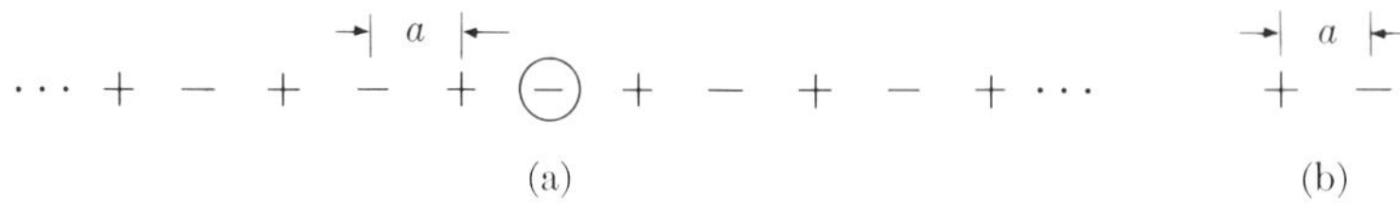

FIG. 2–9. Madelung constant for linear array. (See Example 2–3.) It would take 1.385 times as much energy to remove the single encircled ion from (a) as from the pair (b).

Electronic repulsion energy. Energy is required to bring the electronic fields of two atoms together because of the resulting repulsive forces [the lower curve of Fig. 2–7(a)]. This repulsive energy between the electron shells of adjacent atoms may be expressed by

$$E_R = +b/a^n, \tag{2–8}$$

in which b is a constant and n has a value of approximately 9, as determined experimentally.

The sum U of the net coulombic attractive and electron repulsive energies is $E_C + E_R$; that is

$$U = -Ae^2/a + b/a^n. \tag{2–9}$$

Therefore, at the interatomic distance a', as shown in Fig. 2–7(b), the following relationships may be developed:

$$dU/da = 0 = Ae^2/a^2 - nb/a^{n+1},$$
$$Ae^2 = nb/a^{n-1},$$

or

$$b/a^n = Ae^2/na.$$

Therefore, from Eq. (2–9),

$$U = -Ae^2/a + Ae^2/na$$
$$= [-Ae^2/a]\,[1 - 1/n]. \tag{2–10a}$$

This is the potential energy (net coulombic plus electron repulsive) of one monovalent ion within an ionic crystal, as compared to its energy at infinite separation. The energy per mole is

$$U = -N_0Ae^2\,\frac{1 - 1/n}{a}, \tag{2–10b}$$

where N_0 is Avogadro's number, 6.02×10^{23}.

Since n is approximately 9, the sum of these two energies for NaCl is 184 kcal/gm-mole. This is exceedingly close to the experimental value of 182.4 kcal/gm-mole, as determined through the Born-Haber cycle, which represents a stepwise conversion from the ionic solid to a vapor formed of its ions. This correspondence indicates that the other attraction and repulsion forces such as dipole–dipole, dipole–quadrupole, etc., are either minor or self-canceling. The agreement between calculation and experiment is best for strongly ionic solids. Solids with weaker ionic characteristics usually exhibit less important coulombic energies and more significant covalent effects, thus negating the above calculation procedures.

2–6 Covalent bonds

This type of bond involves mutual sharing of two electrons by adjacent atoms, and is frequently encountered in ceramic materials. Because the theoretical basis for the covalent bond is more complicated than that for the ionic bond, the former has received mathematical and experimental corroboration only in its simplest form, H:H. For this reason, a detailed model of the covalent bond will not be presented here. We shall, however, point out that the filled quantum level (with two electrons of opposite spins between two atoms) apparently produces an energy state lower than that of unpaired electrons in separate atoms. From our current knowledge of the covalent bond, we can utilize several useful qualitative relationships. (1) A covalent bond always involves two electrons. (2) The number of covalent bonds, N, is

$$N = 8 - V, \tag{2–11}$$

where V is the number of valence electrons. (3) No more than four covalent bonds may be associated with any particular atom. (4) Single, double, and triple bonds are possible, involving two, four, and six electrons between an adjacent pair of atoms.

Covalent bonds can be very strong. Evidence of this is found in diamond, and in refractory and abrasive ceramic compounds which depend extensively

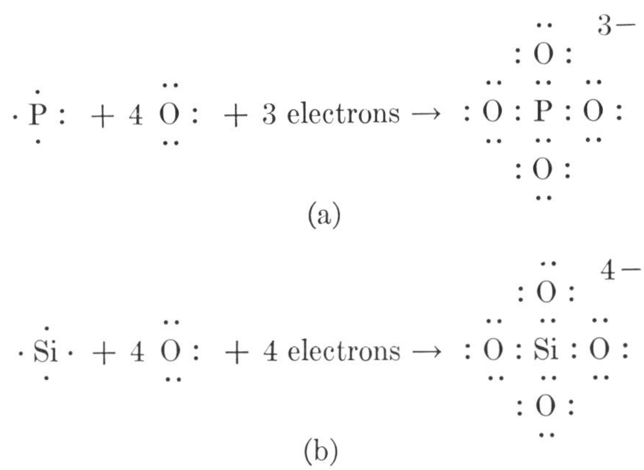

Fig. 2–10. Polyatomic ions. The $(PO_4)^{3-}$ and the $(SiO_4)^{4-}$ ions are tetrahedral, with the oxygen atoms at points of maximum separation. Both ions act as individual units in ceramic materials.

on covalent bonding. Likewise, in a diatomic gas, the rupture of the covalent bonds to give atoms in their nascent state requires a large thermal or electrical activation.

Polyatomic groups. Covalent bonds are associated almost exclusively with the semimetallic and nonmetallic atoms which are located in the upper right-hand corner of the periodic table as it is normally presented (Fig. 1-1). Since these elements can share electrons, it is not surprising to find them clustering into larger polyatomic units. If the polyatomic units satisfy the valence shells of all of the atoms, an independent molecule is formed. Polyatomic groupings may be formed which do not have enough electrons to fill the outer-shell requirements. Such units become more stable if they receive additional electrons from external sources. When this occurs, the unit becomes a polyatomic anion. Figure 2-10 shows the bonding for two ions containing oxygen atoms; these types of ions are commonly encountered in ceramic materials.

Polymerization. When a ceramic material is so composed that the polyatomic groups may be extended by a continued sharing of electrons by additional atoms, polymerization occurs. Since the principles of polymerization involving ceramic materials are identical to those which involve organic materials, it is to be expected that we shall also encounter chains, cross-linking, framework structures, etc., in ceramic materials. But in contrast to organic polymers, which usually contain primary carbon-to-carbon bonds, the polymers most commonly found in ceramics involve alternating silicons and oxygens (Fig. 2-11).

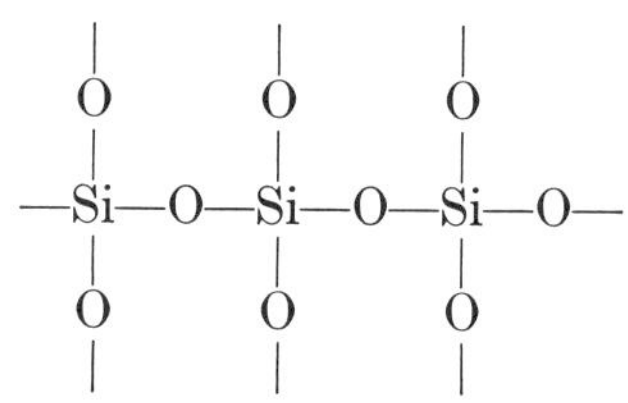

FIG. 2-11. Polymerization in ceramic compounds. The oxygen atoms are shared by adjacent (SiO_4) units.

2-7 Van der Waals bonds

These bonds arise from secondary forces of attraction which originate through electrical dipoles inside the material. In an electrical dipole, the centers of positive and negative charges are displaced relative to each other, and it is therefore possible for coulombic forces to become operative. With the displaced center of charges, the positive end of one dipole can be attracted to the negative end of another dipole, because the negative-to-negative or positive-to-positive distances may be slightly greater than the distances between the unlike charges. Nevertheless, the difference between the attractive and repulsive forces is normally slight, since the displacement of the centers of positive and negative charges can be only a small fraction of an angstrom unit. Consequently, the bonding energy for van der Waals bonds rarely is more than a few kcal/mole, and the presence of van der Waals bonds is normally masked by the presence of stronger primary forces. These van der Waals bonds are observed only in structures which lack local primary bonds. Since van der Waals bonds are weak, we would expect them to be overcome by small thermal, electrical, or mechanical forces. For example, the cleavage of talc is due to the fact that its *inter*layer bonds are chiefly secondary. (The *intra*layer bonds are covalent, and produce a strong two-dimensional sheet.) Van der Waals bonds are also evident in compounds that form true molecules, such as water, SiF_4, and CO_2. They have strong intramolecular forces, but only weak intermolecular attractions; therefore these compounds liquefy and gasify as independent molecules.

Electrical dipoles are inherent in asymmetrical molecules such as HF (Fig. 2-12). Also, dipoles may be induced by an electric field that is applied externally, or by a field which results from the presence of neighboring atoms, molecules, or ions. Thus even symmetrical structures may be induced to develop weak van der Waals attractions to one another. Normally, the attractions of induced dipoles are weaker than attractions occurring inherently, as they do in asymmetrical molecules.

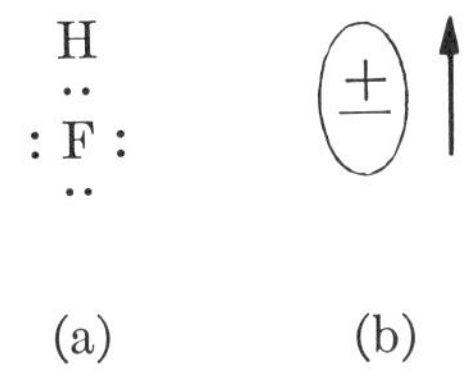

FIG. 2-12. Electrical dipoles. The centers of positive and negative charges are not coincident.

In ceramic materials, there are two main structural and behavioral consequences of the van der Waals forces of attraction: (1) they serve as the loci for plastic slip (e.g., in clays); and (2) they serve to promote physical adsorption of atoms, ions, or molecules onto solid (and liquid) surfaces (e.g., silica gels). These consequences will be detailed later (Sections 5-2 and 8-8).

Dipole moments. The product of the dipole charge Z and the distance d between charges is the dipole moment μ:

$$\mu = Zd. \tag{2-12}$$

The dipole moment is expressed in debyes (1 debye $= 10^{-18}$ esu·cm $= Zd$). Since the dipole moment has direction and is therefore a vector quantity, the moments of polyatomic molecules are calculated by vectorial addition of the moments of individual atom pairs (Fig. 2-13). The values of Table 2-4 are

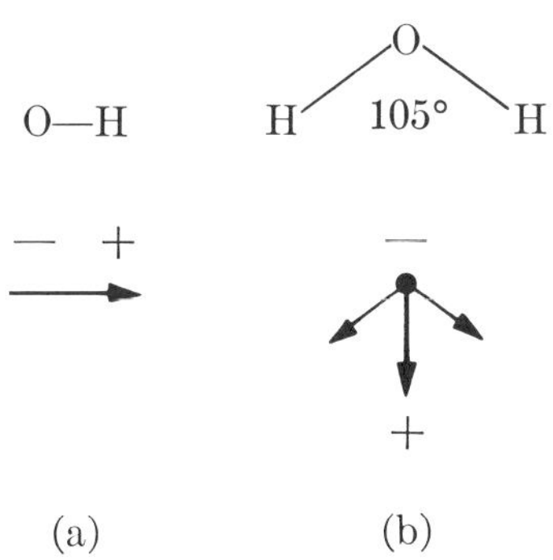

FIG. 2-13. Dipole moments for H_2O. The net dipole moment is a vectorial sum of the component moments.

TABLE 2-4

DIPOLE MOMENTS OF ATOM PAIRS*

Pair	Dipole moment, debyes
C—H	0.4
O—H	1.6
N—H	1.7
C—O	0.7
NO	3.2
NC	1.7

* Adapted from Sinnott, M. J., *The Solid State for Engineers,* New York: J. Wiley and Sons (1958), p. 190.

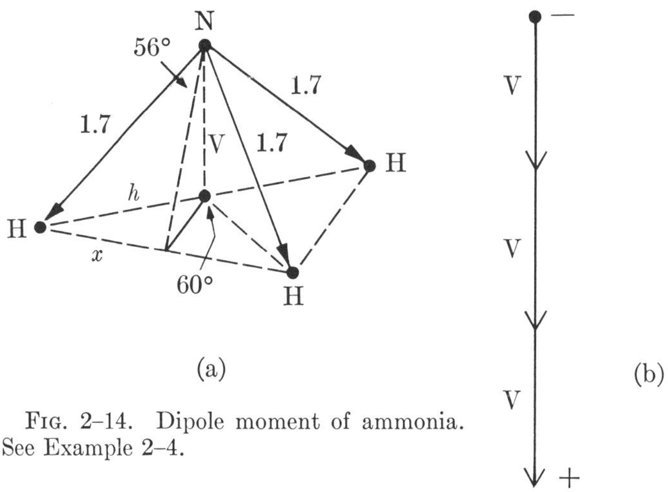

FIG. 2-14. Dipole moment of ammonia. See Example 2-4.

typical, and may be used to make calculations of simple problems, such as Example 2-4. It should be noted, however, that neighboring atoms or ions may induce distortion in any particular dipole; therefore a precise calculation is not always possible.

EXAMPLE 2-4. The H—N—H bond angle in ammonia is 112°, and the H—N couple has a dipole moment of 1.7 debyes. What is the dipole moment of NH_3?

Answer: See Fig. 2-14. Let V be the vertical component and h the horizontal component of each vector. Then

$$\sin 56° = x/1.7, \qquad \sin 60° = x/h;$$
$$h = 1.7 \sin 56/\sin 60 = 1.63;$$
$$V = [(1.7)^2 - (1.63)^2]^{0.5} = 0.5;$$
$$\text{dipole moment of } NH_3 = 3V = 1.5 \text{ debyes.}$$

The stronger the dipole moment, the greater the bonding between structures. Also, for larger molecules with additional polar sites, there is an increase in attraction between adjacent molecules. These facts may become apparent when we consider such properties as melting points with respect to molecule size and structure. Also, characteristics of adsorption onto surfaces of ceramic materials become more pronounced when the dipole moments of the molecules and the substrates are both high. The dielectric constant ϵ and the index of refraction n of the material are also related to the dipole moment, because the displaced centers of positive and negative charges react differently within the electromagnetic field (see Chapter 10).

References for Further Reading

2–1. ADDISON, W. E., *Structural Principles of Inorganic Compounds.* New York: John Wiley & Sons, 1961 (paperback). Chapter 1 introduces electronic theory of the atom, and chemical bonding. Background of freshman chemistry desirable.

2–2. CHALMERS, B., *Physical Metallurgy.* New York: John Wiley & Sons, 1959. The electron structure of atoms, as applied to metals, is presented briefly in Chapter 2.

2–3. KITTEL, C., *Introduction to Solid State Physics,* 2nd edition. New York: John Wiley & Sons, 1956. Advanced textbook on the origin of properties of materials. Chapter 3 presents additional information on the Madelung constant.

2–4. PAULING, L., *The Nature of the Chemical Bond,* 2nd edition. Ithaca, N. Y.: Cornell University Press, 1948. Advanced-level presentation of the chemical bond. This book has for some time been the standard reference on the subject.

2–5. SCARLETT, A. J., and J. GOMEZ-IBANEZ, *General College Chemistry.* New York: Henry Holt and Co., 1954. Introduction to the chemical bond.

2–6. SIENKO, M. J., and R. A. PLANE, *Chemistry,* 2nd edition. New York: McGraw-Hill, 1961. Chapter 4 considers the chemical bond in sufficient detail to provide a background for physical ceramics.

2–7. SINNOTT, M. J., *The Solid State for Engineers.* New York: John Wiley & Sons, 1958. Advanced undergraduate presentation of ionic solids in Chapter 7. Attention is given to ion formation, Madelung constants, and lattice energies. Covalent solids are discussed in Chapter 8.

2–8. SORUM, C. H., *Fundamentals of General Chemistry.* Englewood Cliffs, N. J.: Prentice-Hall, Inc., 1955. Chapter 7 provides an elementary introduction to atomic structure.

Problems

2–1. Give the electronic notations for:
(a) Si^{4+} (b) Zr^{4+} (c) Cr^{3+} (d) Mn^{3+}

2–2. Determine the number of Bohr magnetons for each of the following ions:
(a) Ti^{3+} (b) Ti^{4+} (c) V^{3+} (d) V^{5+} (e) Cr^{2+}
(f) Cr^{3+} (g) Cr^{6+} (h) Mn^{2+} (i) Mn^{3+} (j) Mn^{4+}

(k) Mn^{6+} (l) Mn^{7+} (m) Fe^{2+} (n) Fe^{3+} (o) Co^{2+}
(p) Co^{3+} (q) Ni^{2+} (r) Ni^{3+} (s) Cu^{+} (t) Cu^{2+}
(u) Zn^{2+}

2–3. Calculate the electron repulsion energy for NaCl. The interatomic spacing is 2.8 A. Cite any necessary assumptions.

2–4. Calculate the Madelung constant for each of the following ions in a small cube of MgO which contains only 27 ions. [MgO has the same structure as NaCl, which is shown in Fig. 3–9(b).]

(a) The central ion (b) The face-centered ion
(c) The edge-centered ion (d) The corner ion

2–5. A two-dimensional array of positive and negative ions forms a square pattern, with alternate charges at each adjacent position. Calculate the Madelung constant. [*Hint:* Select an area with boundaries "through" the outermost ions. Use the fraction of these ions which is shared by this area, i.e., one-half of the edge ions and one-fourth of the corner ions.]

2–6. Show the covalent bonding in a sketch of a molecule of ethyl silicate.

Chapter 3

CRYSTALLINE CERAMIC MATERIALS

3–1 Introduction

The internal structure of every ceramic material has a coordinated arrangement of atoms. These arrangements vary from highly repetitive crystal patterns to amorphous and glassy materials with only first-neighbor coordination (see Chapter 4). In this chapter, we shall consider the nature of crystalline ceramic materials.

Ceramic crystals are generally more complex than metallic crystals, because ceramic materials are always compounds of metallic *and* nonmetallic elements. Often several varieties of atoms are present. Since ceramic crystals can be so complex, let us first examine them on the basis of the first neighbors, or short-range order that is common to all ceramic materials; then we shall present details of the long-range order which is repetitive within ceramic crystals. After we consider the more perfect structures, we shall deal with the imperfections which are invariably present within crystals, and which have an important effect on the behavior and properties of the various materials. Finally, we shall discuss the fact that compounds may have more than one crystal structure (polymorphism), depending upon their surrounding environments.

3–2 First neighbors (short-range order)

There is a spatial coordination among the atoms in all types of materials, with the exceptions of monatomic and ionized gases. In polyatomic gases such as Cl_2 and C_2H_6, the coordination involves relatively few atoms. An example is methane, CH_4, which has the coordination shown in Fig. 3–1(a). The coordinations in these molecules are governed by the way in which covalent bonds are shared. In a metallic liquid, each metal atom is surrounded by at least 8 and no more than 15 other metal atoms. Although the neighboring atoms in a liquid metal are constantly being interchanged, it has been verified by x-rays that the coordination distance of nearest neighbors is fairly definite. In contrast to first-neighbor coordination in an unalloyed metal (which involves only *like* atoms), the nearest-neighbor coordination in crystalline ceramic compounds involves *unlike* atoms.

Coordination number. The number of neighbors which an atom has is referred to as its coordination number (CN). Since covalent bonds coordinate specific

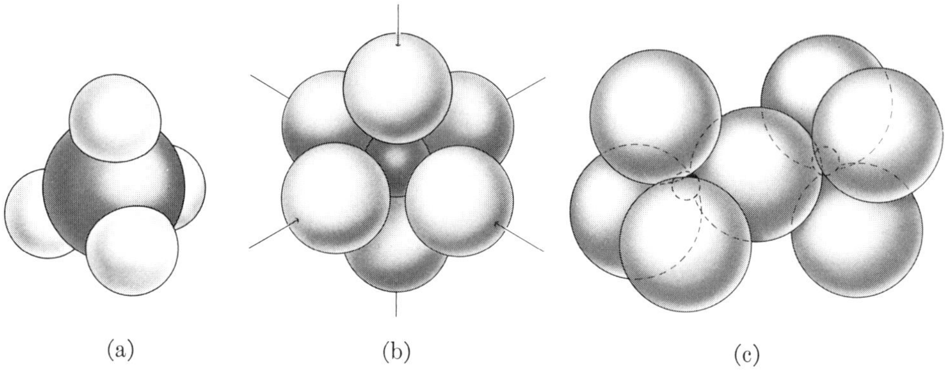

FIG. 3-1. Atomic coordination: (a) Methane. Although there are vibratory movements, each carbon atom is associated (coordinated) with four neighboring hydrogen atoms in the manner shown. (b) MgO. Each magnesium ion is surrounded by six adjacent oxygen ions. Likewise, each oxygen ion is coordinated with six first-neighbor magnesium ions. (c) SiO_2. Each silicon atom is coordinated with four oxygen atoms. However, each oxygen atom is between only two silicon atoms.

atoms, the coordination number in covalently bonded crystals is determined by the relationship of Eq. (2-11). Silicon carbide (β) provides the best example of a covalent ceramic crystal; each atom (silicon and carbon) is coordinated with four neighboring atoms (Fig. 3-2). Each silicon atom has four carbon neighbors, and each carbon atom has four silicon neighbors.

Because ionic bonds are not directional and energy is given off when unlike ions approach each other (Fig. 2-7b), an atom with a high CN is generally more stable than an atom with a low one. However, this statement must be qualified;

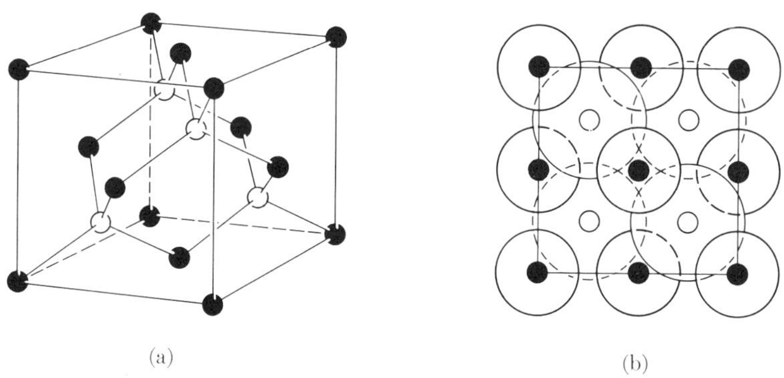

FIG. 3-2. Coordination in β-SiC. Each silicon atom is surrounded by four carbon atoms, and each carbon atom is coordinated with four silicon atoms. The coordination number in a covalent solid depends on the number of shared electron pairs. The two different presentations shown here are equivalent.

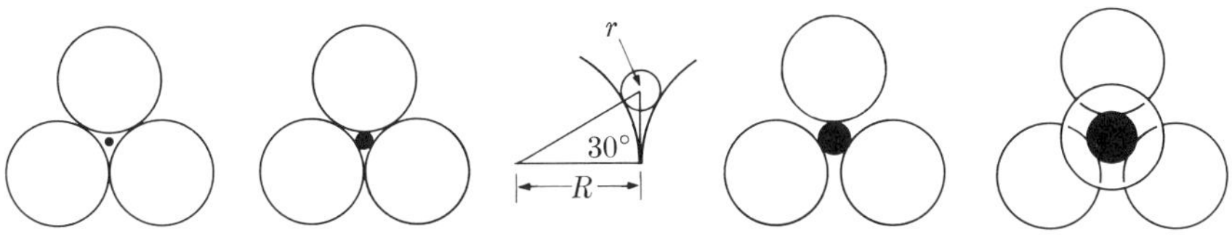

FIG. 3-3. Coordination number versus atomic size ratio. The CN cannot equal 3 when the ratio of r/R is less than 0.155, because the repulsion between like ions is greater than the attraction between unlike ions. If r/R is greater than 0.225, it is possible for four negative ions to surround each positive ion; therefore the CN may be as large as 4.

TABLE 3-1

COORDINATION NUMBER VERSUS ION SIZE

r/R*	Maximum CN
<0.155	2
0.155–0.225	3
0.225–0.414	4
0.414–0.732	6
0.732–1.00	8
1.00	12

* r = radius of small ion; R = radius of large ion.

for if too many like ions are clustered around an individual ion of the opposite charge, the surrounding ions begin to interfere with one another. This leads to the ion-size ratio which may be used to predict the maximum possible coordination in ionic crystals. As shown in Fig. 3–3 and Example 3–1, it is possible to predict a coordination number of 3 when the size ratio lies between 0.155 and 0.225. The range of ratios for other coordination numbers is shown in Table 3–1. Although is is possible to find exceptions to the ranges listed, they are exceedingly rare below the lower limit because of the mutual interference of like ions. Exceptions above the indicated range are somewhat more common, because there is no interference from repulsive forces. Furthermore, there may be other factors, such as partial covalency which favors a specific number of neighbors, although partial covalency does not lead to maximum packing. Zinc sulfide is an example; its bonds are partially covalent and a 4-fold coordination is favored.

EXAMPLE 3–1. Determine the lowest ion-size ratio which will permit a 3-fold coordination (that is, CN = 3).

Answer: Refer to Fig. 3–3(b), where:

$$\begin{aligned} \cos 30^\circ &= R/(R + r), \\ 0.866r &= R - 0.866R, \qquad (3\text{–}1) \\ r/R &= 0.134/0.866 = 0.155. \end{aligned}$$

Large negative ions are never coordinated with as many small positive ions (Table 2–2) as space will allow, primarily because of the requirements of charge balance; for example, each of the large oxygens of Fig. 3–1(c) is coordinated with only two silicons (see also Fig. 3–11), because a Si-to-O atom ratio of one to two must be maintained within SiO_2. However, as predicted, the silicon has a coordination number of 4 because the ion size ratio is 0.3. In fluorite (CaF_2), the Ca^{2+}/F^- size ratio is 0.80, and each Ca^{2+} ion has a CN of 8, as predicted. However, each F^- is coordinated with only four immediately adjacent positive ions because the Ca-to-F atom ratio of one to two must exist to maintain electrical neutrality. This results in a structure which actually contains holes with sufficient space for additional cations (Section 3–4).

Second-neighbor and other longer-range coordinations occurring in those materials which possess repetitive crystalline structures will be discussed in the following sections.

3–3 Crystals (long-range order)

Because it can be shown conclusively that the atoms of crystals are arranged in a repetitive pattern, we use various ways to describe crystal structures. Three-dimensional topology permits the division of space into no more than seven systems (by the use of planes), so that the volume of each individual division is identical.

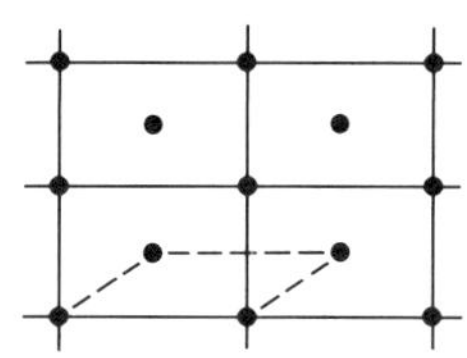

FIG. 3–4. Area-centered rectangle. This two-dimensional lattice could have been drawn as a parallelogram (dashed lines). However, the rectangle is preferred because it better reveals the symmetry.

In all the space systems described in Table 3–2, each position within a given volume or cell has surroundings comparable to an equivalent position in any other cell; thus, if we define one cell, we define them all. In crystals, the most usable repeating volume is called the *unit cell* by crystallographers. The unit cell usually contains a relatively small number of atoms; however, one need not use the minimum number to define the system, because it may be preferable to select cells which have a high degree of symmetry rather than minimum volume. This situation may be illustrated in a two-dimensional analog (Fig. 3–4). If there are equivalent positions at the indicated locations, it is possible to have a

TABLE 3-2

GEOMETRY OF CRYSTAL SYSTEMS
(Compare with Fig. 3-5.)

System	Axes	Axial angles
Cubic	$a_1 = a_2 = a_3$	All angles $= 90°$
Tetragonal	$a_1 = a_2 \neq c$	All angles $= 90°$
Orthorhombic	$a \neq b \neq c$	All angles $= 90°$
Monoclinic	$a \neq b \neq c$	2 angles $= 90°$; 1 angle $\neq 90°$
Triclinic	$a \neq b \neq c$	All angles different; none equal 90°
Hexagonal	$a_1 = a_2 = a_3 \neq c$	Angles $= 90°$ and 120°
Rhombohedral	$a_1 = a_2 = a_3$	All angles equal, but not 90°

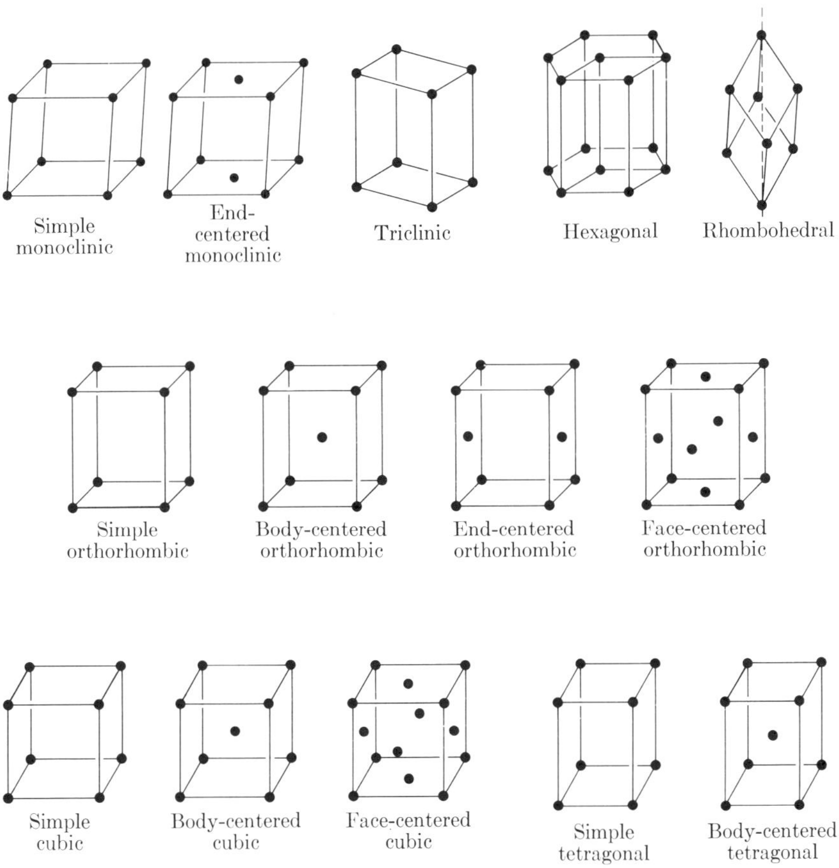

FIG. 3-5. These space lattices continue into three dimensions. Each indicated point has identical surroundings; these points do not necessarily indicate location of atoms. Compare with Table 3-2.

plane group described either by a rectangle or by a parallelogram. In this case, the rectangle can be described as an area-centered rectangle, which is preferable to a parallelogram because it lends greater symmetry to the general structure. Likewise, in three-dimensional atomic arrays we give preference to patterns with greater symmetry, although such a pattern may again involve more than a minimum volume. Cubic lattices may include a simple cubic, a body-centered cubic, or a face-centered cubic repetition (Fig. 3–5). When the other space systems of Table 3–2 are expanded to include modifications of the simple cells, we find that there are 14 space lattices. These are called the *Bravais lattices.*

At this point we need to emphasize that the locations of the Bravais lattice indicate equivalent positions within the crystal structure. For instance, if we use the end-centered orthorhombic lattice as an example, the surroundings at the corner of the unit cell with respect to atom locations, symmetry, etc., are identical with the surroundings for the center of the end of the unit cell. Furthermore, the choice as to which of these two positions serves as the corner of the cell is entirely arbitrary. In the simple orthorhombic structure there is only one equivalent position per cell; both the body-centered orthorhombic and the end-centered orthorhombic lattices have two equivalent positions per unit cell, while the face-centered structures (orthorhombic or cubic) have four equivalent positions per unit-cell volume. A quick examination will indicate why there is no end-centered cubic or end-centered tetragonal structure: the cells of each of these could be redefined to give a simple tetragonal structure.

Example 3–2. Two-dimensional areas can be divided into 5 "area lattices," as contrasted to the 14 space lattices of three dimensions. Cite these 5 "area lattices."

Answer: (a) simple square, (b) simple rectangle, (c) area-centered rectangle, (d) parallelogram, (e) hexagon.

[*Note:* An area-centered square is equivalent to a simple square of smaller dimensions.]

Crystal directions. When we have chosen the unit cell, we may arbitrarily assign one corner as the origin, and let it serve as a reference point for the crystal structure. Both crystal directions and crystal planes may be related to this origin and to the dimensions of the unit cell. Thus we may define a direction within a unit cell as a ray starting at the origin and passing through another location which is a specified number of unit-cell dimensions away from the origin. Thus a [122] direction involves a ray passing from the origin through a position which is one unit cell dimension along the x-axis, and two unit cell dimensions along the y- and z-axes, as illustrated in Fig. 3–6. (It is shown only as far as the $\frac{1}{2}$, 1, 1 position, but would continue to the 1, 2, 2 position.)

Miller indices. The repetitive pattern of atoms within a crystal produces planes which are also repetitive. It is often desirable to identify individual planes within a crystal, and Miller indices (hkl), based on the reciprocals of the

FIG. 3-6. Crystal direction: The [122] direction passes through the points $\frac{1}{2}$, 1, 1 (and also 1, 2, 2).

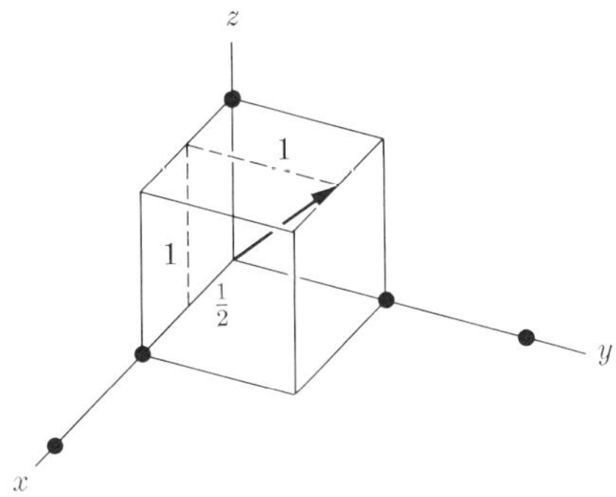

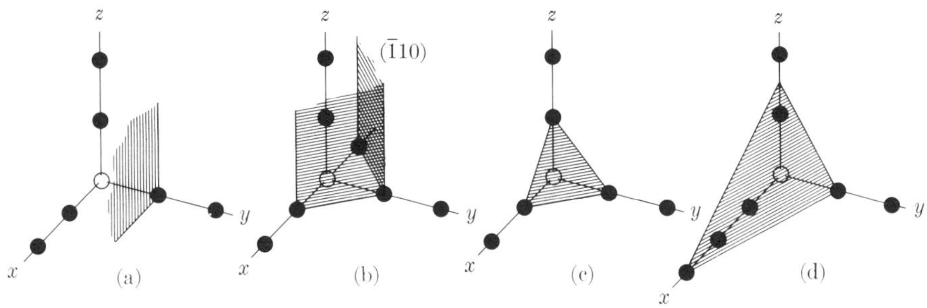

FIG. 3-7. Miller indices (orthorhombic), using reciprocals of the intercepting unit distances: (a) (100), (b) (110) and ($\bar{1}$10), (c) (111), (d) (132).

axial intercepts, are universally used because of their convenience for x-ray calculations. Thus, if a plane cuts the three reference axes at 1, 2, ∞, it is identified by the reciprocals of these intercepts. That is,

$$\frac{1}{1}, \frac{1}{2}, \frac{1}{\infty} = (1\,\tfrac{1}{2}\,0), \tag{3-2a}$$

and clearing fractions, we have

$$(1\,\tfrac{1}{2}\,0) = (210). \tag{3-2b}$$

Figure 3-7 illustrates some of the more commonly encountered planes. Let us note several facts.

(1) The axial distances are expressed as the number of repeating unit-cell distances between the origin and the intercepts of the plane of interest. These units may, or may not, be the same dimensions along the three axes (see Table 3-2).

(2) Parallel planes are given the same notation. Thus, a (113) plane and a (226) plane are equivalently oriented, with the smaller index preferred. (This does not apply to interplanar spacings.)

(3) Planes with negative intercepts are appropriately identified by an overhead bar (Fig. 3-7b).

Symmetrical crystals have comparable directions and planes in which the (*hkl*) notations differ only in the arbitrary choice of the axis label or the axis sign. These related directions and planes are often grouped together for convenience. Thus, in a cubic crystal, which is highly symmetric, the following directions can be grouped together:

$$\left.\begin{matrix}[110], [1\bar{1}0], [\bar{1}10], [\bar{1}\bar{1}0]\\ [101], [10\bar{1}], [\bar{1}01], [\bar{1}0\bar{1}]\\ [011], [01\bar{1}], [0\bar{1}1], [0\bar{1}\bar{1}]\end{matrix}\right\} = \langle 110\rangle.$$

By contrast, since the three crystal axes are not equal in an orthorhombic crystal, only the following can be grouped together:

$$[110], [1\bar{1}0], [\bar{1}10], [\bar{1}\bar{1}0] = \langle 110\rangle.$$

In a similar manner, comparable planes are grouped together with the symbols { }.* Thus there are six planes in the {100} form of a cubic crystal. These are (100), ($\bar{1}$00), (010), (0$\bar{1}$0), (001), and (00$\bar{1}$). The {100} form in a monoclinic crystal, because of a lack of symmetry, involves only (100) and ($\bar{1}$00) planes.

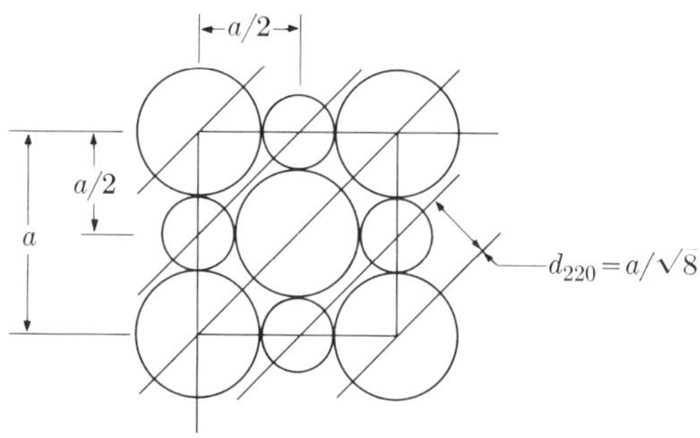

FIG. 3–8. An example of interplanar spacing: d_{220} of a face-centered cubic crystal. The third dimension (not shown) is vertical.

Interplanar spacings. The spacings d_{hkl} between adjacent crystalline planes may be calculated on the basis of geometry, as summarized in Table 3–3. The spacing d_{hkl} is the perpendicular distance of separation between two planes that cross the three axes at points which are separated by unit distances of $1/h$, $1/k$, and $1/l$. Thus, d_{220} for a face-centered cubic crystal is equal to $a/\sqrt{8}$, and adjacent planes cross the three axes at linear separations of $a/2$, $a/2$, and ∞ (Fig. 3–8).

* Since {*hkl*} has been the standard symbol for related (*hkl*) planes, that symbol will be used in this book. However, some crystallographers prefer to use ((*hkl*)) in place of {*hkl*}, because the double parentheses indicate multiple related (*hkl*) planes. Likewise, certain preferences can be cited for using [[ ]] for ⟨ ⟩ to indicate related crystal directions. See M. J. Buerger, *Elementary Crystallography*, New York: John Wiley & Sons (1956), p. 114; and L. Azároff, *Introduction to Solids*, New York: McGraw-Hill (1960), p. 15.

TABLE 3-3
INTERPLANAR SPACINGS

System	Spacings d_{hkl}
Cubic	$\left[\dfrac{h^2+k^2+l^2}{a^2}\right]^{-1/2}$
Tetragonal	$\left[\dfrac{h^2+k^2}{a^2}+\dfrac{l^2}{c^2}\right]^{-1/2}$
Orthorhombic	$\left[\dfrac{h^2}{a^2}+\dfrac{k^2}{b^2}+\dfrac{l^2}{c^2}\right]^{-1/2}$
Hexagonal	$\left[\dfrac{4(h^2+k^2+kh)}{3a^2}+\dfrac{l^2}{c^2}\right]^{-1/2}$
	or $\left[\dfrac{2(h^2+k^2+i^2)}{3a^2}+\dfrac{l^2}{c^2}\right]^{-1/2}$

3-4 Ceramic crystals

Even though ceramic materials are usually compounds, it is advantageous initially to examine structures of metal crystals with a single type of atom, because a knowledge of these will serve in a later analysis of the simpler ceramic crystals.

In single-component metal crystals, if the radii of atoms are considered to be one-half their interatomic distances, atomic packing factors (APF) may be calculated. These factors are 52, 68, 74, and 74%, respectively, for the simple cubic, body-centered cubic, face-centered cubic, and hexagonal close-packed structures. If other things are equal and if the bonding requires no special neighbor selection, larger packing factors provide greater stability because the average interatomic distance is reduced to a value more nearly approaching the separation for the energy trough (Fig. 2-7b). We do not find simple cubic structures among metals because of the low packing factor; however, we do find many metal crystals which possess face-centered cubic or hexagonal close-packed structures. Although the body-centered cubic structure does not have the maximum atomic packing, we find a number of metals with this structure, due to the fact that many metal-bonded crystals exhibit a preference for eight neighbors.

Differences between face-centered cubic and hexagonal close-packed structures are less marked than might be initially surmised. Not only are the packing factors for the atoms identical (74%), but the atomic arrangement on the {111} planes in the face-centered cubic is identical to the atomic arrangement on the (0001) planes of hexagonal close-packed structures. But when these planes are stacked one over another, a difference becomes apparent. In the hexagonal

TABLE 3-4

PACKING IN SELECTED A_mX_n COMPOUNDS

Structure	Anion arrangement	Cation interstitial location	Locations filled	Other examples
CsCl	sc	8-fold	All	
NaCl	fcc	6-fold	All	MgO, MnS, LiF
NiAs	hcp	6-fold	All	FeS
Zincblende (ZnS)	fcc	4-fold	$\frac{1}{2}$	β-SiC, CdS, AlP
Wurtzite (ZnS)	hcp	4-fold	$\frac{1}{2}$	BeO, ZnO, AlN
Corundum (Al_2O_3)	hcp	6-fold	$\frac{2}{3}$	Cr_2O_3, Fe_2O_3, $MgTiO_3$
γ-Al_2O_3	fcc	6-fold	$\frac{2}{3}$	γ-Fe_2O_3
Fluorite (CaF_2)	sc	8-fold	$\frac{1}{2}$	(See below)
Structure	**Cation arrangement**	**Anion location**	**Locations filled**	**Other examples**
Fluorite (CaF_2)	fcc	4-fold	All	ZrO_2, UO_2

close-packed structure, the superposition of alternate (0001) planes produces the following stacking pattern:

$$\ldots \text{A B A B A B A B A B} \ldots \tag{3-3}$$

In the face-centered cubic structures, the stacking pattern is

$$\ldots \text{A B C A B C A B C} \ldots, \tag{3-4}$$

so that every third (111) plane has the identical position. Thus the differences between face-centered cubic and hexagonal close-packed structures are restricted to second-neighbor (and more distant) atoms within the repetitive long-range pattern.

Interstitial positions. Each of the above metal structures, including our hypothetical simple cubic structure, possesses interstitial holes among the atoms. The largest interstitial hole in the simple cubic structure has eight neighboring atoms, and a diameter identical to the minimum value given in Table 3-1 for 8-fold coordination. The largest hole within a face-centered cubic or hexagonal close-packed structure has six neighboring atoms, and a diameter equal to the minimum ratio for 6-fold coordination (Table 3-1). Also within a face-centered cubic and hexagonal close-packed structure, there are 4-fold holes, the radius of which is equivalent to the minimum ratio for 4-fold coordination. In a body-centered cubic structure, the largest interstices are not regular in shape; thus, although the packing factor is slightly lower than in a face-centered cubic structure, there is less opportunity for the inclusion of small atoms.

AX compounds. The simpler ceramic structures may be viewed as combinations of the simple cubic, face-centered cubic, and hexagonal close-packed structures and their interstitial positions (Table 3–4). Generally the anion will take the reference position, and the cation, which is smaller, will take the interstitial position. Figure 3–9 illustrates several examples. Cesium chloride, with a large ion size ratio (0.9), develops an 8-fold coordination for each cesium ion, so that the chlorine has a simple cubic pattern, and the cesium is located in the large interstitial position at the center of the unit cell. In periclase (MgO), the ion-size ratio is only 0.6; hence there can be no more than six oxygen ions around each magnesium ion. The pattern which results is a face-centered cubic structure of oxygen ions, with a magnesium ion in each 6-fold interstice [*cf.* NaCl in Fig. 3–9(b)]. This 6-fold arrangement is also shown in Fig. 3–1(b). Because the resulting pattern is octahedral, i.e., eight-sided, ions with this 6-fold coordination are sometimes considered to be in an octahedral site. The structures of Fig. 3–9 (c) and (d) show cations in the 4-fold interstices of the face-centered cubic and hexagonal close-packed structures. Such crystal structures can occur with AX compounds which have a small ion-size ratio (e.g., BeO), and also with compounds such as ZnS, where the partial covalency of the bonding requires four neighbors for each ion present. In the latter case, the structure is expanded, and its stability, usually due to a high packing factor, is now caused by stronger covalent bonds.

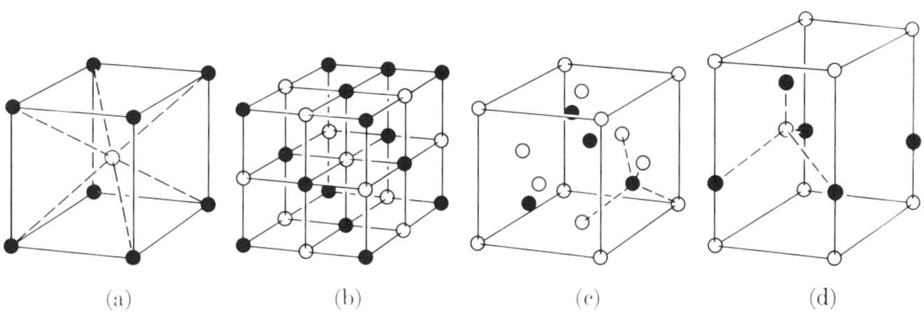

Fig. 3–9. AX-type structures: (a) CsCl has a simple-cubic coordination unit; (b) NaCl has an octahedral unit with six nearest neighbors (cf. Fig. 3–16); (c) and (d) are sphalerite and wurtzite, respectively. Both are ZnS, and are tetrahedrally coordinated with four nearest neighbors.

In the simple cubic structure, there is one 8-fold interstice within each unit cell; therefore, when every interstice is filled with a cation among anions, we have an AX-type compound. Likewise, in a face-centered cubic structure, an AX-compound fills every 6-fold interstice, because there are four equivalent anion positions and four 6-fold sites within each unit cell. The simplified hexagonal close-packed unit cell has two equivalent anion sites and two 6-fold interstices. Pyrrhotite (FeS) and NiAs are typical of this structure. In the former, the sulfur ions maintain a hexagonal close-packed array, and the iron cations are in the 6-fold interstices.

In face-centered cubic and hexagonal close-packed structures, there are twice as many 4-fold interstices as there are equivalent sites per unit cell. Thus the structures of AX type compounds shown in Fig. 3–9 (c) and (d) have only half of the 4-fold sites filled.

EXAMPLE 3–3. Calculate the atomic packing factor for CsCl. [*Note:* When the CN = 8, the radii are increased about 3% over those values of Table 2–2.]

Answer: From Fig. 3–9(a), the lattice constant a_0 for the unit cell is

$$a_0 = (1.03)(2)(1.65 + 1.81)/\sqrt{3} = 4.1 \text{ A}.$$

There is one Cs^+ and one Cl^- ion per unit cell. Therefore, since

$$\text{APF} = \text{volume atoms/volume unit cell}, \tag{3–5}$$

$$\text{APF} = [4\pi/3][(1.65 \times 1.03)^3 + (1.81 \times 1.03)^3]/(4.1)^3 = 0.68.$$

A_mX_n-type crystals. Perhaps the simplest AX_2 compound is fluorite (CaF_2), which may be considered to have a simple cubic structure of anions with calcium cations in the 8-fold interstices (Fig. 3–10b). However, let us note that the

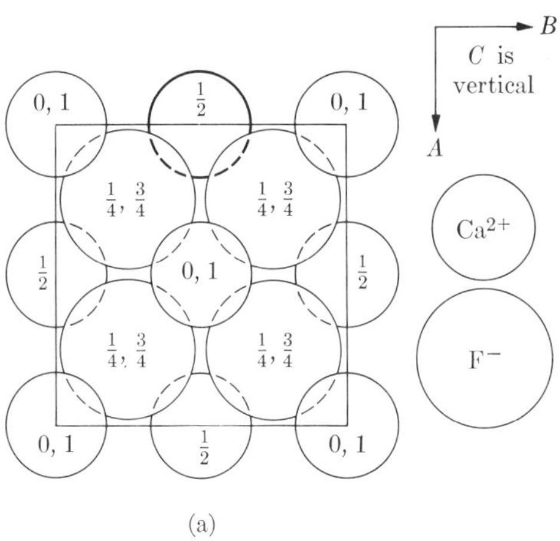

FIG. 3–10. Unit cell of fluorite (CaF_2). These two presentations are equal, but that shown at the left is more commonly used. The fractions indicate the vertical positioning. As an illustration, the heavily outlined Ca^{2+} in the left figure is at an a, b, c position of 0, $\frac{1}{2}$, $\frac{1}{2}$. The a, b positions are indicated in the two dimensions of the paper. The c position is indicated by the inserted number. In the right-hand figure, there are superpositioned F^- ions shown at $\frac{1}{2}$, $\frac{1}{2}$, 0 and at $\frac{1}{2}$, $\frac{1}{2}$, $\frac{1}{2}$. (The $\frac{1}{2}$, $\frac{1}{2}$, 1 position is identical with $\frac{1}{2}$, $\frac{1}{2}$, 0.) The Ca^{2+} ions cause the F^- ions to be separated very slightly (~0.1 A).

calcium-to-fluorine ion ratio must be one to two. Therefore, since there are as many 8-fold interstices as there are equivalent positions in a simple cubic structure, the divalent calcium ions can fill only one-half of the 8-fold positions. [Although it is easy to consider the fluorite structure as simple cubic with only half of the 8-fold interstices filled, the structure is not truly simple cubic because the regular pattern of vacancies makes two types of F^- cubes. An alternative description should be used; namely, that the calcium ions possess a face-centered cubic structure with fluorine ions at each of the 4-fold sites (Fig. 3–10a).]

Alumina, Al_2O_3, and several other compounds have ion-size ratios which favor 6-fold coordination. This is achieved in crystal structures by either a face-centered cubic or hexagonal close-packed anion pattern in which cations are found at only two-thirds of the 6-fold interstices (a consequence of the ion charges). In general, the hexagonal close-packed modification is slightly more stable because the minimum cation-to-cation distance is slightly greater.

EXAMPLE 3–4. Indicate the stacking arrangement of O^{2-} layers and Al^{3+} layers in corundum (Al_2O_3). By themselves, the oxygens form a hexagonal close-packed pattern, and the aluminum ions are located in two-thirds of the 6-fold interstices. [*Note:* The vacancies are so ordered that they have a *DEFDEF* ... pattern. This maintains the maximum average separation of positive ions.]

Answer: The above requirements produce the following sequence for the (0001) stacking:

$$\ldots O_A \quad O_B \quad O_A \quad O_B \quad O_A \quad O_B \quad \ldots$$
$$Al_D \quad Al_E \quad Al_F \quad Al_D \quad Al_E \quad Al_F$$

A complete repetition occurs at every sixth oxygen layer. Therefore the unit cell has a c_0 which includes all of the above layers.

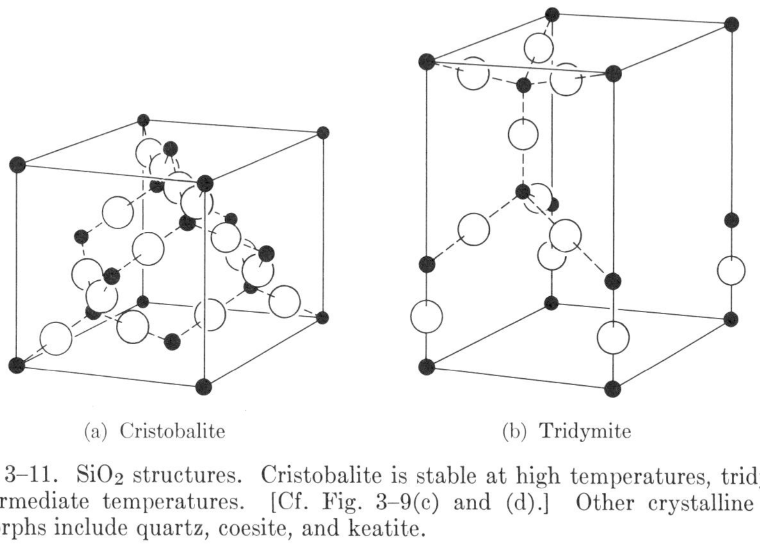

(a) Cristobalite (b) Tridymite

FIG. 3–11. SiO_2 structures. Cristobalite is stable at high temperatures, tridymite at intermediate temperatures. [Cf. Fig. 3–9(c) and (d).] Other crystalline silica polymorphs include quartz, coesite, and keatite.

It is suggested that the reader make a sketch so that he will see the relative locations of ions in successive layers.

In ceramic technology, the most common AX_2-type crystals are the SiO_2 compounds. Among these, cristobalite (Fig. 3–11a) may be compared directly with sphalerite (Fig. 3–9c), in which each zinc and sulfur atom is replaced by a silicon atom, and the oxygen atoms are located between silicon atoms. Cristobalite may also be compared with diamond or with metallic silicon structures whose atoms are located at positions similar to those occupied by silicon atoms in cristobalite; although here the oxygen atoms of cristobalite are absent. Tridymite (Fig. 3–11b) may similarly be compared with wurtzite (Fig. 3–9d). Further discussion of the various silica polymorphs will be given in Section 3–6; however, let us note now that each of these polymorphs possesses a small silicon ion surrounded by four larger oxygen atoms, and each oxygen atom is located between two silicon atoms. In effect, this produces tetrahedra with center silicons and apex oxygens.

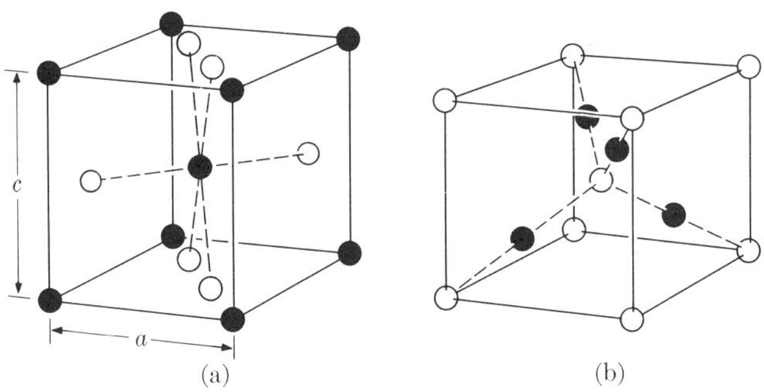

FIG. 3–12. AX_2-type structures: (a) rutile (TiO_2); (b) cuprite (Cu_2O). Also see Fig. 3–10.

Other AX_2 structures of importance in ceramics are TiO_2 and Cu_2O (Fig. 3–12), which are generally more complicated than those structures described previously. However, a structure such as TiO_2 (rutile, shown in Fig. 3–12a) may be analyzed on the basis of interstitial positions. Here the titanium-to-oxygen size ratio is such that a 6-fold coordination is preferred; but the charge ratio requires two oxygens for each titanium. If the denser face-centered cubic or hexagonal close-packed structures of oxygen anions are developed and if only half of the 6-fold interstices are filled, the crystal is distorted and loses its cubic or hexagonal symmetry. Thus one of the less symmetrical Bravais lattices is developed (Fig. 3–5).

AB_mX_n-type compounds. Of rapidly growing importance in ceramic technology are the oxides, sulfides, etc., which have more than one type of cation. Some, such as the spinels ($R^{2+}R_2^{3+}O_4$) and the silicates, have been used many

years but are finding new applications. Others, typified by $BaTiO_3$ (barium titanate) have come to the forefront only since World War II.

For a double oxide, barium titanate possesses a relatively simple structure. Above 120°C (248°F), the unit cell is cubic, with each Ti^{4+} ion octahedrally coordinated among six neighboring oxygen ions (Fig. 3–13). The space lattice is simple cubic, since there is only one equivalent position per unit cell. At room temperature this cube becomes slightly distorted into a tetragonal structure, a fact which has major ferroelectric consequences. More attention will be given to $BaTiO_3$ in Chapter 10.

Fig. 3–13. $BaTiO_3$. This is the structure of perovskite ($CaTiO_3$). At room temperature, the c/a ratio of the tetragonal unit cell is 1.04, and the Ti^{4+} ion has a choice of two locations. (Cf. Fig. 10–17.) Above 120°C, the c/a ratio drops to 1.00 and the structure is cubic.

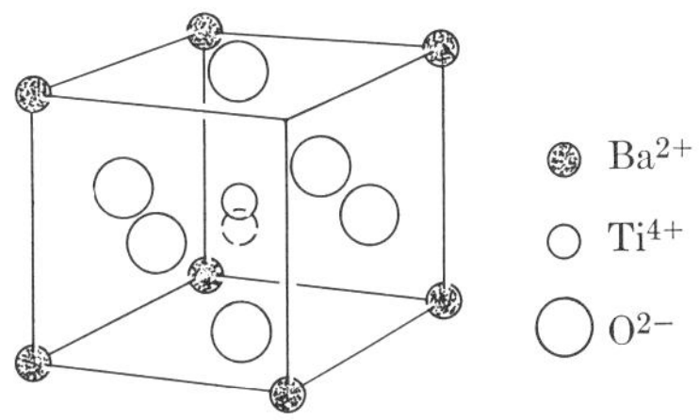

Example 3–5. Assume radii of 1.43 and 1.32 A respectively for Ba^{2+} and O^{2-}. If these ions are located at corner and face-centered positions of a cubic unit cell, what is the radius of the body-centered interstice into which the Ti^{4+} of $BaTiO_3$ must be placed (Fig. 3–13)?

Answer:

$$a_0 = 2(1.43 + 1.32)/\sqrt{2} = 3.89 \text{ A};$$
$$a_0 = 2(1.32 + r_\square),$$
$$r_\square = (3.89/2) - 1.32 = 0.625 \text{ A}.$$

[*Note:* This radius is slightly smaller than that of the Ti^{4+} ion (0.64 A); this causes a distortion within the $BaTiO_3$ at lower temperatures. See above and Section 3–6.]

Calcite ($CaCO_3$) represents a compound structure in which only two types of ions are present, Ca^{2+} and CO_3^{2-}. The four atoms in CO_3^{2-} are so strongly as-

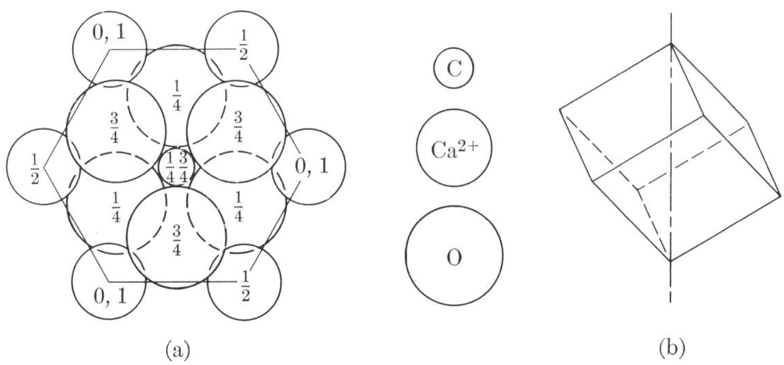

Fig. 3–14. Calcite ($CaCO_3$): (a) atom positions; (b) cleavage planes.

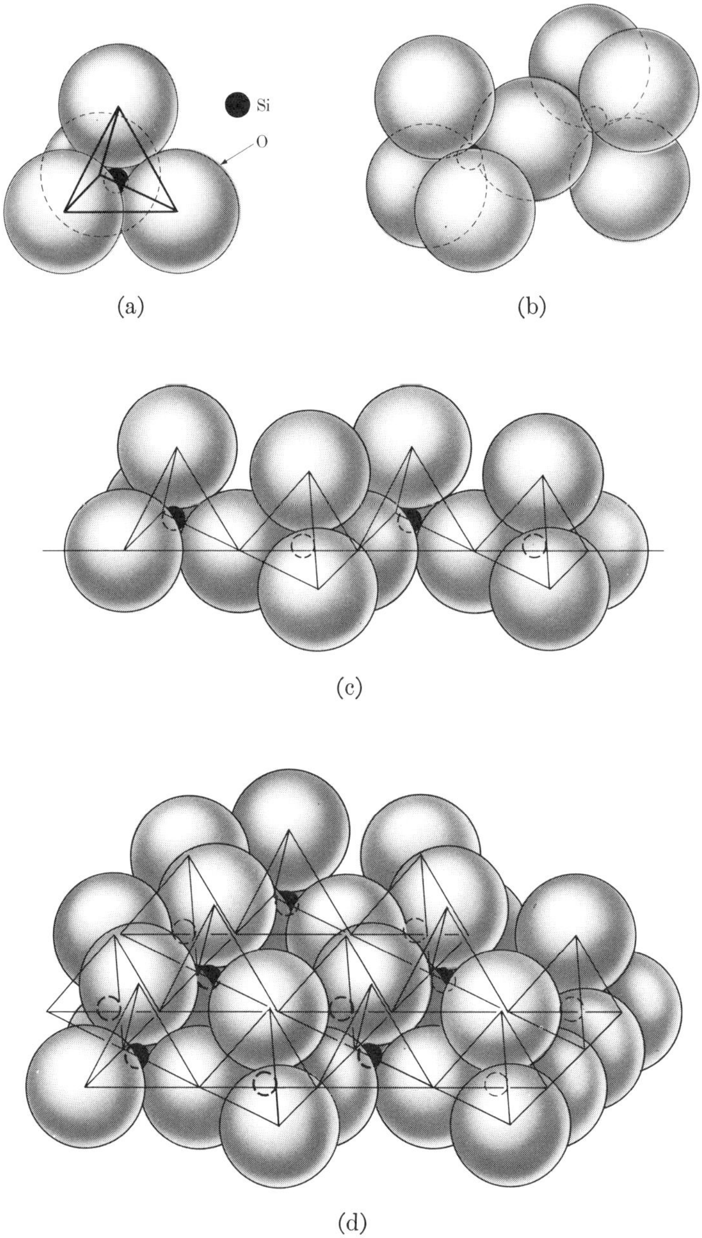

Fig. 3-15. Silicate polymerization: (a) SiO_4^{4-} ion; (b) $Si_2O_7^{6-}$ ion; (c) $(SiO_3^{2-})_x$ chain; (d) $(Si_2O_5^{2-})_y$ sheet. See Fig. 3-11 for framework polymerization. The silicon atoms outline hexagons.

sociated that they act as a unit, and the resulting calcite crystal is very similar to the FeS structure previously described, in that the Ca^{2+} ions possess a hexagonal close-packed pattern, with the CO_3^{2-} ions in the 6-fold positions or interstices. The situation is slightly more complicated than in FeS because the CO_3^{2-} ion is not spherical but planar. In reality, the superpositioning of these ions within the structure requires that each succeeding ion be rotated 60° from the previous one (Fig. 3–14). Thus the structure is rhombohedral, rather than hexagonal close-packed. This increases the minimum distance between the like-charged oxygen ions, and accounts for the typical rhombohedral cleavage within large calcite crystals. The basic crystal structure of dolomite [$CaMg(CO_3)_2$] is very similar to that of calcite, except for a different ordering in the arrangement of the calcium and magnesium cations.

If we are to understand the family of magnetic ceramics, we must also consider the spinel structure (Section 10–6). The prototype of this structure is the mineral spinel ($MgAl_2O_4$), which is typical of a broad range of compositions in which the magnesium may be replaced by other divalent ions of similar size, such as Fe^{2+}, Ni^{2+}, Zn^{2+}, and Mn^{2+}. Likewise, there can be extensive isomorphous replacement of the aluminum by trivalent ions of comparable size, such as Fe^{3+}, Cr^{3+}, and occasionally Mn^{3+}. In addition to the above familiar ions, many of the lanthanide elements (rare earths) may be incorporated into the spinels at positions and in amounts consistent with their sizes and valences. In brief, the spinel structure contains a basic pattern of face-centered cubic oxygen ions. The O^{2-} ions are sufficiently limited in their approach to one another so that space is left in the interstices for the cations. Various distributions of the divalent and trivalent cations among the 4-fold and 6-fold interstitial sites become significant when we consider those details of the spinel structure which control the magnetic properties. (This aspect will be discussed later, in Section 10–6.)

Silicate structures. Although the silicates present a variety of structures, they do have some basic features which can be correlated. The silicons are surrounded tetrahedrally with four oxygens. A single SiO_4^{4-} tetrahedron acts as an ion in those compounds which have a sufficient number of metal atoms to supply the necessary electrons. One such example is forsterite (Mg_2SiO_4) in which there are two Mg^{2+} ions for each SiO_4^{4-} ion. With fewer metal atoms to supply electrons, polymerization occurs to produce repeating tetrahedral units. Figure 3–15(b) shows two tetrahedra sharing one oxygen at their apexes to produce an $Si_2O_7^{6-}$ ion. Akermanite ($Ca_2MgSi_2O_7$) is the prototype for this structure.

An extension of the polymerization into a chain (Fig. 3–15c) produces the metasilicates, such as enstatite ($MgSiO_3$). There is no anion in the usual sense. However, there are two negative charges along the chain for each SiO_3^{2-} (mer). Cations of metal atoms are present in the classical sense, with coulombic attraction to all adjacent negative charges.

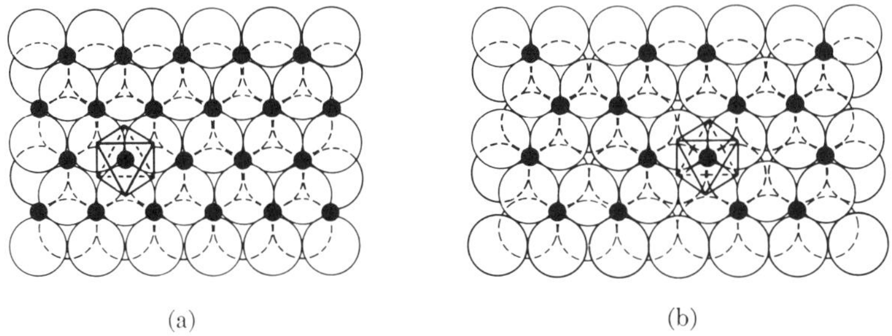

Fig. 3-16. Octahedrally coordinated sheets: (a) brucite, $Mg_3(OH)_6$, trioctahedral; (b) gibbsite, $Al_2(OH)_6$, dioctahedral. All cations have a 6-fold coordination.

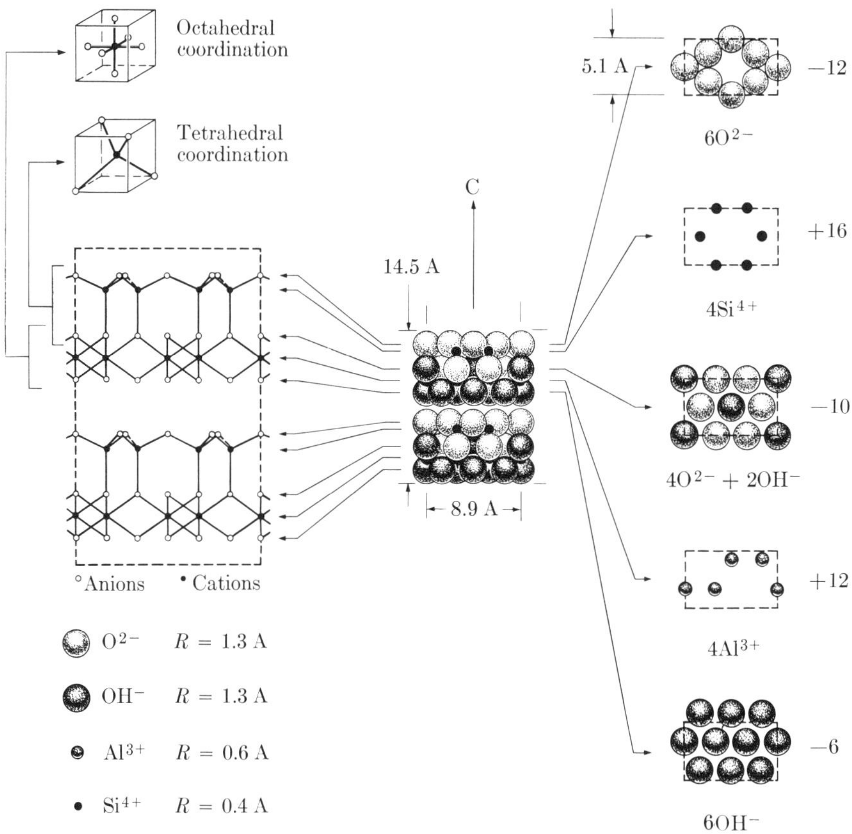

Fig. 3-17. Kaolinite, $Al_2Si_2O_5(OH)_4$. The unsatisfied oxygens of the tetrahedrally coordinated silicate layer are also octahedrally coordinated in the gibbsite layer. (F. H. Norton, *Elements of Ceramics*. Reading, Mass.: Addison-Wesley, 1952, page 9.)

Sheet structures may be obtained by polymerization of the tetrahedral units into two dimensions (Fig. 3–15d). Similarly, framework structures result from polymerization into three dimensions. When three-dimensional polymerization is allowed to continue until it produces a crystalline array, we find silica phases, such as those shown in Fig. 3–11. If the polymerization produces a noncrystalline framework structure, a glass results.

Sheet or layered structures are not limited to silicates. Brucite [$Mg(OH)_2$] has a two-dimensional structure with an octahedral (6-fold) coordination (Fig. 3–16). Because the ratio of ionic radii in gibbsite [$Al(OH)_3$] also requires 6-fold (or octahedral) coordination units, its structure is similar to that of $Mg(OH)_2$, with one exception: in gibbsite, only two out of the three possible cation positions are filled. This is another consequence of the charge-balance limitations. Octahedral structures containing R^{3+} ions are called *di*octahedral because only two of three positions are filled. In *tri*octahedral layers, all three positions are filled by R^{2+} ions.

The above octahedrally coordinated layered structures are self-contained, and are independent of primary bonds with adjacent layers (although van der Waals bonds do exist). In contrast, the tetrahedrally coordinated layered silicates have unsatisfied primary-bond requirements on one surface of the sheet, which means that these layered silicates must be associated with other structural units. The simplest relationship is between the structures shown in Figs. 3–15(d) and 3–16(b), in which the incompletely bonded oxygens of the tetrahedral silicate layer replace some of the hydroxyl ions on one side of the octahedral layer (Fig. 3–17). The resulting structure is that of kaolinite, the most common of several clay minerals, which has a combined octahedral-tetrahedral layer requiring no external primary bonds on its two plane surfaces. In Chapter 8, we shall relate this structure to the plastic characteristics of the clay, particularly wet clay.

Composite layered structures exhibit a number of possible variations: (1) The octahedral layer may contain either divalent or trivalent ions, such as Mg^{2+} versus Al^{3+}. (2) A triple structure of tetra-octa-tetra may occur, in talc and in other phases. (3) Some of the Si^{4+} ions in the tetrahedral sites may be replaced by an Al^{3+} ion and an alkali ion (to balance the charge) which lie at the outside surface of the structure; these structures are the micas. Table 3–5 summarizes additional possibilities.

3–5 Imperfections

Although long-range order is characteristic of all crystals, we have observed that crystals do not always exhibit a perfect array of atoms. Imperfections that exist merit attention beyond their relative numbers, because they have profound influences on the properties of crystalline materials. We may divide these imperfections into three main groups: point, linear, and surface defects.

TABLE 3–5

LAYERED SILICATES AND RELATED STRUCTURES

Structure	Composition	Sequence	Coordination units
Brucite	$Mg(OH)_2$	3 OH^-	
		3 Mg^{2+}	Trioctahedral
		3 OH^-	
Gibbsite	$Al(OH)_3$	3 OH^-	
		2 Al^{3+}	Dioctahedral
		3 OH^-	
Kaolinite	$[Al_2Si_2O_5(OH)_4]_2$	3 OH^-	
		2 Al^{3+}	Dioctahedral
		2 O^{2-}, 1 OH^-	
		2 Si^{4+}	Tetrahedral
		3 O^{2-}	
Serpentine	$[Mg_3Si_2O_5(OH)_4]_2$	3 OH^-	
		3 Mg^{2+}	Trioctahedral
		2 O^{2-}, 1 OH^-	
		2 Si^{4+}	Tetrahedral
		3 O^{2-}	
Pyrophyllite	$[Al_2Si_4O_{10}(OH)_2]_2$	3 O^{2-}	
	$n = 2\ Al^{3+}$	2 Si^{4+}	Tetrahedral
		2 O^{2-}, 1 OH^-	
		-n-	Di- or trioctahedral
		2 O^{2-}, 1 OH^-	
Talc	$[Mg_3Si_4O_{10}(OH)_2]_2$	2 Si^{4+}	Tetrahedral
	$n = 3\ Mg^{2+}$	3 O^{2-}	

Micas	Muscovite	$[KAl_3Si_3O_{10}(OH)_2]_2$	0.5 K^+	CN = 12
		m = 2 Al^{3+}	3 O^{2-}	
			1.5 Si^{4+}, 0.5 Al^{3+}	—Tetrahedral
	Phlogopite	$[KAlMg_3Si_3O_{10}(OH)_2]_2$	2 O^{2-}, 1 OH^-	
		m = 3 Mg^{2+}	-m-	—Di- or trioctahedral
			2 O^{2-}, 1 OH^-	
	Biotite	$[KAl(Mg, Fe)_3Si_3O_{10}(OH)_2]_2$	1.5 Si^{4+}, 0.5 Al^{3+}	—Tetrahedral
		m = 3 $(Mg, Fe)^{2+}$	3 O^{2-}	
	Other micas—other substitutions		0.5 K^+	CN = 12
Montmorillonite		$[R^+_{.33}Al_{1.67}Mg_{.33}Si_4O_{10}(OH)_2]_2$	0.17 R^+ or 0.08 R^{2+}	Exchangeable ions
			3 O^{2-}	
			2 Si^{4+}	—Tetrahedral
			2 O^{2-}, 1 OH^-	
			1.67 Al^{3+}, 0.33 Mg^{2+}	—Dioctahedral
			2 O^{2-}, 1 OH^-	
			2 Si^{4+}	—Tetrahedral
			3 O^{2-}	
			0.17 R^+ or 0.08 R^{2+}	Exchangeable ions
Illite		Combination of substitutions found in the micas and montmorillonite above		
Chlorite			3 OH^-	
		Cf. Brucite	3 Al^{3+}	—*Tri*octahedral (Excess positive)
			3 OH^-	
	Alternate layers		3 O^{2-}	
			2 (Al^{3+}, Si^{4+})	—*Tetra*hedral (Excess negative)
			2 O^{2-}, 1 OH^-	
		Cf. Mica	2 Al^{3+}, (or 3 Mg^{2+})	—Di- or trioctahedral
			2 O^{2-}, 1 OH^-	
			2 (Al^{3+}, Si^{4+})	—Tetrahedral (Excess negative)
			3 O^{2-}	

Point defects. These involve the presence of excess atoms, or the absence of atoms that normally should be present, within the crystal structure. The excess atoms, commonly called *interstitialcies*, are illustrated by Fig. 3–18(a). It should be noted that a point defect involves a randomly placed defect, rather than a repetitious pattern of vacant sites or interstitial sites, such as occurs when calcium ions take alternate interstitial sites in fluorite (Section 3–4).

FIG. 3–18. Examples of point defects: (a) interstitialcy; (b) vacancy (Schottky defect); (c) displacement (Frenkel defect). In each case ceramic crystals require electrical neutrality.

A vacancy is known as a *Schottky defect* (Fig. 3–18b). It may be surmised, correctly, that such defects greatly increase the diffusion rates of atoms within crystalline solids. Either interstitialcy or vacancy defects are more common in compounds whose composition is not strictly stoichiometric. They are thus favored by the presence of cations which maintain multiple valence states.

A pair of defects involving an interstitialcy and a vacancy is sometimes categorized separately as a *Frenkel defect* (Fig. 3–18c), and is considered as simply a displacement of atoms or ions. Therefore, over-all electrical balance is maintained within the crystal. Defects such as these are often responsible for the process of self-diffusion in stoichiometric crystals.

Linear defects (dislocations). The most common type of one-dimensional defect within a crystal structure is a dislocation which is a line imperfection (Fig. 3–19). An *edge dislocation* is most readily described as the edge of an extra plane of atoms inserted within the crystal structure; thus a displacement vector (Burgers) is introduced perpendicular to the defect. A second type of linear defect is a *screw dislocation*, in which the displacement is parallel to the defect line. These imperfections are somewhat complex in ceramic crystals, because severe coulombic restrictions impose requirements that do not apply in metals. For example, an edge dislocation is limited to those planes which do not produce neighboring like charges. The effect of dislocations on the mechanical properties of ceramic crystals will be discussed in Chapter 8.

Surface defects (boundaries). A structure may have an imperfection which extends in two dimensions throughout a crystal; a rather common type of this defect is a stacking fault. An example may be cited in the face-centered cubic structures which were described in Section 3–4 as having an

$$\ldots A\ B\ C\ A\ B\ C\ A\ B\ C \ldots \tag{3-4}$$

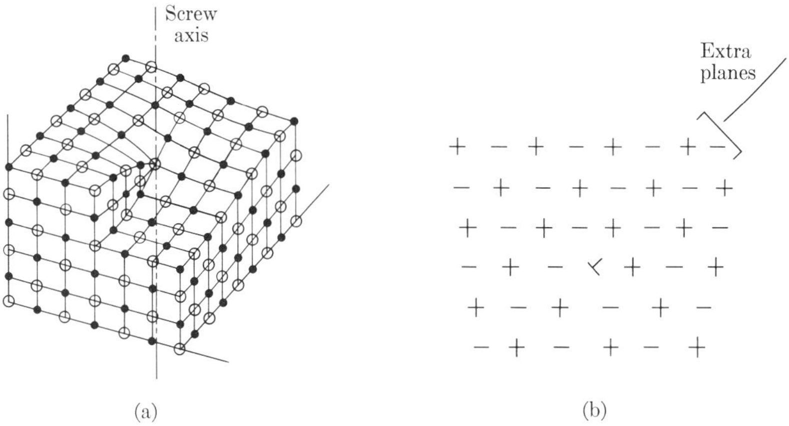

FIG. 3-19. Examples of linear defects (dislocations). (a) Screw dislocation: the crystal "plane" normal to the linear defect forms a helix around the defect. (b) Edge dislocation: the edge is a terminating line of an extra crystal plane, normal to the plane of the page. The restrictions placed on these imperfections are heavier in ceramic crystals than in metal crystals.

repetition of the {111} planes. It is not uncommon to find that the stacking sequence in crystals like cristobalite (SiO_2) gives a repeating pattern such as

$$\ldots \mathrm{A\ B\ C\ A\ B}\ \overset{\downarrow}{\underset{\uparrow}{\,}}\ \mathrm{A\ B\ C\ A\ B\ C} \ldots \qquad (3\text{-}6)$$

in which, in effect, a plane is missing.

A second type of irregularity in the long-range repetitive structure of crystals occurs at a twin boundary where the stacking pattern is reversed, such as

$$\ldots \mathrm{A\ B\ C\ A\ B}\ \overset{\downarrow}{\underset{\uparrow}{\mathrm{C}}}\ \mathrm{B\ A\ C\ B\ A} \ldots \qquad (3\text{-}7)$$

Stacking faults and twin boundaries are generally low-energy defects, since they do not involve changes in first-neighbor coordination. Rather, they produce longer-range, second-neighbor alterations.

A third type of two-dimensional defect found in most solid materials occurs because solids usually contain grains. These grains are separate crystals with different orientations, so that the boundary between the grains is a region of transition or crystalline mismatch. The normal type of grain boundary with a large angle of mismatch may also be classed as an interface, and will be discussed in Chapter 5. However, boundaries with a small angle of "tilt" will be included here because they represent an aligned sequence of edge dislocations (Fig. 3-20). Small-angle grain boundaries are of specific interest in mechanical deformation, in that these boundaries represent the sites at which dislocations become stabilized, and their further movement is restricted.

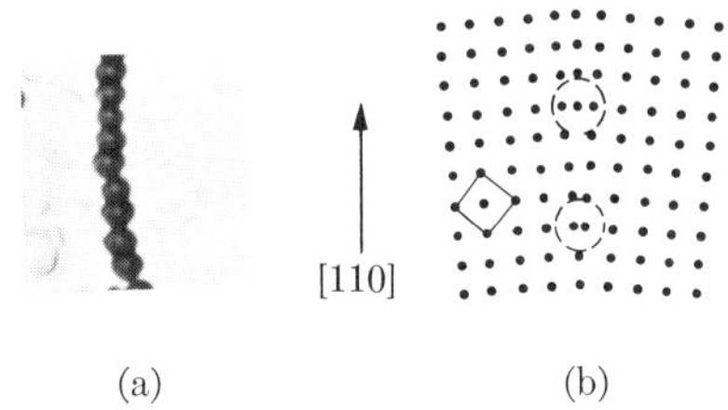

FIG. 3-20. Small-angle boundary: (a) LiF crystal etched to show the ends of edge dislocations (×300); (b) schematic representation of the above. Only the cations (or anions) are shown; exaggerated angle. (Photomicrograph courtesy of A. S. Keh.)

EXAMPLE 3-6. The displacement vector for LiF is in the [110] direction and therefore is 2.97 A. Determine the angle of this small-angle boundary. (Refer to Fig. 3-20.)

Answer: Since Fig. 3-20(a) is enlarged (× 300), the dislocation pits are about 0.0006 cm (60,000 A) apart; hence

$$\sin\theta = 2.97/60{,}000 = 0.00005$$
$$\theta = 10''.$$

Volume defects. A fourth category of lattice imperfection may be indicated for crystals which have every position correctly filled with the right type of ion, but for which there are small variations in interatomic spacings throughout the volume of the crystal. This may occur, for example, as a result of thermal vibrations of the atoms when the crystals are at temperatures above absolute zero. Because of these thermal vibrations, the lattice constants of the unit cells are changed. Furthermore, the resistance to transfer of thermal and electrical energy is increased. These factors will be discussed further in Chapter 9.

Solid solution. Another crystalline imperfection may arise when several types of cations or anions are present to give a solid solution. When a second element dissolves in a solid, the resulting phase is called a solid solution. If the size of the atoms of the solute is comparable to the size of the atoms of the solvent, the solute atom will substitute for one of the matrix atoms in the crystal lattice. This is called a substitutional solid solution. This solid solution is limited by geometric and electronic relationships. The limitation of geometric size in ceramic solid solutions is similar to the size limitation in metals; that is, there can be extensive isomorphous replacement in a solid solution which has a close-packed structure if the radius of the substituting ion is within 15% of that of the replaced ion. Such substitutional solid solutions have already been mentioned for the spinels (Section 3-4), and also exist for many other types of ceramic crystals, such as Fe^{2+} for Mg^{2+} in periclase (MgO), Cr^{3+} for Al^{3+} in corundum (Al_2O_3) to produce ruby, and F^- for OH^- in the synthetic micas. Size limitations are not so rigid in structures which are not close-packed, so that the substitution, for instance, of Ca^{2+} for Mg^{2+} in enstatite ($MgSiO_3$) is possible.

Each of the examples just cited maintains the original electronic balance by substituting ions with identical valences. This electronic requirement places certain limitations on what may be substituted in solid solutions, but does not completely exclude the possibility of replacing ions of certain valences by ions of other valences. For example, a Ca^{2+} Al^{3+} ion pair in anorthite ($CaAl_2Si_2O_8$)

may be replaced by a $Na^{+}Si^{4+}$ pair. Likewise, it is possible to substitute some Li^{+} for Mg^{2+} in periclase (MgO) if there is a simultaneous substitution of Fe^{3+} for Mg^{2+}, *or* of OH^{-} for O^{2-}. In other words, the factor that must be maintained is the charge balance.

Ion exchange represents another type of solid solution in which ions of one type may be replaced by ions of another type, depending on the activity of the various ions in the surrounding environment. This phenomenon will be discussed further in Chapters 5 and 8, but let us note here that such alterations in the crystal structure occur most readily in the open-type crystals which have a low atomic packing factor, and between ions with a low valence.

Defect structures. Charge balances can be maintained in some crystals by defect structures. For example, wüstite (FeO), which has the same structure as periclase (MgO), contains an equilibrium amount of ferric iron (Eq. (3–8)):

$$2Fe^{2+} + \tfrac{1}{2}O_2 \rightleftarrows 2Fe^{3+} + O^{2-}. \qquad (3\text{–}8)$$

A cation vacancy must exist for every two Fe^{3+} ions present (Fig. 3–21). In wüstite, the Fe^{3+}-to-Fe^{2+} ratio may be as high as 0.7, and therefore up to 17% of the cation sites may be vacant. Beyond this ratio, magnetite (Fe_3O_4) is formed; however, it is interesting to note that both magnetite and wüstite have the same face-centered arrangement of oxygen ions. In magnetite, there is an ordering of the cations and interstitial positions to give a spinel (Sections 3–4 and 10–6).

Fe^{2+}	O^{2-}	Fe^{2+}	O^{2-}	Fe^{2+}	O^{2-}	Fe^{3+}
O^{2-}	Fe^{2+}	O^{2-}	Fe^{2+}	O^{2-}	□	O^{2-}
Fe^{2+}	O^{2-}	Fe^{2+}	O^{2-}	Fe^{3+}	O^{2-}	Fe^{2+}
O^{2-}	Fe^{2+}	O^{2-}	Fe^{2+}	O^{2-}	Fe^{2+}	O^{2-}
Fe^{3+}	O^{2-}	Fe^{2+}	O^{2-}	Fe^{2+}	O^{2-}	Fe^{2+}
O^{2-}	□	O^{2-}	Fe^{2+}	O^{2-}	Fe^{2+}	O^{2-}
Fe^{2+}	O^{2-}	Fe^{3+}	O^{2-}	Fe^{2+}	O^{2-}	Fe^{2+}

FIG. 3–21. Defect structures: the iron/oxygen ratio varies with the oxygen pressure through the introduction of cation vacancies. This defect structure permits increased diffusion rates.

EXAMPLE 3–7. Wüstite (FeO) contains 24 w/o oxygen. What are the numbers of vacancies, Fe^{2+} ions, and Fe^{3+} ions per 100 O^{2-} ions?

Answer: Basis, 100 gm = 24 gm O^2 and 76 gm ($Fe^{2+} + Fe^{3+}$);

$$24/16.00 = 1.500 = 100\ O^{2-}\ \text{ions},$$

$$76/55.84 = 1.362 = \underline{91}\ \text{Fe ions}.$$

Therefore 9 vacancies.

Hence there are 9 vacancies, 18 Fe^{3+} and 73 Fe^{2+}; $Fe^{3+}/Fe^{2+} = 0.247$.

3-6 Polymorphism

Alternate structures with identical compositions are variously called isomers, allotropes, and polymorphs. The latter term will be employed here, since it is in more general use among crystallographers. Examples have already been cited of ZnS and SiO_2 (Figs. 3-9 and 3-11). Note that the structures of polymorphs are usually closely related, and that their coordination numbers and first-neighbor atoms are often similar. Differences arise mainly in the second-neighbor and stacking patterns. Therefore, it is not surprising to find that the several structures possible for a polymorphic compound do not vary markedly in stability and physical properties. Nevertheless, if one polymorph is to change to another, such as tridymite to cristobalite (SiO_2), bonds must be broken for a *reconstructive* transformation to occur.

Some compounds permit a more subtle type of structural change, which involves bond lengths or bond angles, but no change in either first or second neighbors. For example, there are oxygens between the silicons in the cristobalite shown in Fig. 3-11(a). At lower temperatures (below 220°C) a *displacive* transformation reduces the Si—O—Si bond angles slightly (Fig. 3-22), which means, of course, that the volume decreases during cooling. Furthermore, there is a reduction of symmetry within the crystal.

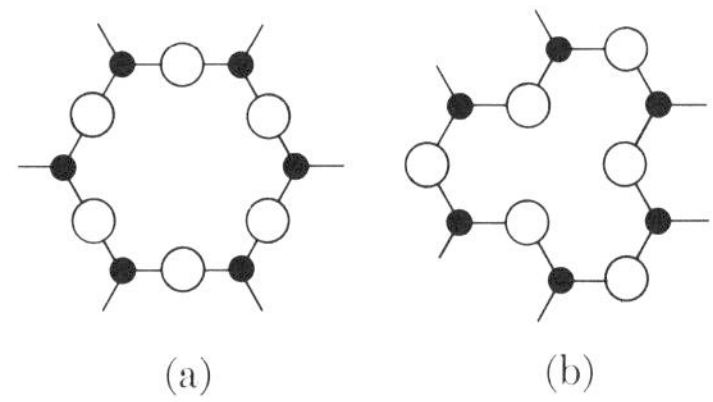

FIG. 3-22. Displacive transformation: (a) high-temperature cristobalite; (b) low-temperature cristobalite (schematic diagram).

A second example of displacive transformation with only a minor coordination change occurs in $BaTiO_3$ (Section 3-4). Above the Curie temperature the structure is cubic, a factor which will be discussed in Chapter 10. Below this temperature, with the reduction in entropy, the cations shift with respect to the oxygen anions, and their atomic packing factor increases.

Two general rules may be formulated concerning polymorphic changes. (1) Lower temperatures usually mean greater order and lower symmetry, as illustrated by comparing cristobalite with tridymite and quartz; also cristobalite at a high temperature versus cristobalite at a low temperature; or cubic $BaTiO_3$ versus tetragonal $BaTiO_3$. (2) Polymorphism occurs more readily when no bonds are broken, as in displacive transformations between high- and low-temperature forms of quartz, cristobalite, and tridymite (Fig. 3-22). In contrast, the cristobalite-to-tridymite-to-quartz reconstructive transformations (Fig. 3-11) are extremely sluggish because they require significant activation energy in order to break bonds and realign the SiO_4 coordination units.

References for Further Reading

3–1. ADDISON, W. E., *Structural Principles in Inorganic Compounds*. New York: John Wiley & Sons, 1961. Recommended for the student who has had a general college chemistry course with only limited emphasis on the structure of compounds.

3–2. AZÁROFF, L. V., *Introduction to Solids*. New York: McGraw-Hill, 1960. An exceptionally good presentation of atomic packing within crystals is presented in Chapters 3 and 4. Imperfections are presented in Chapter 5. Advanced undergraduate and graduate levels.

3–3. BRINDLEY, G. W., editor, *X-ray Identification and Crystal Structures of the Clay Minerals*, London: Mineralogical Society, 1951. The most authentic compilation of material about clay structures.

3–4. CULLITY, B. D., *Elements of X-ray Diffraction*. Reading, Mass.: Addison-Wesley, 1956. General textbook on X-ray diffraction. Introductory level.

3–5. EVANS, R. C., *Crystal Chemistry*. Cambridge, England: The University Press, 1948. Provides a detailed systemization of crystal chemistry. Advanced undergraduate level.

3–6. GRIM, R. E., *Clay Mineralogy*. New York: McGraw-Hill, 1953. A detailed monograph on the structure and properties of clay.

3–7. HAUTH, W. F., "Crystal Chemistry in Ceramics," *Bull. Amer. Ceram. Soc.*, **30**, 1951. A series of eight articles on the subject. The silicate structures are presented in considerable detail.

3–8. KINGERY, W. D., *Introduction to Ceramics*. New York: John Wiley & Sons, 1960. The structure of crystals is presented in Chapter 4. Structural imperfections receive a thorough discussion in Chapter 6. Advanced undergraduate and graduate levels.

3–9. SINNOTT, M. J., *The Solid State for Engineers*. New York: John Wiley & Sons, 1958. Chapter 7 discusses the structure of ionic solids. Advanced undergraduate level.

3–10. VAN VLACK, L. H., *Elements of Materials Science*, second edition. Reading, Mass.: Addison-Wesley, 1964. The concepts of crystal structure are introduced in Chapter 3; Chapter 6 gives additional attention to metallic crystals.

Problems

3–1. Determine the origins of the figures 0.225, 0.414, and 0.732 in Table 3–1.

3–2. From Table 2–2, (a) calculate the lattice constant for CsCl. (b) What is the distance between the centers of nearest Cs^+ ions? Cl^- ions? (c) What is the distance between the "surfaces" of nearest Cs^+ ions? Cl^- ions? (d) Calculate the density of CsCl. (e) What are the interplanar spacings between the following planes: (110), (320)?

3–3. Derive the tetragonal relationship of Table 3–3.

3–4. Calculate the density of MgO.

3–5. (a) From Table 2–2, calculate the density of CaF_2. (b) Account for any discrepancy from the handbook value of 3.18 gm/cm^3.

3–6. The density of $BaTiO_3$ is 6.1 gm/cm^3. (a) If a cubic structure is assumed, what is the lattice constant? (b) If the c/a ratio is 1.01, what are the lattice constants?

3–7. (a) What are the first-neighbor, second-neighbor, and third-neighbor ions for the Ti^{4+} of $BaTiO_3$? (b) How many are there of each?

3–8. (a) Calculate the size of the largest hole (for a spherical atom) in cristobalite. (b) What is the largest passageway for diffusion? (c) What assumptions are necessary for the above calculations?

3–9. The small-angle boundary in MnS is 5′. What is the dislocation spacing along the boundary? The boundary approximates the (110) plane. (MnS has the same structure as MgO.)

3–10. Magnesiowustite [(Mg, Fe)O] has an Mg/Fe ratio of 0.33.
(a) What is the w/o Mg? (b) MgO? (c) FeO?

3–11. Jette, E. R., and F. Foote [*J. Chem. Phys.*, **1**, p. 29 (1933)] give the following data for $Fe_{<1}O$ (or $FeO_{>1}$).

Edge of unit cell, A	Density, gm/cm^3	Fe/O
4.202	5.613	0.84
4.285	5.624	0.91
4.292	5.658	0.92
4.301	5.724	0.95

(a) What is the w/o Fe in each? (b) How many cation vacancies are there per cm^3 in each?

3–12. Assume that the oxygen atoms are placed directly between adjacent silicons in high cristobalite. A displacive transformation produces a 1% linear contraction. What is the Si—O—Si bond angle?

Chapter 4

GLASSES AND OTHER NONCRYSTALLINE PHASES

4–1 Introduction

Glasses are formed from supercooled liquids; however, glasses are not typical of the liquids that we normally encounter. Though they are solids at normal temperatures, glasses do not have the crystalline structures that characterize most of the other solid engineering materials that we use. All in all, the vitreous (glassy) materials have unusual characteristics which provoke considerable scientific curiosity; they also have many unique and useful engineering properties arising from their atypical nature.

Glass has been given several definitions, varying somewhat with the goal of the writer; but two features are usually mentioned: (1) a glass is formed from an extremely viscous supercooled liquid, and (2) the liquids which form glasses possess a polymerized network structure with short-range order. A third point is sometimes added, namely: the composition is inorganic. This last point is inserted when specific attention is given to the major commercial glasses, and has led to the ASTM definition of glass as *an inorganic product of fusion which has cooled to a rigid condition without crystallizing.* It must be acknowledged, however, that some organic and semiorganic compounds have vitreous characteristics. This relationship is especially emphasized nowadays, with organic chemists extending their developments toward higher-temperature polymers and considering boron-based and aluminum-based polymers as well as silicones. There appears to be no categoric delineation between inorganic and organic glasses.

There are numerous metal-nonmetal compositions which cannot form glasses. These include slags and mattes, which will be considered briefly, and fused salts which will not be considered.

4–2 The structure of glass

Although a glass lacks the high order of a crystal, it is not devoid of structure. All glasses have two characteristic structural features: (1) first-neighbor, or short-range order, and (2) a continuous (usually three-dimensional) framework of strong primary bonds.

The short-range order is illustrated schematically in Fig. 4–1(a) for a B_2O_3 glass. Each boron is coordinated with three neighboring oxygens (and each oxygen joins two borons). The long-range order in the crystalline compound of Fig. 4–1(b) is lacking in the glass.

Silica glass (sometimes inappropriately called quartz) is the most common of the glass structures having a single oxide. Its structure is identical to that of Fig. 4–1(a), but it has a fourth oxygen above or below the plane of the paper. The SiO_4 tetrahedral unit provides the short-range order in fused silica. Each tetrahedron is connected to another through the jointly shared apex oxygen (cf. Fig. 3–15), thus producing the three-dimensional framework of strong Si—O—Si bonds.

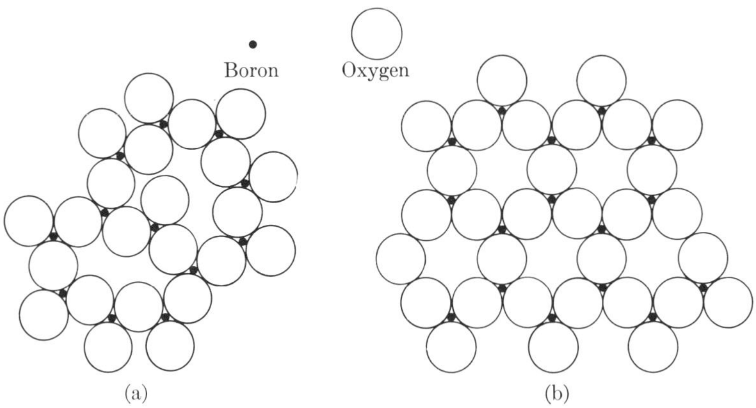

Fig. 4–1. Glassy versus crystalline structures: B_2O_3. (a) The glass has only first-neighbor or short-range order. (b) Crystalline compounds have long-range as well as short-range order.

Laboratory proof for the above conclusions is attained by x-ray diffraction. Silica glass produces a rather definite diffraction band, which indicates an average interatomic spacing of 1.62 A. This corresponds to the value of 1.60 A for the closest spacing (Si—O bonds) in crystalline silicas. There is a little more variation in this spacing in vitreous silica than there is in cristobalite or other silicas, which is to be expected, since the absence of long-range order in a glass introduces stresses and strains within the tetrahedral units; and further, the bond angles have greater variability. These facts will become important later (Chapter 8) when the mechanical characteristics of glasses will be compared with the properties of crystalline ceramics.

Example 4–1. A silica glass has a density of 2.3 gm/cm^3. Assuming that the atom sizes are the same as in crystalline silica, calculate the atomic packing factor (APF) for the glass.

Answer:

$$\rho = 2.3\ \text{gm/cm}^3 = \frac{(\text{number of SiO}_2/\text{cm}^3)\ (60.02)}{6.02 \times 10^{23}},$$

$$\text{number of SiO}_2/\text{cm}^3 = 2.28 \times 10^{22},$$

$$\text{SiO}_2/\text{A}^3 = 0.0228.$$

$$\text{Atom volume within 1 A}^3 = (0.0228)\ (4\pi/3)\ [(0.39)^3 + 2(1.32)^3]$$

$$= 0.45\ \text{A}^3;$$

$$\text{APF} = 0.45.$$

Polycomponent glasses. Most glasses contain more than one oxide component, since in the absence of long-range crystalline order, the structure does not have severe geometric limitations placed on its components. Of course, the requirement remains that the electronic charge balance must be maintained.

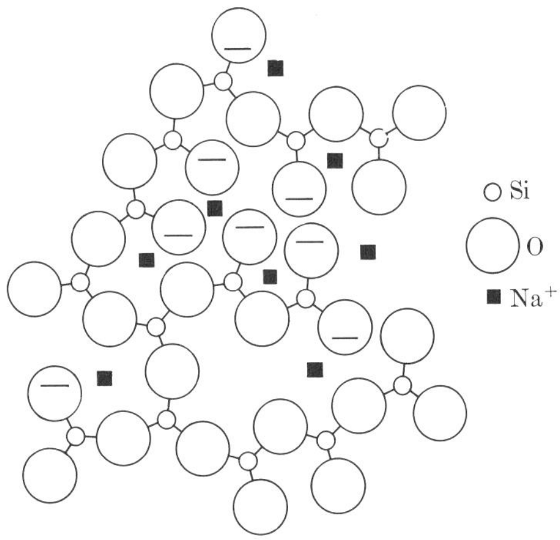

FIG. 4–2. Alkali-silicate glass. The O/Si ratio is higher than in a pure silica glass, and the bonding is more ionic in character. [The fourth oxygen (not shown) is coordinated with each silicon.]

In a glass structure, the component oxides lose their identity. If the components are B_2O_3 and SiO_2, the glass is composed of both triangular and tetrahedral coordination units. If Na_2O, K_2O, CaO or similar oxides are added, the framework structure is depolymerized (see Fig. 4–2). In effect, the addition of Na_2O to a glass breaks into the framework to produce two O^- branches and two Na^+ ions. Addition of CaO leads to an analogous situation, i.e., it produces one Ca^{2+} ion.

There is evidence that polycomponent glasses have a complexity of structure greater than those described above. For example, the cation positioning within the glass cannot be entirely random, since these ions must occupy sites that are interstitial to the polymerized network. Further, their mutual repulsion leads to geometric separation from other cations.

EXAMPLE 4–2. The silicate polymerization factor is defined as the ratio of oxygen atoms shared by two tetrahedra to those associated with only one tetrahedron. What is the polymerization factor for a Na_2SiO_3 glass?

Answer: Let

$$X = \text{number of oxygens joining two tetrahedra;}$$
$$Y = \text{number of oxygens associated with only one tetrahedron;}$$
$$N = X + Y = \text{total oxygens;}$$
$$2X/4 = \text{number of silicon atoms with shared oxygens;}$$
$$Y/4 = \text{number of silicon atoms with unshared oxygens;}$$
$$M = (2X + Y)/4 = \text{total silicons.}$$

For Na_2SiO_3,

$$M = 1 = (2X + Y)/4, \qquad N = 3 = X + Y;$$

therefore $X = 1$ and $Y = 2$. Polymerization factor $= X/Y = 0.5$.

Immiscible glasses. Not all components of a glass mixture exhibit complete solubility. When they do not, two separate liquids form, each with a somewhat different structure. The most readily cited example of immiscibility is SiO_2 and CaO. At 1700°C (3090°F), a silica glass may dissolve up to 5% CaO (much more if either Al_2O_3 or Na_2O is also present). Additional CaO produces a second glass with lower SiO_2 content, which is less polymerized and more ionic than that of the high-silica glass. At higher temperatures, the miscibility gap between these two glassy liquids is reduced as the randomness is increased.

Other examples of immiscible glasses are found in most of the $RO\text{-}SiO_2$ systems; also in the opaline glasses and in certain $Na_2O\text{-}B_2O_3\text{-}SiO_2$ glasses. In the latter (commercial) example, the original glasses may be thermally separated into two phases after shaping and the SiO_2-lean phase is removed by leaching, leaving only the SiO_2-rich (96%) phase.

Ion exchange. The cations within a glass are subject to diffusion and exchange just as they are in a crystalline silicate. This fact becomes particularly important in glass fibers, where the surface-to-volume ratio is large. The most common exchange is between Na^+ ions in the glass and H^+ ions in the aqueous solutions or from surface-adsorbed atmospheric moisture:

$$Na^+_{\text{glass}} + H^+_{\text{surface}} \rightleftarrows H^+_{\text{glass}} + Na^+_{\text{surface}}. \tag{4-1}$$

For this reason, glass fibers for electrical insulation must be low in Na_2O, since the surface Na^+ ions lead to adsorption of moisture and thence to subsequent electrical breakdown.

4-3 The composition of glass

The structures of the vast majority of commercial glasses are based on a framework of SiO_4 tetrahedra. However, glasses are not limited to the silicates. Other compounds, such as the aforementioned B_2O_3, find considerable application in glasses designed for high-temperature utilization. In addition, P_2O_5-type glasses are increasing in technical importance, particularly because of their spectral absorption–transmission characteristics. A variety of other major types of glasses is also available.

Zachariasen rules. Why is SiO_2 the most commonly used glass former? And why do some oxides, even with rapid cooling, fail to form glass? Considerable attention has been given to these questions. After detailed analysis, Zachariasen* proposed the following empirical rules for glass formation.

Rule 1: Each oxygen should be coordinated with not more than two metal atoms. (This rule singles out those oxides which develop network structures.)

Rule 2: The metal atoms should have a small coordination number. (This rule singles out those oxides with stronger bonds.)

Rule 3: The oxygen polyhedra share corners only, not edges or faces.

Rule 4: Three or more corners of each polyhedral coordination unit must be shared. (This rule also ensures polymerization of a three-dimensional network structure.)

Relatively few oxides meet all of these requirements. It is impossible for oxides of monovalent or divalent metals to do so because they lack the required number of oxygens to form a network. Further, the R_2O_3 oxides can form a network only if the metal atom is small enough to permit triangular coordination units. Boron is the only Group III element which can form a 3:2 coordination with oxygen, i.e., three oxygens surround each boron and two borons are associated with each oxygen. Under favorable conditions, aluminum with its octahedral ionic coordination can enter into a glass structure. The RO_2 and R_2O_5 oxides are the most likely candidates for glass formation; SiO_2, GeO_2, P_2O_5, and As_2O_5 actually form glassy structures. Other elements of Groups IV and V have a coordination number which is too high. (Both 6-fold and 8-fold coordination are common.) Also, Groups VI and VII oxides either have too high a coordination number or they form nonpolymerized polyatomic ions, such as ClO_4^- and SO_4^{2-}.

In addition to the oxides B_2O_3, SiO_2†, GeO_2, P_2O_5, and As_2O_5, which are known to be *network formers*, several oxides can, to a limited extent, enter an existing network. Alumina is the most commonly used example, because aluminum ions are small and can proxy for some of the silicons in the tetrahedral positions. Also, it is probable that a network of tetrahedrally coordinated units can contain a limited number of octahedral units. Other ions which behave similarly to aluminum are zirconium, titanium, vanadium, antimony, and lead. On the basis of the Zachariasen rules, each of these elements is considered to be a marginal possibility for network formation.

---

* Zachariasen, W. H., "The Atomic Arrangement in Glass," *Journ. Amer. Chem. Soc.*, **54**, 3941 (1932).

† BeF_2 has valence and size ratios almost identical to those of SiO_2; therefore it is also a network former.

Table 4-1
Bond Strengths of Oxide Glass Components*

Oxide	Dissociation energy, kcal/mole	Coordination number of cations	Bond strength, kcal/mole
Network formers			
B_2O_3	356	3	119
GeO_2	431	4	108
SiO_2	424	4	106
V_2O_5	449	4 (5)	112–90
P_2O_5	442	4 (5)	111–88
Al_2O_3	402–317	4	101–79
B_2O_3	356	4	89
As_2O_5	349	4 (5)	87–70
Sb_2O_5	339	4 (5)	85–68
ZrO_2	485	6	81
Intermediates			
TiO_2	435	6	73
PbO	145	"2"	73
ZnO	144	"2"	72
Al_2O_3	402–317	6	67–53
ThO_2	516	8	64
BeO	250	4	63
ZrO_2	485	8	61
CdO	119	"2"	60
Network modifiers			
Sc_2O_3	362	6	60
La_2O_3	406	7	58
Y_2O_3	399	8	50
SnO_2	278	6	46
Ga_2O_3	267	6	45
In_2O_3	259	6	43
ThO_4	516	12	43
PbO_2	232	6	39
MgO	222	6	37
Li_2O	144	4	36
PbO	145	4	36
ZnO	144	4	36
BaO	260	8	33
CaO	257	8	32
SrO	256	8	32
CdO	119	4	30
Na_2O	120	6	20
CdO	119	6	20
K_2O	115	9	13
Rb_2O	115	10	12
HgO	68	6	11
Cs_2O	114	12	10

*Adapted from K. H. Sun, "Fundamental Conditions of Glass Formation," *Amer. Ceram. Soc.*, **30**, 277–281 (1947).

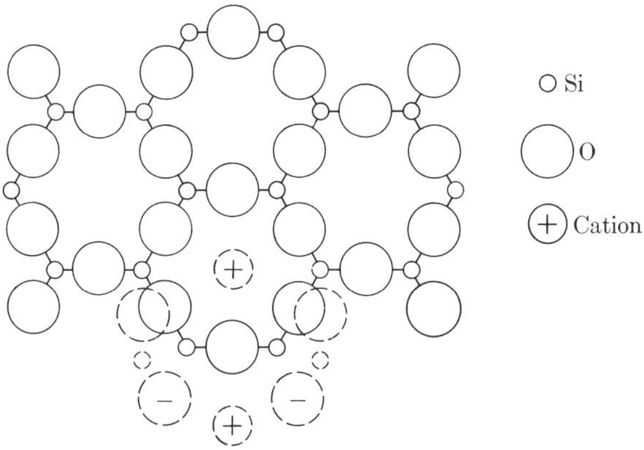

FIG. 4-3. Depolymerization by cations. Solid lines: former positions. Dashed lines: position after Na_2O addition.

Other oxides are widely used in glasses, including Na_2O, K_2O, CaO, BaO, MgO, PbO, and the rare-earth oxides, to name but a few. The term *network modifier* has been given to these oxides. In effect, they open up the three-dimensional framework structure, as shown in Fig. 4-3, and have a very definite purpose in glass. Not only do they contribute to the optical or thermal properties that are often desired, but they also increase the workability of a glass by increasing its fluidity. This increase in fluidity is a direct consequence of the depolymerizing effect. There is a limit to the amount of network modifier which can be added; the addition of these network modifiers increases the ionic character of the glass, and crystallization occurs more readily as the polymerization ratio is decreased below 1.0 (see Example 4-2.) Slags from blast furnaces offer a ready illustration: those slags with a ratio of 1.0 are vitreous when cooled at a rate of 10°F/sec. A slag with a polymerization ratio of 0.5 must be cooled even faster than that if it is to remain glassy. It is almost impossible to avoid devitrification (crystallization) in an orthosilicate slag (which has a polymerization ratio, or P.R., equaling zero).

Bond strength (Table 4-1) is the dissociation energy of the oxide divided by the coordination number, and this factor can be used as a qualitative index of the glass-forming tendencies of oxides. The network-forming oxides have bond strengths of more than 80 kcal/bond. In contrast, the network modifiers have bond strengths of less than 60 kcal/bond. The oxides with bond energies between these two values are called intermediates. It is immediately apparent that some oxides, such as Al_2O_3, have two bond energies depending on the coordination of their cations within the glass. This factor accounts for the variation in the glass-forming abilities of these oxides.

Commercial glasses. Individual commercial glass compositions are extremely numerous, since incremental variations are made to provide specific properties,

TABLE 4–2

COMMERCIAL GLASS TYPES*

Type†	Major Components, %								Comments
	SiO_2	Al_2O_3	CaO	Na_2O	B_2O_3	MgO	PbO	Other	
Fused silica	99								Very low thermal expansion, very high viscosity
96% silica (vycor)	96				4				Very low thermal expansion, high viscosity
Pyrex type	81	2		4	12				Low thermal expansion, low ion exchange
Containers	74	1	5	15		4			Easy workability, high durability
Plate	73	1	13	13					High durability
Window	72	1	10	14		2			High durability
Lamp bulbs	74	1	5	16		4			Easy workability
Lamp stems	55	1		12			32		High resistivity
Fiber	54	14	16		10	4			Low alkali
Thermometer	73	6		10	10				Dimensional stability
Lead glass tableware	67			6			17	K_2O 10	High index of refraction
Optical flint†	50			1			19	BaO 13 K_2O 8 ZnO 8	Specific index and dispersion values
Optical crown†	70			8	10			BaO 2 K_2O 8	Specific index and dispersion values

* Data adapted from various sources, but primarily from A. K. Lyle, "Glass Compositions," *Handbook of Glass Mfg.* (F. V. Tooley, editor). New York: Ogden Publishing Co., 1953, p. 5.

† See Table 11–2 for additional optical glasses.

such as index of refraction, color, dispersion, viscosity, etc. However, commercial glasses may be divided into relatively few general categories, as indicated in Table 4-2. The choice of a glass depends on striking a balance between the requirements for processing and the properties of the material. For example, a fused silica glass has both a very low thermal expansion and a very high viscosity. Thus, even though there are many applications where the low thermal expansion would be desirable, costs of processing of fused silica glass may exclude its use because of the difficulty involved in working it.

4-4 Transformation range

Changes in the density of glass during cooling reveal some of the characteristics of the glass structure. When glass is in the form of a supercooled liquid, its density increases rapidly as the temperature is lowered, because the atoms rearrange themselves into a more tightly packed structure. However, at much lower temperatures there is less increase in density (Fig. 4-4). In effect, below the "transition" temperature, the atoms cannot rearrange themselves into denser packings, and any decrease in volume is simply a reflection of the decrease in mean interatomic spacing encountered in various crystalline silicates.

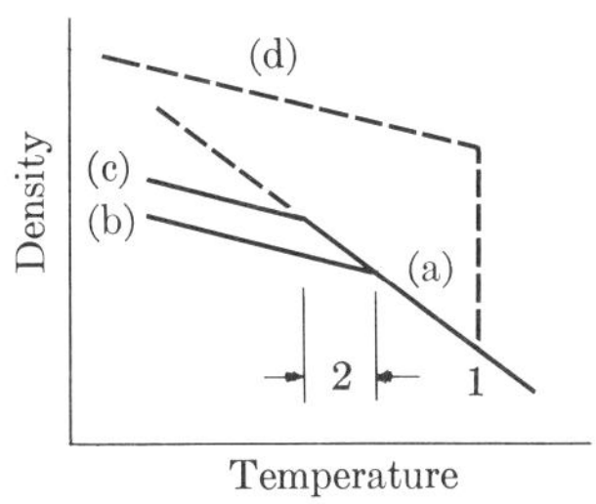

FIG. 4-4. Density versus temperature during cooling: (a) liquid silicate; (b) rapidly cooled glass; (c) slowly cooled glass; (d) comparable crystalline silicate; (1) crystallization temperature; (2) transformation range of fictive temperatures. Above the fictive temperature the atoms can rearrange themselves for more efficient packing during cooling. Only the interatomic distance can be reduced when the glass is below this temperature.

This transition temperature, called the *fictive* temperature, is considered to be the point at which glasses are transformed from a supercooled liquid to a vitreous solid. This temperature is not unique, but varies with the rate of cooling. The slower the cooling process, the more time there is for atoms to rearrange themselves and densify. Furthermore, a rapidly cooled glass may undergo further densification if it is reheated slowly, permitting it to reach the transformation range.

EXAMPLE 4-3. The two glass samples (one cooled rapidly, the other slowly) of Fig. 4-4 are reheated very slowly. Indicate the volume changes qualitatively.

Answer: see Fig. 4-5.

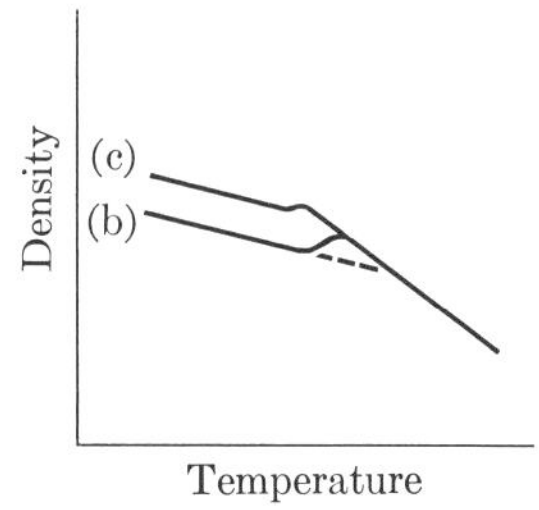

FIG. 4-5. Density versus temperature during slow heating (see Example 4-3): (b) glass which has been cooled rapidly; (c) glass which has been cooled slowly. When there is sufficient time for a slow reheat, atoms can rearrange themselves below the original fictive temperature.

4-5 Crystallization (devitrification) of glass

We have discussed the fact that a glass forms from a supercooled liquid, which implies that a glass is only metastable and will eventually crystallize if conditions are favorable. Four factors affect the devitrification: (1) time, (2) temperature, (3) nucleation, and (4) internal structure.

The time required for devitrification is simply a measure of the rate at which it occurs. The longer the time, the more extensive the arrangement of the atoms into a crystalline structure. The fact that early historic glasses are found intact indicates that under certain conditions the time is exceedingly long. However, we must expect that crystallization will eventually take place, because at ambient temperatures the free energy of glass is higher than the free energy of the corresponding crystalline phases.

Higher temperatures produce faster rates of crystallization, as we can observe in various glasses that are overheated in use. We can expect the following activation energy relationships:

$$\text{Rate} = Ae^{-E/kT}, \qquad (4\text{-}2a)^*$$

or

$$\log_e 1/t = A' - E/kT. \qquad (4\text{-}2b)$$

In this relationship, t is the time it takes to achieve any given state of reaction; E is the activation energy for the individual step; k is Boltzmann's constant, T is the absolute temperature, and $A' = \log_e A$ is a constant. Higher temperatures favor faster devitrification because there is a greater probability that any particular network bond may receive sufficient energy to be ruptured, and thus permit the formation of the lower-energy, longer-range crystalline structure. However, a maximum rate is reached, so that with still higher temperatures the

* This Arrhenius equation is an approximation for any reaction. Therefore we should not expect it to precisely define rates of reaction. However, it is a very useful relationship.

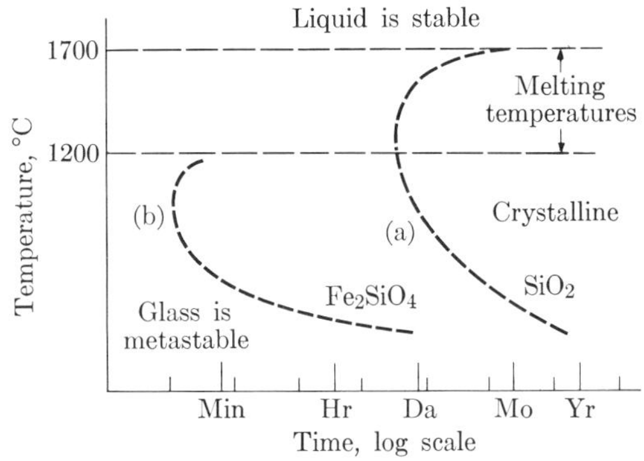

FIG. 4-6. Glass devitrification (schematic drawing). (a) Network glass (silica). (b) Partially ionic glass [fayalite (Fe_2SiO_4) liquid].

rate falls, because crystallization depends not only on the probability of bond rupture and long-range ordering, but also on the formation of a new surface between the glassy and crystalline phases. The characteristics of such surfaces will be considered in the following chapter. It should be noted here, however, that such a surface requires additional energy which must be obtained from the free energy released by crystallization. Near the true crystallization temperature, the value for ΔF (i.e., $F_{gl} - F_{cryst}$) is so small that there is very little "driving force" to promote nucleation and new surfaces. This means that the transformation time just below the true crystallization temperature can be very long.

Figure 4-6 summarizes the above points by illustrating longer devitrification times for lower temperatures, and slightly delayed crystallization just below the true freezing temperature, where nucleation becomes a problem. In addition, a comparison is made between two types of glass. The SiO_2 glass has an excellent network structure. Consequently, it devitrifies more slowly than a Fe_2SiO_4 liquid, which has a large percentage of discrete SiO_4^{4-} ions and relatively little polymerization. The relationship between the devitrification time and the internal structure points up the fact that devitrification depends to a large degree on composition. We shall give more attention to crystallization in Section 14-5.

4-6 Slag and mattes

At high temperatures, slags and mattes are true liquids of metal-nonmetal compounds, and both have metallurgical significance. They differ from each other in that a slag is essentially an oxide liquid and a matte is essentially a sulfide liquid. They are not completely miscible in each other, a fact which has a commercial application as a separation procedure in some smelting processes.

Their immiscibility probably results from the size difference between the S^{2-} and O^{2-} ions ($r_S = 1.74$ A, $r_O = 1.32$ A).

Metallurgical mattes are ionic in behavior and crystallize even during a severe quench. Also, with some exceptions (see Section 4–3), the basicity of metallurgical slags is high enough to produce a liquid with extensive ionization but only limited polymerization. The composition of an open-hearth slag, for example, is that of a liquid structure with anions similar to those shown in Fig. 3–15(a) and (b). The thermodynamic activities of the slag components, as well as the fluidity of the slag itself, are directly related to the degree of polymerization. Therefore, the slag structure provides us with an excellent procedure for studying the reactions which take place during metallurgical refining processes.

4–7 Inorganic linear polymers

In two previous locations (Sections 1–3 and 4–1) we have called attention to the fact that there is no sharp delineation between ceramic and organic materials. An area in which there is a considerable overlap is that of the inorganic polymers. Here we shall briefly consider those inorganic polymers that do not customarily form three-dimensional framework structures. One- and two-dimensional polymers, which are characterized by strong covalent intramolecular bonds and weak van der Waals intermolecular bonds (Section 2–7), possess many properties which are noticeably different from those of glasses or crystalline compounds.

The best known of the inorganic polymers are the silicones, of which there are as many varieties as there are of the carbon-based compounds. It is commonly assumed that the silicones are simply silicon analogs of the carbon compounds. This is true to a certain extent, but there are definite differences. The hydrosilicon (i.e., silane) chains are much less stable than the comparable hydrocarbon chains. In fact, the longest silane known contains only six silicons (Si_6H_{14}).

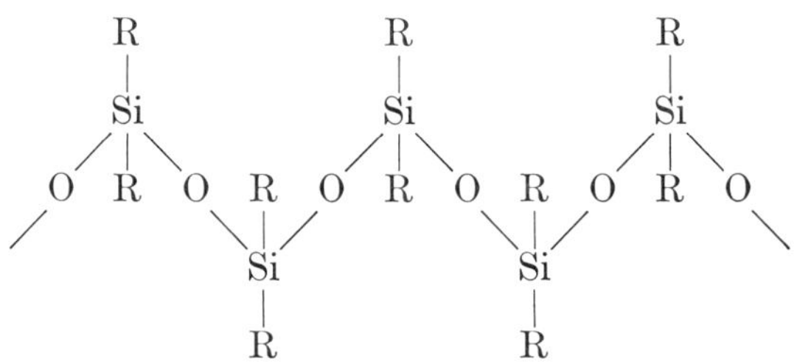

R = H, Cl, OH, CH_3, etc.

FIG. 4–7. Siloxane silicones. The strong Si-O bonds give thermal and chemical stability to the chain.

The commonest silicones are the siloxanes, which consist of a central chain, quite stable, of silicon-oxygen bonds (Fig. 4–7). The siloxanes are more widely utilized as engineering materials than comparable (C—O) polymers, such as the polyformaldehydes. Since the bonds of the central chain of a silicone are essentially identical to those in the previously discussed silicates (Fig. 3–15c), these silicones possess a chemical and thermal stability comparable to many ceramics. The mechanical properties, on the other hand, are markedly different, because adjacent molecules behave quite independently, and plastic deformation readily occurs unless cross-linking is introduced. Also, many of the linear silicones may be highly amorphous; consequently, their initial modulus of elasticity is low, and they exhibit rubberlike characteristics.

References for Further Reading

4–1. Davies, T. H., "The Physics of Glass," *The Science of Engineering Materials* (J. E. Goldman, editor). New York: John Wiley & Sons, 1957, pp. 489–510. Excellent summary of the interrelationship of structure and property.

4–2. Douglas, R. W., "The Properties and Structure of Glass," *Progress in Ceramic Science*, Vol. 1 (J. E. Burke, editor). New York: Pergamon Press, 1961, pp. 200–223. A concise summary, for simple glasses, of the relationship of properties to structure.

4–3. Hauth, W. F., "Crystal Chemistry in Ceramics," *Bull. Amer. Ceram. Soc.*, **30**, 1951. Part VIII concerns the structure of glass.

4–4. Kingery, W. D., *Introduction to Ceramics*. New York: John Wiley & Sons, 1960. Noncrystalline solids is the subject of Chapter 5. Primary attention is given to glass.

4–5. Morey, G. W., *The Properties of Glass*, second edition. New York: Reinhold Publishing Co., 1954. This is the classical monograph on glass.

4–6. Phillips, C. J., *Glass: Its Industrial Applications*. New York: Reinhold Publishing Co., 1960. Semitechnical introduction to glass and its uses.

4–7. Rochow, E. G., *Chemistry of the Silicones*, second edition. New York: John Wiley & Sons, 1951. Brief presentation of silicone chemistry. Previous course in organic chemistry is desirable, but not mandatory.

4–8. Stanworth, J. E., *Physical Properties of Glass*. Oxford: Clarendon Press, 1950. The properties are correlated with the structure. A physical chemistry background is assumed.

Problems

4–1. A plate glass has a composition of 14 Na_2O-13CaO-72SiO_2, and a density of 2.5 gm/cm^3. Compare its atomic packing factor with that of a silica glass (Example 4–1).

4–2. A vycor glass has 96 w/o SiO_2 and 4 w/o B_2O_3 as its principal constituents, and an atomic packing factor of 0.44. Calculate its density.

4–3. Determine the silicate polymerization factor for a glass containing (a) $Na_2Si_2O_5$, (b) $CaSiO_3$, (c) 5 w/o Na_2O, 95 w/o SiO_2, and (d) 14 w/o Na_2O, 13 w/o CaO, 72 w/o SiO_2.

4–4. Following the procedure of Example 4–2, determine a polymerization factor for anhydrous borax ($Na_2B_4O_7$).

4–5. The plate glass of Problem 4–1 is to be made from soda ash, calcite, and silica sand. What weight ratios of each should be used?

4–6. Ethyl silicate solutions may be used to form an amorphous silica bond for refractory materials. Indicate the probable chemical reactions.

Chapter 5

SURFACES AND INTERFACES

5–1 Introduction

The behavior of an atom within a crystal (or glass) is governed by all of its adjacent atoms. Since the atoms located at the surface of a crystal have surroundings different from the atoms of the interior lattice, they behave differently.

This difference in behavior may be traced to differences in energies possessed by surface and lattice atoms. If we calculate a Madelung constant (Section 2–5) for a surface atom, we obtain a lower value than that of the Madelung constant for an interior atom, which indicates that surface atoms have higher energy than interior atoms.

There are a number of consequences of this differential energy relationship, three of which are of prime importance in ceramic materials: first, surface adsorption occurs. Second, surface areas tend to decrease in order to reduce surface energy. Third, certain ion exchanges may be promoted. In this chapter we shall consider each of these factors.

5–2 Surface adsorption

To gain an understanding of adsorption, let us consider the nature of fractured surfaces. Figure 5–1 shows three examples of fractured surfaces, each of which possesses sites where surface attractions may adsorb ions or molecules. There are two types of adsorption: chemical and physical. Parts (a) and (b) of Fig. 5–1 show how chemical adsorption may take place where there are positive or negative ions resulting from the fractures. Atoms which are chemically adsorbed become a part of the surface structure, since they are held by strong ionic or covalent bonds (approximately 100 kcal/mole, as indicated in Section 2–4). Thus we say that, in essence, there is a chemical reaction between the exposed crystal surface and the chemically adsorbed atom.

On the other hand, physical adsorption is a result of secondary van der Waals attractions, and this adsorption depends on these weak dipole interactions, which have a bonding energy of only a few kcal/mole.

The type of adsorption varies with the crystal structure, and can be either or both. For example, with NaCl (Fig. 5–1a), either chemical or physical

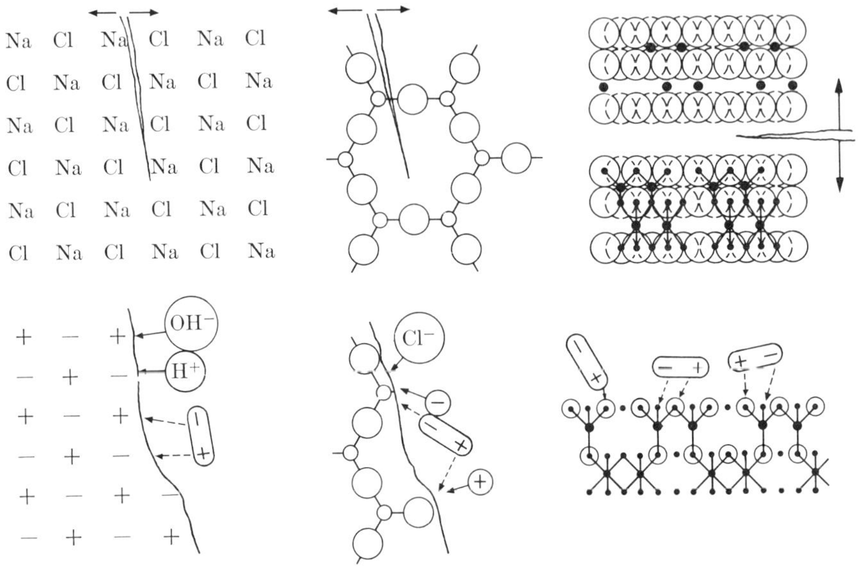

(a) (b) (c)

FIG. 5-1. Schematic drawing of fractured surfaces. Weak van der Waals attractions (dashed lines) may be set up between polar molecules, (− +), and positive or negative sites on the fractured surface. If fractured primary bonds are exposed, an actual chemical (ionic or covalent) bond may be established between the adsorbed atom and the surface atom. Parts (a) and (b) show chemical and physical adsorption, but only physical adsorption can occur on the surface shown in (c). Chemical adsorption can occur around the edges of the layered structure.

adsorption can occur. However, since bonds which are formed by chemical adsorption are more stable than those formed by physical adsorption, it is usual for any physically adsorbed atom or molecule to be replaced subsequently by chemically adsorbed atoms, which produce primary bonds and release more energy.

Kaolinite crystals, as seen in Fig. 5-1(c), display two types of surfaces: edges, which exhibit positions capable of chemical as well as physical adsorption; and sheet surfaces, which, on the other hand, do not have ruptured primary bonds; thus in their case only physical adsorption may occur.

EXAMPLE 5-1. Assume one $(OH)^-$ adsorption site per surface sodium ion on cubes (one-micron edge) of NaCl. How many $(OH)^-$ ions will be adsorbed per gram of NaCl?

Answer: Determine surface Na^+ ions per gm, given that

$$\text{Surface area/cm}^3 = 6d^2/d^3 = 6/d \text{ cm}^2/\text{cm}^3,$$
$$\text{Surface area/gm} = 6/\rho d \text{ cm}^2/\text{gm}.$$

Since NaCl is fcc,

$$\begin{aligned}\text{Surface Na}^+\text{ ions/cm}^2 &= (\sqrt{2}/a_0)^2 = 2/a_0^2\text{ ions/cm}^2,\\ \text{Surface Na}^+\text{ ions/gm} &= (2/a_0^2)(6/\rho d)\\ &= 12/[(5.6\times 10^{-8}\text{ cm})^2(2.16\text{ gm/cm}^3)(0.0001\text{ cm})]\\ &= 1.8\times 10^{19}\text{ ions per gm.}\end{aligned}$$

EXAMPLE 5–2. If a water molecule is adsorbed onto each exposed oxygen on the tetrahedral side of a kaolinite sheet (Figs. 3–17 and 5–1c), how much weight will be gained by the kaolinite?

Answer: There are three such oxygens for every two silicons (Fig. 3–17); therefore

$$(3H_2O)/[Al_2Si_2O_5(OH)_4] = (3\cdot 18)/258 = 21\%.$$

Because adsorption reduces the free energy of atoms or molecules, it also reduces their chemical activity, with consequences which are observable in drying operations (Section 6–2). Normally, the vapor pressure of water is 1 atmosphere at 100°C. However, adsorbed water must be heated to higher temperatures, in order to develop this normal vapor pressure and thus make possible complete removal of water during a drying operation. This water-removal factor assumes great importance when one is dealing with ceramic materials such as clays, which are extremely fine-grained and contain a large fraction of adsorbed moisture. Note, however, that the extent of chemical activity and desorption will vary if the adsorption sites are not equivalent.

TABLE 5–1

SURFACE-TO-VOLUME RATIOS*

Diameter, mm	Thickness, mm	Surface/volume, mm^{-1}
2.0	0.1	5.5
0.02	0.001	550
0.0002 (0.2μ)	0.00001 (100 A)	55,000

* Example: discs with a d/t ratio of 20.

Colloids. When a particle becomes very small, the surface-to-volume ratio becomes very large (Table 5–1). In a particle which has the dimensions of a colloid (100 A to 3μ), the surface-to-volume ratio is so large that the surface properties outweigh the properties of the interior. Adsorbed ions give the particle enough of a charge to attract it to charges of opposite sign and repel it from like charges.

Colloidal particles can have adsorbed layers which are more than one molecule thick, because the adsorbed layers can build up, one on top of another. In colloidal suspensions, the adsorbed layer may become thick enough to effectively alter the solid/liquid ratio, and hence may change the viscosity of the suspension (Section 8–8). Although an adsorbed layer has no sharp external boundary, we can effectively calculate its thickness by experiments. Thicknesses of layers have been found to range from a few angstroms on surfaces which are physically adsorbed to several hundred angstroms on surfaces which are chemically adsorbed.

EXAMPLE 5–3. An effective adsorbed layer is 100 A thick. What is the maximum fraction of solids which could be present in a suspension when the dimension of the particle is (a) 1μ? (b) 0.1μ? (c) 0.01μ? (Assume perfect packing of cubes.)

Answer: Solid volume/total volume $= d^3/(d + 2a)^3$.

For $a = 100$ A, and $d = 1\mu$, fraction $= (10{,}000)^3/(10{,}200)^3 = 0.95$.
For $a = 100$ A, and $d = 0.1\mu$, fraction $= (1000)^3/(1200)^3 = 0.58$.
For $a = 100$ A, and $d = 0.01\mu$, fraction $= (100)^3/(300)^3 = 0.04$.

5–3 Ion exchange

In Sections 3–5 and 4–2, when we spoke briefly of ion exchange, we mentioned only the fact that ions of one type can be replaced by ions of another element. Montmorillonite (Table 3–5), which is the most prevalent mineral of bentonite, provides an example of base exchange. Between layers, its structure contains cations which produce an overall charge balance. If the ions that fill these positions are Na^+, the structure swells because it adsorbs water. But if Ca^{2+} ions fill these positions, the bentonite is nonswelling. Both types of structure are found in nature, and may be changed from one type to another depending on the desired utilization. If we wish to convert a Ca^{2+} bentonite to a Na^+ bentonite, for example, we expose the bentonite to a strong solution of soda, which gives us

$$Ca^{2+}_{\text{bent}} + 2Na^{+}_{\text{sol'n}} \rightleftarrows Na^{+}_{2\,\text{bent}} + Ca^{2+}_{\text{sol'n}}. \tag{5–1}$$

With this reaction, as with other reactions, an equilibrium may be established:

$$K = \frac{(a_{Na^{+}_{2\text{bent}}})(a_{Ca^{2+}_{\text{sol'n}}})}{(a_{Ca^{2+}_{\text{bent}}})(a^{2}_{Na^{+}_{\text{sol'n}}})^2}. \tag{5–2}$$

If the concentration of the soda solution is high, reaction (5–1) moves to the right. The reaction can be reversed by increasing the concentration of the Ca^{2+} ion in the solution.

More quantitative data are given in Fig. 5–2. The two clays, montmorillonite and kaolinite, were both treated with an acid solution so that essentially all exchangeable cations were replaced with H^+ ions. A suspension of each clay was then treated with NaOH, and the pH was measured, as a function of the

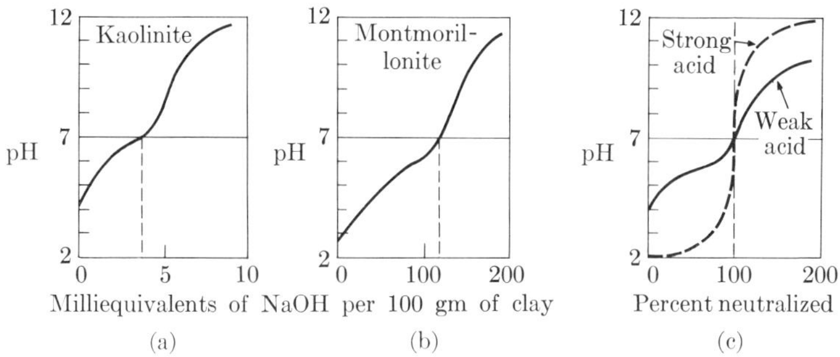

FIG. 5-2. Base-exchange capacity. The neutral point (pH = 7) is the measure used. The clays (and other ion-exchangeable structures) react as very weak acid solutions [Eq. (5-4)]. [After Marshall, C. E., and C. A. Krinbill, "The Clays as Colloidal Electrolytes," *J. Phys. Chem.*, **46,** 1077-90 (1942).]

milliequivalents of NaOH per 100 gm of clay. The amount of NaOH required to neutralize the suspension (pH = 7) is a measure of the base-exchange capacity of the clay. Here the average chemical activity of the H^+ ions with the clay is the same as their chemical activity in pure water. Three features are noteworthy.

1. The following reactions are appropriate:

$$H^+_{clay} \rightleftarrows H^+_{sol'n}, \tag{5-3}$$

$$H^+_{clay} + NaOH \rightleftarrows Na^+_{clay} + H^+ + OH^-. \tag{5-4}$$

Before any NaOH is added, the clay contributes H^+ ions to the solution. When Na^+ ions have been added to replace the adsorbed H^+ ions, the H^+ ion concentration of the solution is lowered (and the pH is higher). The clay acts as a very weak acid (Fig. 5-2c).

2. The neutralization point for clay is not as abrupt as for acids, because the various adsorbed H^+ ions are held with bonding energies of different strengths. This factor is predictable, since there is a variety of adsorption sites.

3. The base-exchange capacity of montmorillonite is much greater than for kaolinite, which is a logical expectation if we consider the structure of the two clays (Table 3-5). Montmorillonite must contain ions that are exchangeable between the layers, but kaolinite does not have this requirement.

The above information is corroborated by x-ray and other data for many of the more complicated compounds. The ions of these structures react with their environments through their surfaces; that is,

$$A^+_{struct} + B_{environ} \rightleftarrows B^+_{struct} + A_{environ}. \tag{5-5}$$

Since structural sites are not all equivalent, chemical activities for components of the above reaction vary (compare with the statement on drying in Section 5-2).

5-4 Interfacial energies

Surfaces of various types may be classified as interfaces. These include external crystal/vapor surfaces, interphase surfaces (e.g., solid/liquid boundaries), and grain boundaries. Atoms at any of these interfaces possess more energy than atoms in the interior (Section 5-1). Therefore, there is a natural tendency for interfacial areas to diminish in extent.

The existence of interfacial energies has several consequences which are of importance in the realm of ceramic materials: (1) Sintering can occur. (2) Microstructures are altered. (3) Properties are affected.

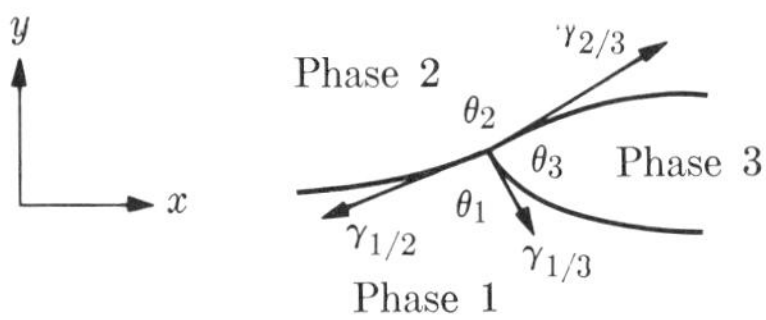

FIG. 5-3. Geometric balance. At equilibrium, a vectorial balance is maintained between the interfacial energies and the dihedral angles at all phase junctions.

Geometric equilibrium. Interfacial energies, expressed as ergs/cm^2, become directional in behavior when one interface is balanced against another (Fig. 5-3). The most stable interfacial configuration is one which has the lowest total free energy. Thus, at equilibrium, there is a vectorial balance,

$$\frac{\gamma_{1/2}}{\sin\theta_3} = \frac{\gamma_{1/3}}{\sin\theta_2} = \frac{\gamma_{2/3}}{\sin\theta_1}, \tag{5-6}$$

where the interfacial energies between adjacent phases are indicated by γ, and the dihedral angle at the junction edge of any phase or grain is indicated by θ. Alternately, the relationship may be expressed as

$$\sum \mathbf{V}_x = 0, \tag{5-7}$$

and

$$\sum \mathbf{V}_y = 0, \tag{5-8}$$

where $\mathbf{V}_x$ is the horizontal component of the surface-tension vectors and $\mathbf{V}_y$ is the vertical component.

There are several examples of dihedral angles in Fig. 5-4. Figure 5-4(a) represents the most general case, where three separate phases are involved. Figure 5-4(b) provides a rather specific but extremely common example where the interfaces are grain boundaries. Since the three interfaces (grain boundaries) are equivalent, the three dihedral angles are 120°.* A two-grain, two-phase junc-

---

* There may be a variation of energy with boundary orientation. This variation is negligible unless: (a) a low-energy twin boundary is involved, (b) a small-angle grain boundary is involved (Section 3-5), or (c) the crystal structure is markedly anisotropic; e.g., enstatite ($MgSiO_3$), which has a chain structure (Fig. 3-15c).

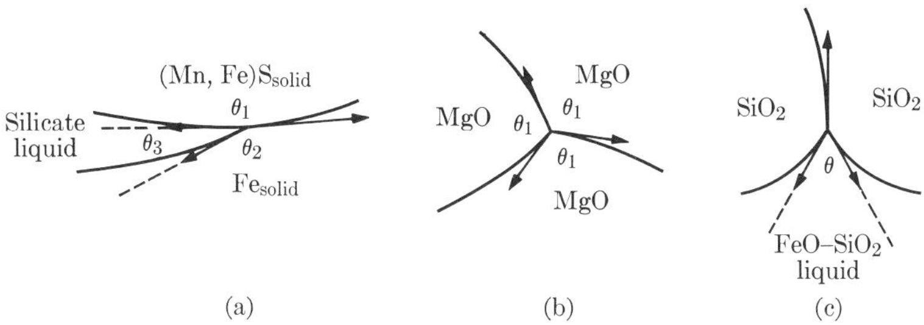

FIG. 5–4. Dihedral angles. The three-boundary junction is perpendicular to the plane of the paper. (a) Three separate phases and three different boundaries. (b) Three separate grains of the same phase, and three equivalent boundaries. (c) Two equivalent boundaries (cf. Fig. 7–9). Equilibrium is reached more rapidly at higher temperatures.

tion is shown in Fig. 5–4(c). This relationship is particularly important when microstructures are considered (Chapter 7).

EXAMPLE 5–4. A silicate liquid penetrates between two periclase (MgO) grains with a dihedral angle of 27°. What is the ratio of the phase-boundary to grain-boundary energies?

Answer:

$$\gamma_{S/S} = 2\gamma_{S/L}\cos(\theta/2), \tag{5-9}$$

$$\gamma_{S/L}/\gamma_{S/S} = 1/(2)(\cos 13.5°) = 0.52.$$

Wetting. The surface of a solid is wetted when the total energy of the several surfaces is reduced. Complete wetting occurs when

$$\gamma_{S/L} + \gamma_{L/G} < \gamma_{S/G}. \tag{5-10}$$

Figure 5–5 shows other interesting relationships, the consequences of which will be discussed in later chapters.

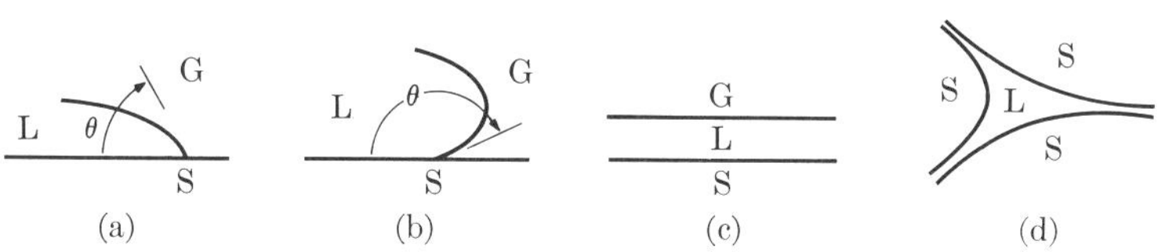

FIG. 5–5. Wetting: (a) $\gamma_{S/G} > \gamma_{S/L}$, partial wetting; (b) $\gamma_{S/G} < \gamma_{S/L}$, nonwetting; (c) $\gamma_{S/L} + \gamma_{L/G} < \gamma_{S/G}$, complete wetting; (d) $2\gamma_{S/L} < \gamma_{S/S}$, complete grain-boundary wetting. [In (a) and (b), the vectorial balance is met partially through the rigidity of the solid. With time, a dihedral angle will be developed in the solid. Cf. Fig. 5–4(a).]

References for Further Reading

5–1. ADAM, N. K., *The Physics and Chemistry of Surfaces*, third edition. London: Oxford University Press, 1941. This monograph presents the basic facts known about surface chemistry up to 1941, and is still standard in its field. Advanced level.

5–2. BRINDLEY, G. W., "Ion Exchange in Clay Minerals," *Ceramic Fabrication Processes* (W. D. Kingery, editor). New York: John Wiley & Sons, 1958, pp. 8–23. Relation of ion exchange to crystal structure.

5–3. ILER, R. V., *The Colloid Chemistry of Silica and the Silicates*. Ithaca, N. Y.: Cornell University Press, 1955. Assumes a physical chemistry background. Chapter 8 deals with adsorption onto silica and the silicate surfaces.

5–4. KINGERY, W. D., *Introduction to Ceramics*. New York: John Wiley & Sons, 1960. Surfaces and interfaces are the topics of Chapter 7.

5–5. MCLEAN, D., *Grain Boundaries in Metals*. London: Oxford, 1957. Summarizes various studies of grain boundaries, their energies, and their effects on properties. Advanced level; however, a metallurgical background is not mandatory.

5–6. SMITH, C. S., "Grains, Phases, Interfaces: An Interpretation of Microstructure," *Trans. A.I.M.E.*, **175**, 15, 1948. Emphasizes the importance of boundaries and interfaces. Requires an introductory familiarity with metallurgy.

Problems

5–1. One gm of quartz increases in weight to 1.02 gm when it is crushed (in humid air) to a powder composed of particles each of which has a dimension of one micron. What is the thickness of the adsorbed moisture film? The density of quartz is 2.65 gm/cm^3.

5–2. Kaolinite platelets average 0.9μ in diameter and 0.1μ thick. It can be shown that the effective adsorbed layer is 30 A thick on the flat surfaces and 200 A around the edges. What is the effective increase in volume of the particle as a result of the adsorption?

5–3. The surface energy of periclase (MgO) is 1000 $ergs/cm^2$. How much energy is required to crush one gm of MgO to a powder whose particles each have a dimension of one micron? The density of MgO is 3.65 gm/cm^3.

5–4. The following dihedral angles were determined for a microstructure containing a solid oxide, a solid sulfide and a liquid silicate: (a) oxide between two sulfide grains, 112°; (b) sulfide between two oxide grains, 100°; (c) liquid between two sulfide grains, 60°; (d) liquid between two oxide grains, 0°; (e) liquid between an oxide and a sulfide grain, 70°. If the energy of the oxide/oxide boundary is 900 $ergs/cm^2$, what are the energies of the other boundaries?

5–5. The surface energy of water is 70 $ergs/cm^2$ and that for a glass/gas surface is 580 $ergs/cm^2$. When a drop of water is on the glass, a contact angle of 55° is realized. (a) What is the energy of the glass/liquid interface? (b) Assume no change in $\gamma_{L/G}$ or $\gamma_{S/L}$. To what value would the gas/solid energy have to be changed to produce complete wetting? (c) To provide nonwetting (that is, $\theta > 90°$)? (d) Assume no change in $\gamma_{S/G}$ or $\gamma_{L/G}$. To what value would the liquid/solid energy have to be changed to provide complete wetting? (e) What is wrong with the general assumptions cited above? (f) Why do the vertical vectors in (a) not balance?

Chapter 6

EQUILIBRIA AND REACTIONS BETWEEN CERAMIC PHASES*

6–1 Introduction (phase rule)

Information on phase equilibria is of prime importance in predicting the behavior of materials, because equilibrium must be attained (or at least approached) in order to ensure chemical stability. Also, knowledge of the phase equilibria may be useful even if equilibrium does not exist, because these equilibria reveal the direction of change in chemical potential, and can also indicate limits for service temperatures and pressures. Further, phase relationships provide a basis for estimates of chemical analyses.

The rule which applies to phase relationships at equilibrium is frequently encountered in ceramic technology:

$$P = C + E - V. \tag{6–1a}$$

Here, P is the number of phases, a number which depends on the number of components C, the number of environmental variables E, and the variance V of the system. We shall now describe each of these.

Phase. A phase is a structurally homogeneous part of a material system. Structural features such as first neighbors, short-range order, and long-range order (or lack of it) are consistent throughout a given phase. The phase composition may be varied by being a solution or by the presence of defects. However, such atom variations are statistically distributed. There are limitations to the compositional variation because the free-energy levels must be at their lowest in order that equilibrium may occur. If excessive changes are made in the composition of the phase, new structures may provide a more stable pattern (see Section 6–3).

Component. The components of a system represent the minimum number of pure substances out of which a given system can be produced. Thus, a component may be elemental, or it may be a compound. Because phase relation-

* Only a limited number of phase diagrams are presented in this book. (See Appendix F.) The reader is referred to *Phase Diagrams for Ceramists,* Columbus, O., American Ceramic Society, for other ceramic phase diagrams.

ships become more complicated with each additional component, it is desirable to use a compound as a component if the compound is not altered. For example, when considering ceramic materials, it is usually preferable to deal with a single component such as Al_2O_3, rather than with the *two* elemental components, aluminum and oxygen.

Environmental factors. It is reasonable to consider any number of environmental factors that might cause important changes in the system. Thus, if only temperature is taken into account, Eq. (6–1a) becomes

$$P + V = C + 1. \qquad (6\text{–}1b)$$

Or, if both temperature and magnetic field are factors, the resulting equation is

$$P + V = C + 2. \qquad (6\text{–}1c)$$

This form of the phase rule is also used when temperature and pressure are the essential environmental factors.

Variance. The degree of freedom or variance of a system refers to the number of unassigned or independent variables. Since we know that the final component is the remainder from the other components, the total number of variables in an equilibrated system is $(C - 1) + E$. A completely unrestricted system will have only one phase; that is, when $V = (C - 1) + E$, each variable may be changed independently and $P = 1$. By contrast, if $V = 0$, there is no choice among the variables, and the number of phases will be at a maximum for the system at hand. In such a case, at least one of the phases must disappear before any of the variables may be changed.

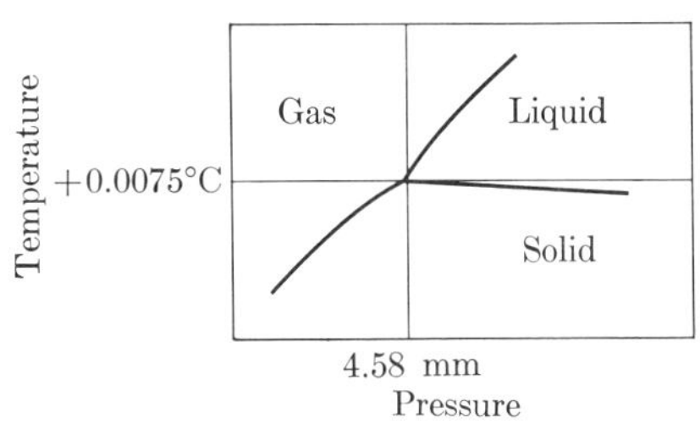

FIG. 6–1. The H_2O system. When three phases are present, there is no choice of temperature and pressure. For two phases to coexist, one may either choose temperature *or* pressure; then the other environmental factor is fixed.

6–2 Single-component systems

Phase relationships in a one-component system depend only on environmental factors. Figure 6–1 shows the phase relationships for H_2O when temperature and pressure are varied.

The enthalpy H of a phase will change as the temperature T is varied at constant pressure. Hence

$$\Delta H = \int_{T_1}^{T_2} C_p \, dT, \qquad (6\text{–}2)$$

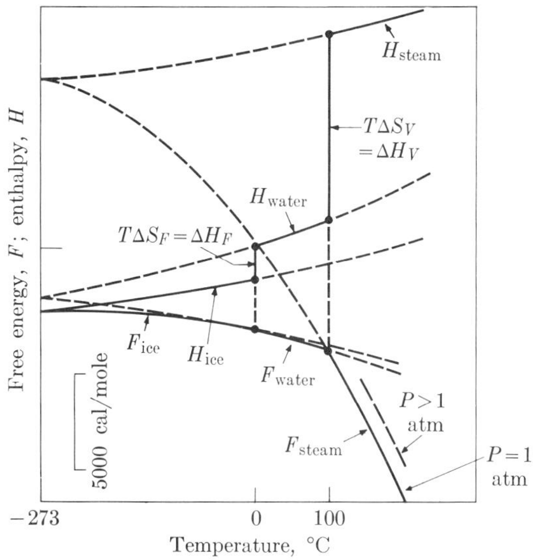

FIG. 6–2. Energies of H_2O (at atmospheric pressure, except as indicated). The stable form is the phase with the lowest free energy.

where ΔH is the change in enthalpy and C_p is the specific heat at constant pressure. This is shown for the three most common phases of H_2O in Fig. 6–2, as is the free energy F (which equals $H - TS$) for each phase. It may be noted that

$$dF/dT = -S. \tag{6–3}$$

Since the entropy S is always positive, and tends to increase at higher temperatures, the free energy of a substance decreases (at an accelerating rate) as the temperature is increased.

Transformation temperatures. At a transformation temperature, two phases of a given substance are in equilibrium with each other, meaning that they have equal free energies (Fig. 6–2). Thus, any heat of transformation ΔH_{tr} is simply a reflection of a change in entropy ΔS_{tr}. Thus

$$\Delta F = F_1 - F_2 = 0 = (H_1 - TS_1) - (H_2 - TS_2);$$

therefore

$$\Delta H_{tr} = T\,\Delta S_{tr}. \tag{6–4}$$

Energy relationships for more orthodox ceramics may also be represented by the type of F-T plots presented in Fig. 6–2. Figure 6–3 shows an F-T plot for silica. It is significant to note that the "driving force," ΔF, is never very great among the various silica polymorphs. This plot also indicates the metastable ranges of the several phases.

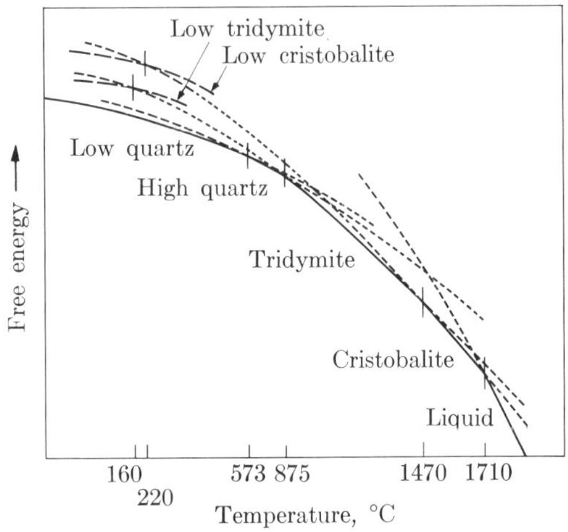

FIG. 6-3. Free energy versus temperature of SiO_2 (schematic), at atmospheric pressure. The stable form at any temperature is the phase with the lowest free energy.

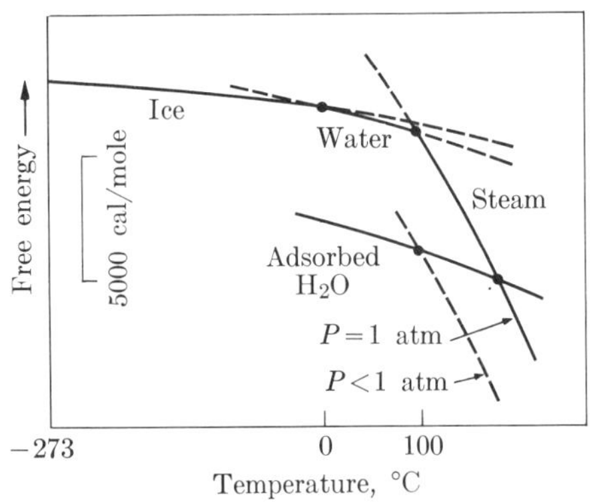

FIG. 6-4. Free energy of adsorbed H_2O (schematic). Atmospheric pressure, except as indicated.

Adsorption energy. At this point, it may be advantageous to consider the free-energy relationships for adsorbed components. The example chosen is H_2O adsorbed onto a crystalline surface (such as the kaolinite layer shown in Fig. 3-17). The free energy for H_2O at one of the adsorption sites is indicated by a separate curve in Fig. 6-4.

Our previous knowledge permits us to note the following points: (1) The temperature must be raised above 100°C (or the pressure reduced below one atm) for desorption to occur. (2) The adsorbed H_2O has a high degree of order (low entropy) such as that found in the crystalline substrate. In this respect, some mineralogists and ceramists compare the adsorbed H_2O with polymorphs of ice. (3) The free-energy curves for adsorbed H_2O are different at different adsorption sites. Specific contrasts may be drawn between H_2O at (a) the upper surface, (b) the lower surface, and (c) the edges of kaolinite (cf. Fig. 3-17), and for water adsorbed as a second layer over the first layer. These contrasts provide a range of equilibrium curves on the F-T plot, and also a range of drying temperatures or pressures (Section 5-2).

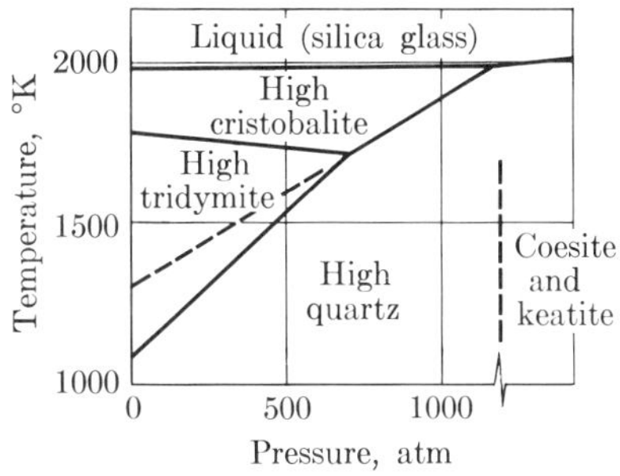

FIG. 6–5. The SiO_2 system. The denser phases are the higher-pressure phases. (A more complete phase diagram would include the low-temperature modifications of each phase. See Section 3–6.)

Transformation pressures. Phase changes can occur at constant temperature if the pressure is changed. Figure 6–1 and the free-energy curves in Fig. 6–2 indicate that the gas phase becomes more stable as the pressure is reduced. Conversely, the denser phase is brought on by higher pressures (LeChatelier's principle). The following (Clausius-Clapeyron) relationship exists:

$$dT/dP = T\,\Delta V_{\mathrm{tr}}/\Delta H_{\mathrm{tr}}, \qquad (6\text{–}5)$$

where ΔV_{tr} is the volume change accompanying the transformation, and ΔH_{tr} is the corresponding heat of transformation. The ice-to-water transformation will therefore have a negative value of dT/dP, since the liquid is denser (Fig. 6–1).

Figure 6–5 shows the phase diagram for the SiO_2 system. The boundaries which define the stabilities of the four low-pressure phases have been rather definitely established and are consistent with Eq. (6–5). As we mentioned above, the higher-pressure phases are denser; however, the stability limits of coesite and keatite have not been fully ascertained.

6–3 Polycomponent systems (binary)

Probably many readers have a general familiarity with the two-component phase diagram, so we shall present only a summary of it here. However, the presentation will include certain details that are not usually encountered, so that the binary diagram can serve as a basis for later ternary systems (Section 6–4). For these considerations, the pressure is regarded as fixed at one atm, and hence is eliminated as the environmental factor in the application of the phase rule (Eq. (6–1b)).

General. A T-X binary phase diagram shows the types of phases as a function of temperature T and composition X. Thus, in Fig. 6–6, a composition containing 80%* Al_2O_3 has a single liquid phase at 2200°C (3990°F), and a

---

* Weight percents are used for condensed phases unless mole, volume, or atom percent are specifically stated. However, volume percents are used for gaseous phases, unless another basis is specifically stated.

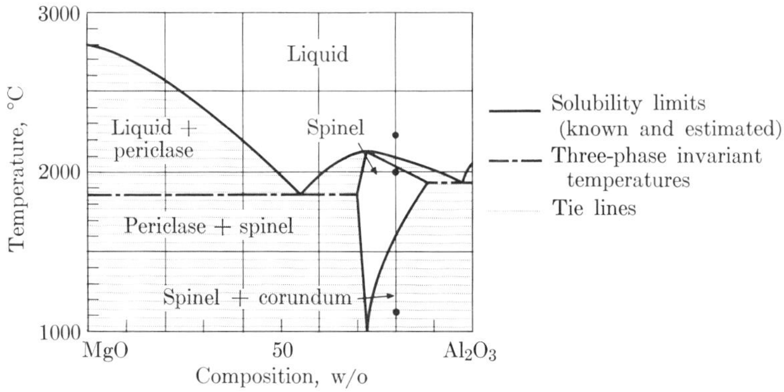

FIG. 6–6. The MgO-Al_2O_3 system.

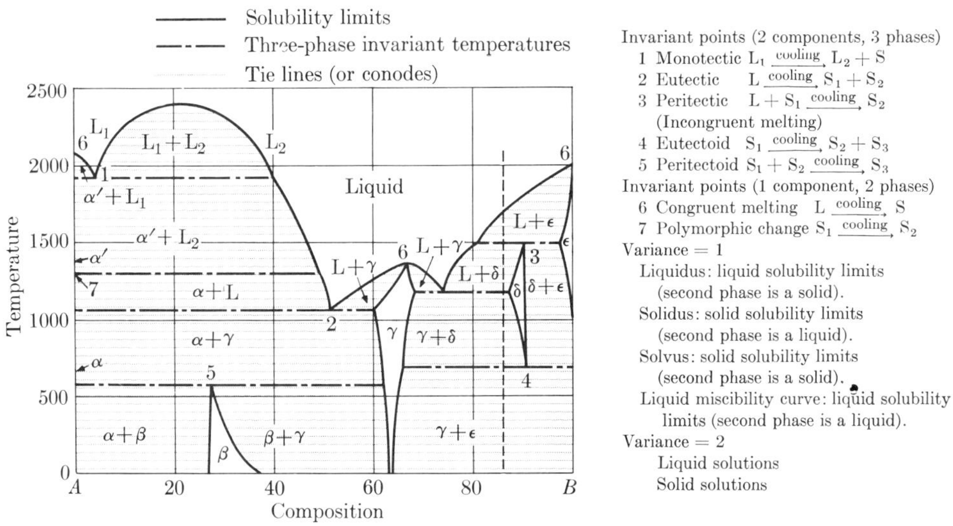

FIG. 6–7. Binary phase relationships (hypothetical) at constant pressure.

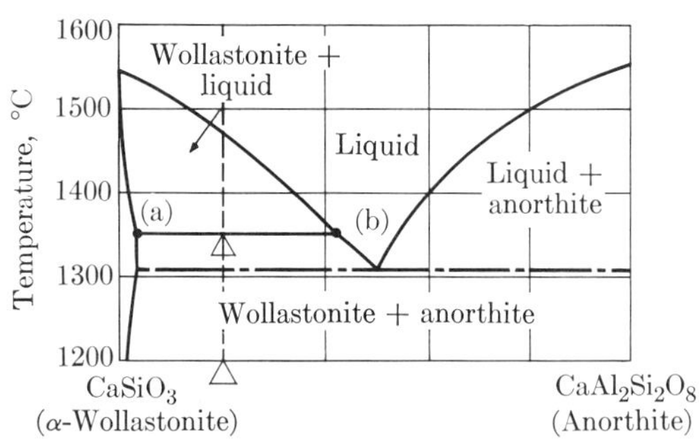

FIG. 6–8. Binary quantitative relationship: $CaSiO_3$-$CaAl_2Si_2O_8$ (α-wollastonite-anorthite).

single solid phase at 2000°C (3630°F). At 1100°C (2010°F), there are two coexisting solid phases: corundum and spinel. The corundum is nearly pure Al_2O_3. The spinel (basically $MgAl_2O_4$) dissolves additional Al_2O_3 into its structure. The solubility limit is shown by the boundary between the spinel area and the spinel-plus-corundum field.

Variance. In single-phase areas, both temperature and composition may vary independently without engendering a new phase. Solubility lines or limits, which constitute the boundary between one- and two-phase fields, define T-X relationships in which the variance is limited to one degree of freedom. If either the composition or the temperature is specified, the other is thereby established. At the points where two solubility lines cross, there are three phases in equilibrium with one another; whereupon both temperature and composition become invariant. At least one phase must disappear before either variable can be changed. Figure 6–7 summarizes a number of variance possibilities.

Tie lines (conodes). The compositions of equilibrated phases are determined by the curves of the solubility limits. [See points (a) and (b) of Fig. 6–8.] Tie lines or conodes are used to join the two equilibrated compositions, and hence they are horizontal (isothermal) lines extending across the two-phase fields.

Center-of-gravity principle. The total composition of the system must lie somewhere between the compositions of two equilibrated phases. In fact, the overall composition, Δ, of solid and liquid in Fig. 6–8 lies at the center of gravity relative to the amounts of the two phases and their respective compositions:

$$\frac{\text{Amount of } L}{\text{Amount of } W} = \frac{\Delta - (\text{a})}{(\text{b}) - \Delta} = \overline{\text{a}\Delta}/\overline{\text{b}\Delta}, \qquad (6\text{–}6\text{a})$$

where L is the liquid and W is wollastonite ($CaSiO_3$). Equation (6–6a) is known as the *lever rule*. Other forms of the lever rule as applied to the above examples include:

$$\frac{\text{Amount of } L}{\text{Total}} = \frac{\Delta - (\text{a})}{(\text{b}) - (\text{a})} = \overline{\text{a}\Delta}/\overline{\text{ab}}; \qquad (6\text{–}6\text{b})$$

$$\frac{\text{Amount of } W}{\text{Total}} = \frac{(\text{b}) - \Delta}{(\text{b}) - (\text{a})} = \overline{\text{b}\Delta}/\overline{\text{ab}}. \qquad (6\text{–}6\text{c})$$

Sequence of crystallization. Observation of the sequence of crystallization makes it possible to obtain considerable information about compositions and equilibrium changes. The sequence of crystallization is usually relatively simple in a binary system, because there are usually only primary and secondary crystalline phases. For example, in Fig. 6–8, any composition on the left side of the eutectic has α-wollastonite ($CaSiO_3$) as a primary phase and anorthite ($CaAl_2Si_2O_8$) as a secondary phase.

Resorption. Occasionally, more complicated situations arise. A liquid of composition 85% B (15% A), as shown in Fig. 6–7, will on cooling first solidify to ϵ as the primary phase. However, at the peritectic temperature, a resorption or peritectic reaction occurs, with secondary δ replacing the primary phase. Eventually a tertiary phase, γ, is generated. A characteristic of the resorption reaction is that it is slow, because it requires diffusion of the atoms through the δ phase which forms at the interface between the liquid and the ϵ phase. Therefore it is not uncommon to find remnants of phases which would have disappeared if equilibrium conditions had prevailed.

Subsystems. Binary systems may be divided into subsystems if the end members of a subsystem behave congruently over the range of specific interest. For example, the MgO-Al_2O_3 system (Fig. 6-6) may be divided into two binaries; (a) periclase-spinel (MgO-$MgAl_2O_4$), and (b) spinel-corundum ($MgAl_2O_4$-Al_2O_3).

6-4 Polycomponent systems (ternary and quaternary)

Many ceramic materials contain three or more components. No new principles are involved in these polycomponent systems; however, a complication arises because paper is limited to two dimensions and our minds to three. These are not enough dimensions to accommodate the $(C - 1) + E$ variables in an equilibrated system (Section 6-1). There are three procedures which facilitate the handling of the problem: (a) the use of triangular and tetrahedral plots to present compositional data; (b) fixing of variables; and (c) fixing of phases by saturation.

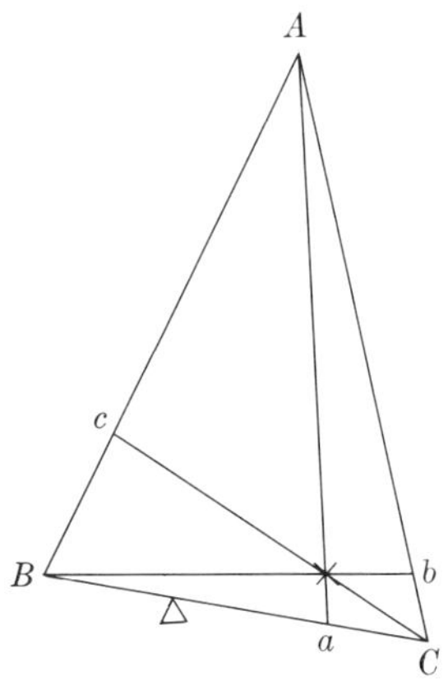

FIG. 6-9. Triangular plot: $\overline{aX}/\overline{aA} + \overline{bX}/\overline{bB} + \overline{cX}/\overline{cC} = 1$. This is a silica-anorthite-wollastonite system, with $A = SiO_2$, $C = CaAl_2Si_2O_8$, and $B = CaSiO_3$. See also Fig. 6-12.

Compositional plots. Since a three-component system has only two independent compositional variables (the third can only be the remainder), ternary compositions can be presented in two dimensions. A triangle provides a convenient plot, and although an equilateral one is commonly used, it is to the reader's advantage to realize that other triangles are also appropriate. Figure 6-9 is used as a specific example.

The center-of-gravity principle is applicable here. The overall compositions are expressed in terms of the corner components, the amount of each being given as follows:

$$\text{Amount of } A = \overline{aX}/\overline{aA}. \tag{6–7}$$

Therefore

$$\overline{aX}/\overline{aA} + \overline{bX}/\overline{bB} + \overline{cX}/\overline{cC} = 1. \tag{6–8a}$$

Three items should be noted. (1) The overall composition must lie within the triangle. (2) The corner compositions may be components, equilibrated phases, raw materials, or any other compositions of interest. (3) When one of the three compositions is zero, a binary relationship exists between the other two.

EXAMPLE 6–1. Refer to Fig. 6–12. (a) Locate the compositional position for 50% SiO_2, 25% Al_2O_3 and 25% CaO.

Answer: It is the point denoted by a cross in Fig. 6–12.

(b) Locate the compositional position for 8% tridymite (SiO_2), 23.5% wollastonite ($CaSiO_3$), 68.5% anorthite ($CaAl_2Si_2O_8$ or CAS_2).

Answer: It is the point denoted by a cross in Figs. 6–10 and 6–12.

A four-component system may be presented as a tetrahedron with a different substance or component at each of the four corners. Equation (6–8b) applies,

$$\overline{aO}/\overline{aA} + \overline{bO}/\overline{bB} + \overline{cO}/\overline{cC} + \overline{dO}/\overline{dD} = 1, \tag{6–8b}$$

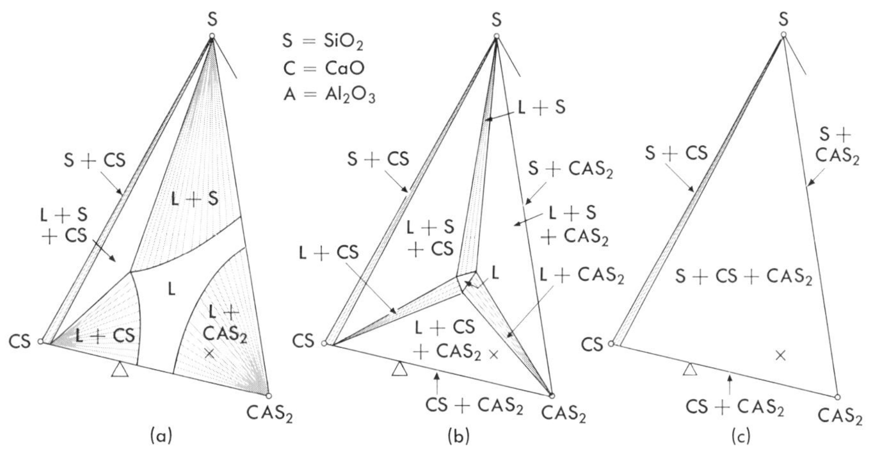

FIG. 6–10. Isothermal plots: SiO_2-$CaSiO_3$-$CaAl_2Si_2O_8$. (a) 1350°C (2460°F). (b) 1200°C (2190°F). (c) 1100°C (2010°F). Compare with items discussed in Section 6–4.

where O is the overall composition. Everything cited in the preceding paragraph can be restated to apply here.

Isothermal plots. Three constant-temperature sections of the silica-wollastonite-anorthite (SiO_2-$CaSiO_3$-$CaAl_2Si_2O_8$) system are shown in Fig. 6-10. A direct comparison may be made between these sections and binary sections (Fig. 6-7 and 6-8) for relationships such as tie lines, variance, and the center-of-gravity principle for any two or three phases (or for the three components).

A three-phase composition in a three-component system at constant pressure has a variance of one (Eq. (6-1b)). This situation is found in each of the isotherms of Fig. 6-10. However, it is evident that there is a specific temperature between 1200°C and 1100°C where four phases coexist (L, S, CS, and CAS_2), and the variance is reduced to zero. The temperature of this invariant equilibrium, as determined by experiment, is 1170°C (2140°F).

EXAMPLE 6-2. Refer to Fig. 6-12. (a) Locate the composition for 33% liquid, 13% wollastonite ($CaSiO_3$) and 54% anorthite ($CaAl_2Si_2O_8$ or CAS_2) at 1200°C.

Answer: It is the point denoted by crosses in Figs. 6-12 and 6-10(b).

(b) What would be the composition of the liquid at 1400°C if the overall composition is 50% SiO_2, 25% Al_2O_3 and 25% CaO?

Answer: 52% SiO_2, 22% Al_2O_3, 26% CaO.

Liquidus plots. The most common of the saturation plots is the liquidus surface (Fig. 6-11). These plots are widely used in ceramic literature because they supply extensive information for determining sequence of crystallization, variance relationships, and quantitative phase relationships.

The liquidus plot shows the isotherms at which the liquid is saturated with solid phases. Figure 6-11 should be compared directly with Figs. 6-8 and 6-10.

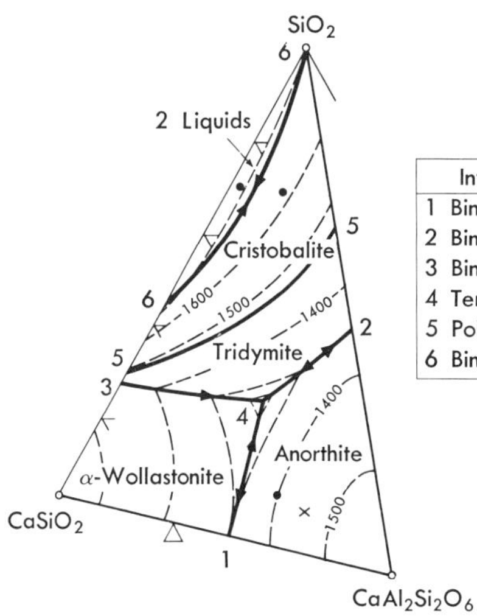

Invariant point	Phases	Temperature, °C
1 Binary eutectic	L-CS-CAS_2	1299
2 Binary eutectic	L-S-CAS_2	1359
3 Binary eutectic	L-S-CS	1436
4 Ternary eutectic	L-S-CS-CAS_2	1170
5 Polymorphic change	L-Cri-Tri	1470
6 Binary monotectic	L_1-L_2-S	1698

FIG. 6-11. Liquidus plot: SiO_2-$CaSiO_3$-$CaAl_2Si_2O_8$. Temperature is plotted perpendicularly to the plane of the paper. (Compare with Figs. 6-8, 6-9, 6-10, and 6-12.)

Further, let us note that Fig. 6–11 is simply a subsystem of the CaO-Al_2O_3-SiO_2 ternary system (Fig. 6–12). If a subsystem involves three equilibrated solid phases, it is called a *compatibility triangle*.

If temperature is the only environmental factor, the invariant points of a liquidus plot have four equilibrated phases, which include the liquid and the three solid phases with which it is saturated. (Or, if a monotectic is involved, the four phases are two liquids and two solids.) Thus there is an invariant point at the junctions of the three solubility surfaces: (1) a four-phase eutectic reaction involves an invariant point which lies *within* the compatibility triangle of the three equilibrated solid phases. (2) A four-phase peritectic reaction involves an invariant point which lies *outside* the corresponding compatibility triangle, and the inappropriate solid is resorbed.

Since one of the four phases in a peritectic reaction must disappear before further cooling can occur under equilibrium conditions, inspection is required to determine whether the liquid can continue to exist at lower temperatures: (2a) Under equilibrium conditions, the liquid will disappear at the peritectic reaction temperature if the total composition lies within the compatibility triangle of the three solid phases present. (2b) The inappropriate solid will be resorbed and disappear if the total composition lies in a compatibility triangle other than that for the three solid phases which are in equilibrium at the peritectic reaction temperature. Cooling will then continue, with the liquid and the two remaining solid phases, to the next invariant temperature.

EXAMPLE 6–3. Refer to Fig. 6–12. (a) Locate the first invariant reaction on cooling for 50% SiO_2, 25% Al_2O_3, 25% CaO. Is it a peritectic or a eutectic?

Answer: Point 16. It is a eutectic reaction involving L, S, CS, and CAS_2.

(b) Locate the first invariant reaction on cooling for 25% SiO_2, 70% Al_2O_3, 5% CaO. Is it a peritectic or a eutectic? What phase must disappear before further cooling under equilibrium conditions?

Answer: Point 3. It is a peritectic reaction involving L, C, A_3S_2, and CAS_2. The liquid disappears.

(c) Locate the first invariant reaction on cooling for 40% SiO_2, 49% Al_2O_3, 11% CaO. Answer the same questions here as in (b).

Answer: Point 3. It is the same peritectic reaction as (b). Corundum (Al_2O_3) disappears.

Sequence of crystallization. Assuming equilibrium conditions, the sequence of crystallization may be followed for a material containing 25% CaO, 25% Al_2O_3, and 50% SiO_2 (point × of Fig. 6–11). The liquid becomes saturated with respect to anorthite ($CaAl_2Si_2O_8$) at 1450°C (2640°F). Therefore, anorthite is the primary phase. As anorthite crystallizes, the remaining liquid becomes depleted of the anorthite composition. Since the overall composition must be one which is intermediate between those of the equilibrated phases, the tie line

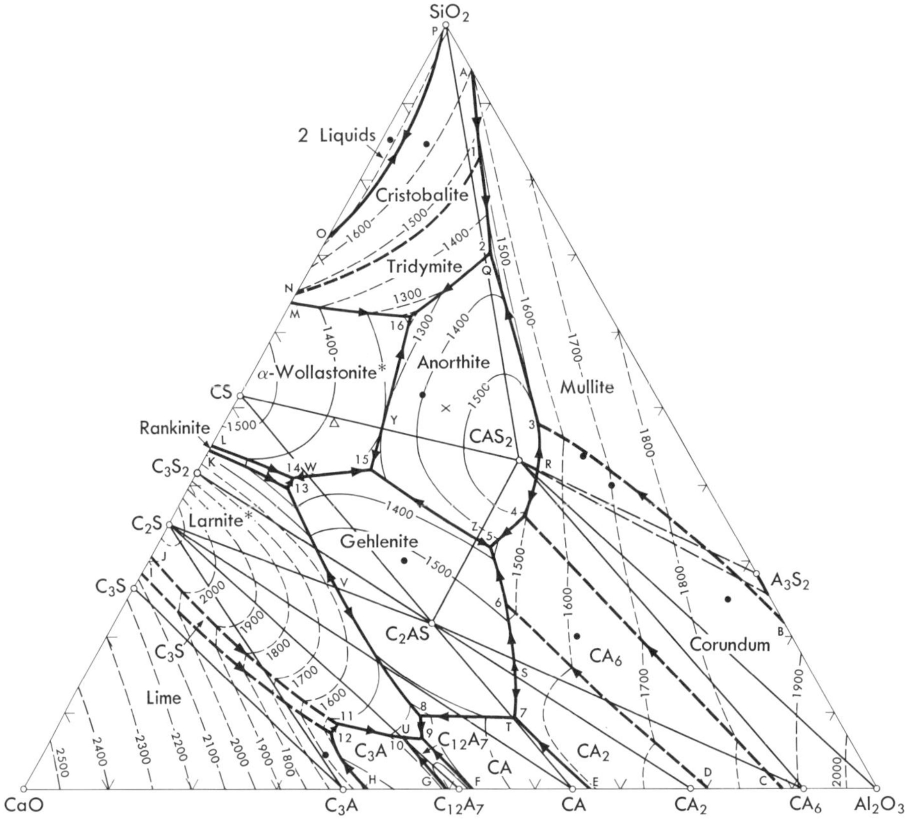

Compound	Melting temp., °C	Phase name	Composition Al_2O_3	Composition CaO	Composition SiO_2
A-Al_2O_3	~2020	Corundum	100	—	—
C-CaO	~2570	Lime	—	100	—
S-SiO_2	1723	Cristobalite	—	—	100
S-SiO_2		Tridymite	—	—	100
S-SiO_2		Quartz*	—	—	100
CS	1544	Wollastonite*	—	48	52
A_3S_2	~1850	Mullite	72	—	28
CAS_2	1553	Anorthite	37	20	43
C_2AS	1593	Gehlenite	37	41	22
C_3S_2	*D*†	Rankinite	—	58	42
C_2S	~2130	Larnite*	—	65	35
C_3S	*D*†	—	—	74	26
C_3A	*D*†	—	38	62	—
$C_{12}A_7$	1455	—	51	49	—
CA	~1605	—	64.5	35.5	—
CA_2	~1750	—	78.5	21.5	—
CA_6	*D*†	—	91.5	8.5	—

*Name of 25°C polymorph
†*D* = decomposes

Invariant Points									
Ternary					Binary				
	Temp., °C	Al_2O_3	CaO	SiO_2		Temp., °C	Al_2O_3	CaO	SiO_2
1	1470	12	5	83	A	~1590	6	—	94
2	1345	20	10	70	B	~1840	78	—	22
3	1512	37	15	48	C	~1850	89	11	—
4	1495	41	23	36	D	~1730	80	20	—
5	1380	39	29	32	E	~1595	66.5	33.5	—
6	1475	45	31	24	F	1400	53	47	—
7	1500	53	37.5	9.5	G	1395	50	50	—
8	1380	42	48.5	9.5	H	1535	41	59	—
9	1335	43.5	49.5	7	I	~2070	—	72	28
10	1335	41.5	52	6.5	J	~2050	—	70	30
11	1455	33	58.5	8.5	K	1464	—	55.5	44.5
12	1470	33	59.5	7.5	L	1460	—	55	45
13	~1315	12	49	39	M	1436	—	36	64
14	~1310	12	47	41	N	1470	—	35	65
15	1266	20	38	42	O	1707	—	27	73
16	1170	15	23	62	P	1707	—	1	99
					Q	1368	19.5	11.5	70
					R	1547	39.5	18.5	42
					S	1552	50	35	15
					T	1512	52.5	38	9.5
					U	1335	41	52.5	6.5
					V	1545	24	50	26
					W	1318	13	46	41
					Y	1307	37	30	33
					Z	1385	19	34	47

FIG. 6–12. The CaO-Al_2O_3-SiO_2 system. [Adapted, with some simplifications, from Osborn, E. F., and A. Muan, "The System CaO-Al_2O_3-SiO_2," *Phase Equilibrium Diagrams of Oxide Systems*, Columbus, O.: American Ceramic Society, Plate I (1960).] This diagram includes some recent revisions, and therefore may not match older diagrams exactly.

at 1350°C (2460°F) of Fig. 6–10(a) locates the liquid's composition. Between 1300°C and 1200°C, the altered liquid also becomes saturated with respect to α-wollastonite, the secondary phase. The composition of the liquid at 1200°C (2190°F) is indicated in Fig. 6–10(b). The final liquid has the ternary eutectic composition shown in Figs. 6–11 and 6–12. At that temperature (1170°C or 2140°F), the liquid becomes saturated with respect to a third solid phase, i.e., tridymite.

A second example (11% CaO, 49% Al_2O_3 and 40% SiO_2) will illustrate the sequence of crystallization when resorption occurs (Fig. 6–12). Corundum (Al_2O_3) is the primary phase. However, the composition is in the compatibility triangle (subsystem) for silica-anorthite-mullite (SiO_2-$CaAl_2Si_2O_8$-$Al_6Si_2O_{13}$); therefore corundum cannot be a final phase. As Al_2O_3 is removed from the liquid, the composition of the latter is shifted until at about 1650°C (3000°F) mullite becomes a secondary phase. From this point, the liquid composition follows the saturation curve between the corundum (Al_2O_3, that is, A) and mullite ($Al_6Si_2O_{13}$, that is, A_3S_2) areas until a third solid phase (anorthite, that is, CAS_2) appears. This produces an invariant condition. Before further cooling can occur, one of the four phases must disappear. Since the three solids, corundum (Al_2O_3), mullite (A_3S_2), and anorthite (CAS_2) cannot comprise the overall 11C-49A-40S composition, one of these three phases is resorbed via a peritectic reaction. When the corundum does disappear, the remaining three phases (liquid, mullite, and anorthite) can be cooled to the ternary eutectic at 1345°C (2450°F) with the liquid composition at point 2 in Fig. 6–12.

We highly recommend that the reader try his hand at determining the sequence of crystallization for other compositions, and also at calculating the amounts of phases at selected temperatures. Some compositions can be easily handled if reasonable care is exercised; for example, 25CaO-55Al_2O_3-20SiO_2 or 40CaO-30Al_2O_3-30SiO_2. Others will require close attention to the phase rule and the center of gravity principle; such as 5CaO-70Al_2O_3-25SiO_2; 12CaO-44Al_2O_3-44SiO_2; 62CaO-33Al_2O_3-5SiO_2; 10CaO-5Al_2O_3-85SiO_2; or 14CaO-1Al_2O_3-85SiO_2.

Other procedures. We can sometimes simplify ternary and quaternary systems by fixing the percentage of one of the components. It would be possible, for

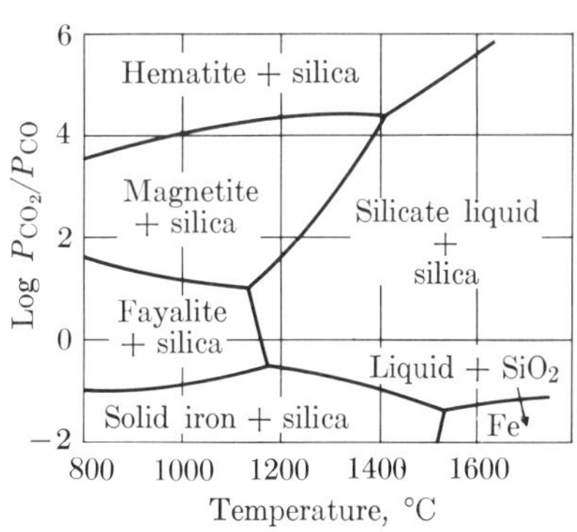

FIG. 6–13. Fe-Si-C-O system: each phase is saturated with silica; gas is present in all areas. [Adapted from Colligan, G. A., "Iron-Silica Sand Interface Reactions," The University of Michigan, Ph.D. dissertation (1959); and Darken, L. S., "Melting Points of Iron Oxides and Silica: Phase Equilibrium in the System Fe-Si-O as a Function of Gas Composition and Temperature," *Jour. Amer. Chem. Soc.*, **70**, 2048 (1948).]

example, to select a 30 percent SiO_2 cut of Fig. 6-12 and make a pseudo-binary diagram of temperature versus the CaO and Al_2O_3 composition. Such a procedure has the advantage of revealing sub-liquidus phase relationships. However, since true binaries are not represented, care must be taken to account for all of the variables before making interpretations.

The ceramist may use other procedures for revealing the effect of the pertinent variables. Figure 6-13 shows a plot of phase relationships in the Fe-C-O-Si system. Since this system always required excess silica, one composition variable was removed. The amount (activity) of the oxygen was varied through a control of the CO/CO_2 ratios. Thus the curves of the figure become univariant lines in which four phases are involved. These four include those shown in each pair of adjacent fields, plus a CO_2/CO gas.

6-5 Rates of reaction

The reactions of ceramics, as a class, occur slowly. This fact is illustrated by the metastability of glass (Section 4-5), but is also observable in other materials through the nonequilibrium microstructures which develop during cooling (see Fig. 7-8). However, the fact that the reaction rates of ceramic materials are slower than those of metals does not mean that these rates may be ignored. In fact, many ceramic materials are used at high temperatures where their rates of reaction are increased. Furthermore, high-temperature reactions are the basis of ceramic processing (Chapter 14).

Reactions can be categorized as homogeneous and heterogeneous. Homogeneous reactions occur within a single phase; the most familiar example is the combustion of a gas such as methane with the oxygen in air and the formation of H_2O and CO_2 (also gases). There are few such reactions in solid ceramic phases unless one considers diffusion as a reaction. This will be discussed in a later subsection.

A heterogeneous reaction involves two or more phases, and is therefore the usual type of ceramic reaction. Included are drying, adsorption, crystallization, sintering, oxidation, reduction, dissociation, solution and precipitation of phases. Because this type of reaction does involve more than one phase, several steps are involved: (a) Nucleation of a new phase (if not already present). (b) Diffusion of the reactants to the reaction site. (c) Actual reaction at the phase boundary. (d) Diffusion of the products of reaction away from the phase boundary. The rate of reaction is limited by the slowest of the above steps.

Nucleation. If a new phase must be formed (e.g., the crystallization of glass), it must start in a very small region with a high interface-to-volume ratio. This requirement is a definite deterrent to nucleation, because energy is necessary in order to form the new phase boundary. And energy is not available in a single-phase material until there has been considerable supercooling, and until the free energy of the metastable phase greatly exceeds that of the new stable phase. It is for this reason that phase transformations occur more slowly immediately below the transformation temperature than they do when additional supercooling is employed (Fig. 4-6).

If no new phase is required to initiate a reaction, these nucleation lags do not occur. Alternatively, if nucleators are present, the lags are decreased. As we recall, nucleators may be impurity phases or structural imperfections (Section 3-5) which provide sites from which growth can continue. In principle, nucleators provide the required surface, so that less supercooling is required. We shall cite specific examples in Section 14-5 for nucleation in glass.

Diffusion. Atomic movements (or diffusion) do occur in ceramic solids as well as in liquids and gases. Of course, movements are much slower in ceramic solids; however, these atoms must move if a reaction is to proceed. The point defects of Fig. 3-18 become important in this situation, as indicated in Fig. 6-14. For an atom to be moved, an energy barrier must be overcome. The energy necessary to do this is called the activation energy. Since, as we know, atom movement increases at higher temperatures, this activation energy is usually supplied from thermal sources.

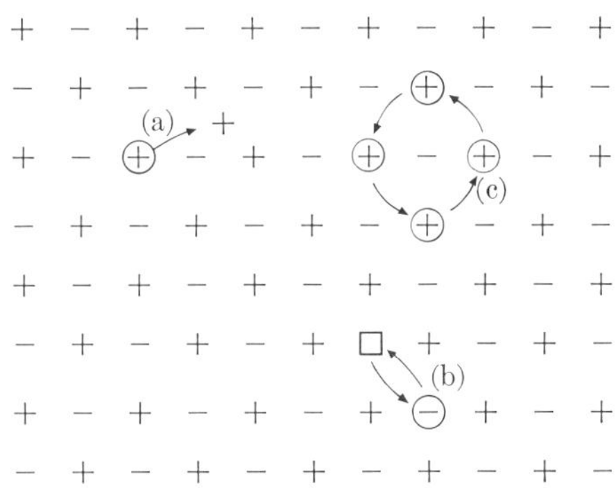

FIG. 6-14. Diffusion mechanisms: (a) interstitialcy diffusion; (b) vacancy diffusion; (c) ring diffusion.

In the absence of a concentration gradient (or more correctly, an activity gradient), atom movements do not bring about a net diffusion of atoms. With a gradient, there *is* a net diffusion, because atoms move from regions of high concentration to regions of low concentration. The flux J of atom movements along a gradient, dC/dx, is

$$J = -D(dC/dx), \tag{6-9}$$

where D is a diffusion coefficient and is expressed in units such as cm^2/sec. The dependence of the diffusion coefficient on temperature may be expressed as follows:

$$D = Ae^{-Q/kT}. \tag{6-10}$$

Among these terms, Q is the previously cited activation energy, k is Boltzmann's

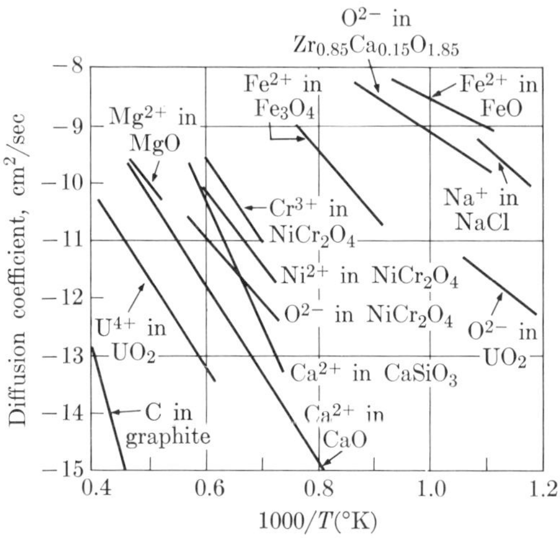

Fig. 6–15. Diffusion coefficients versus $1/T$. (Adapted from Kingery, W. D., *Introduction to Ceramics*. New York: John Wiley & Sons, 1960, page 232.)

constant, and T is temperature. The term A is a constant which varies from material to material.

A summary of diffusion coefficients in several selected crystalline ceramic solids is plotted as a function of $1/T$ in Fig. 6–15. We may make several observations. (1) The diffusion coefficient increases with temperature. (2) The diffusion coefficient is high for those ions which can enter existing lattice vacancies. This accounts for the high coefficients of Fe^{2+} in FeO, and of O^{2-} in $Zr_{0.85}\,Ca_{0.15}\,O_{1.85}$. (3) Compounds which melt at higher temperatures have lower ion diffusion coefficients than those which melt at lower temperatures. Here a comparison may be made between Na^+ in NaCl and Mg^{2+} in MgO. Although glass is not among the examples given in Fig. 6–15, the diffusion coefficients of ions in glass are appreciably higher than in close-packed crystalline solids; e.g., the diffusion coefficient of Na^+ in a sodium silicate glass is 10^{-5} cm^2/sec at 1000°C (1830°F).

Because at lower temperatures the diffusion coefficients are lower, the rates of reaction are markedly reduced, and thus cause a delay in (a) the movement of the reactants to the reaction site, and (b) the movement of the products away from the reaction site.

References for Further Reading

6–1. Eitel, W., *Silicate Melt Equilibria.* New Brunswick, N. J.: Rutgers University Press, 1951. A thorough presentation of relationships encountered in solid–liquid ceramic equilibria.

6–2. Levin, E. M., C. R. Robbins, and H. F. McMurdie, *Phase Diagrams for Ceramists.* 1964 edition. Columbus, O.: American Ceramic Society. This volume is a compendium of all previously published ceramic phase diagrams, and is essential to the library of a ceramist.

6–3. Osborn, E. F., and A. Muan, *Phase Equilibrium Diagrams of Oxide Systems.* Columbus, O.: American Ceramic Society, 1960. A series of ten ternary diagrams, printed on large-scale plates, which give detailed information involving the most common refractory and slag silicates.

6–4. Rhines, F. N., *Phase Diagrams in Metallurgy.* New York: McGraw-Hill, 1956. A detailed presentation of phase relationships encountered in metallurgy, which gives less attention to liquidus and more to subliquidus relationships than the above references.

Problems

(Assume equilibrium unless indicated otherwise. Refer to Appendix F where necessary.)

6–1. Kaolinite is calcined, then heated to just above the eutectic temperature. How much melt and how much mullite result?

6–2. A melt of MgO and SiO_2 is cooled until the liquid, upon analysis, is shown to contain 65% MgO. The liquid comprises 70% of the total. What was the composition of the original melt?

6–3. See Reference 6–2. How much liquid will be present if SiO_2 is equilibrated at 1650°C (3000°F) with 2 w/o of each of the following oxides: (a) CaO? (b) MgO? (c) MnO? (d) FeO? (e) K_2O? (f) Na_2O? (g) Li_2O? (h) Al_2O_3?

6–4. (a) What is the primary phase for a melt containing 50% SiO_2, 30% Al_2O_3, and 20% CaO? (b) The secondary phase? (c) The tertiary phase?

6–5. Indicate the sequence of crystallization of the following compositions.

(a) 25% CaO-55% Al_2O_3-20% SiO_2 (b) 40% CaO-30% Al_2O_3-30% SiO_2
(c) 5% CaO-70% Al_2O_3-25% SiO_2 (d) 12% CaO-44% Al_2O_3-44% SiO_2
(e) 62% CaO-33% Al_2O_3- 5% SiO_2 (f) 10% CaO- 5% Al_2O_3-85% SiO_2
(g) 14% CaO- 1% Al_2O_3-85% SiO_2

6–6. For the seven compositions of Problem 6–5, show the compositions of the phases at each 100°C interval between the liquidus and the solidus temperatures.

6–7. For the seven compositions of Problem 6–5, how much of each phase would be present at each 100°C interval between the liquidus and the solidus temperatures?

6–8. Repeat Problem 6–5, substituting MgO for CaO.

6–9. Repeat Problem 6–6, substituting MgO for CaO.

6–10. Kaolinite and gibbsite (Table 3–5) are mixed and calcined to give equal weights of Al_2O_3 and SiO_2. Then CaO is added to give: (a) 1%, (b) 5%, (c) 10%, (d) 18%, (e) 25%, (f) 50%, and (g) 65% CaO. What phases are present at 1400°C in each case? What are the phase compositions? How much of each phase is present per 100 gm total?

6–11. The reaction between excess kyanite (Al_2SiO_5) and an iron silicate slag (approximating $FeSiO_3$ in composition) is continued until equilibrium is established at 1600°C, 1300°C, 1200°C, and 1100°C. (a) What phases are present? (b) What are their compositions? [*Note:* The final cooled equilibrated product contains 75% mullite ($Al_6Si_2O_{13}$).]

6–12. On the basis of Fig. 7–7, draw the diagram for the binary FeO-Al_2O_3.

Chapter 7

CERAMIC MICROSTRUCTURES

7–1 Introduction

Commercial ceramic materials, except in the case of glass and in certain single-crystal applications, have microstructures. The term "microstructure," as applied here, is used to describe those structural characteristics found in polygranular (polycrystalline or polyphase) materials. Microstructures usually lend themselves to observation and study by microscopic means, and some of them are even visible to the unaided eye (e.g., the microstructure of coarse grinding wheels). On the other hand, there are certain colloidal ceramic materials (e.g., clay casting slips) which are submicroscopic, but whose structural features, like those of the larger microstructures, result from grain and phase composites (Section 7–6). Thus in the discussions in this chapter we shall not group ceramic microstructures by their sizes, but by other characteristics.

Microstructures can be characterized by five factors: (1) grain size, (2) grain shape and distribution, (3) phase quantity, (4) grain orientation, and (5) porosity. Sometimes the fifth factor, porosity, is considered to be a phase of zero composition. As such, it could be incorporated under (3). Each of these factors will be considered; but first, let us briefly give our attention to procedures of micrographic observation.

7–2 Ceramic micrography

Micrographic techniques may be divided into two classifications: (1) phase identification, and (2) microstructural measurements. Both transmitted and reflected light procedures are available for these qualitative and quantitative purposes. However, since there is decreasing emphasis on transmitted and increased emphasis on reflected light procedures, only the latter will be discussed here. (See Reference 7–1 for specific information on transmitted light techniques.) In addition to the optical methods cited above, electron microscopy is becoming very important; we shall therefore consider its advantages and limitations.

Reflected light techniques. In comparison with transmitted light techniques, the use of reflected light has the direct advantage of laboratory simplicity: (1) only one surface has to be polished; and (2) any opacity of phases is relatively unimportant. Also, it is possible to observe fine details of grain and phase

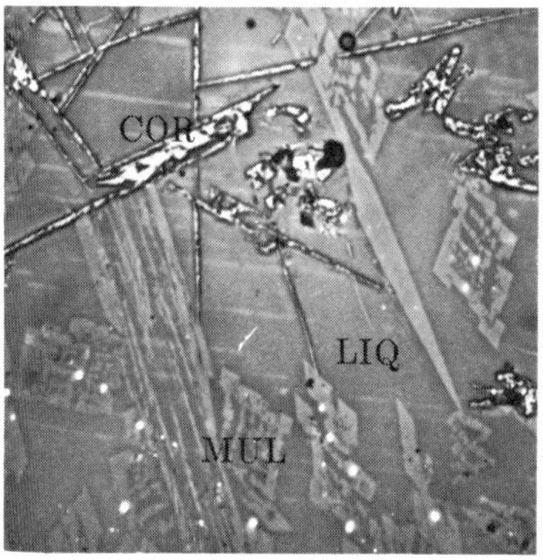

Identifying characteristics of the phases are: CORundum (Al_2O_3): hardness, polishes in relief; shape, cross section of basal plates; reflectivity, bright. MULlite ($Al_6Si_2O_{13}$): hardness, no relief; shape, cross section of elongated prisms; reflectivity, slightly brighter than equilibrated glass. LIQuid silicate (now a glass): reflectivity, low; location, matrix.

FIG. 7–1. Photomicrograph showing phase differences (×500; reflected light; unetched).

juxtaposition. Admittedly, we can generally view only two dimensions; however, internal reflections may be used to determine some of the three-dimensional relationships among nonopaque phases.

Micrographic procedures involving reflected light usually depend on some initial knowledge of the phases to be encountered. This information may come (1) through a knowledge of total composition, (2) from prior analyses of structures by x-ray diffraction, or (3) from prior determinations of optical indices by transmitted light. Identification of specific phases by means of reflected light utilizes properties such as (1) reflectivity, (2) grain shape, (3) grain distribution, (4) relative hardness, and (5) color of internal reflections. These characteristics are compared in Fig. 7–1 for corundum (Al_2O_3), mullite ($Al_6Si_2O_{13}$), and glass. Quantitative procedures are discussed in subsequent sections, and other examples of photomicrographs of ceramic materials made with reflected light are presented at the end of the chapter (Fig. 7–14 to 7–23).

Electron microscope techniques. The major advantage of electron microscopic procedure is that it can resolve submicron microstructural details. Thus Fig. 7–2 shows surface relief characteristics of a magnetic ceramic which would have been missed if the ceramic had been observed by normal reflected or transmitted light optical microscopy, which has a resolution limit of about 0.5 μ. Inasmuch as electrons are readily absorbed, electron microscopic procedures cannot be used for thick sections, except to provide silhouettes of the geometry of the material. Therefore, for electron transmission examination, it is necessary either to prepare extremely thin sections ($<1\mu$) or to prepare collodion replicas of the surface of the ceramic; this was done for Fig. 7–2. In practice, surfaces may be (1) fractured, (2) polished, or (3) polished and etched, depending on the point of interest of the micrographer. Principles and techniques of the electron microscope are presented in References 7–3 and 7–5, as well as in a number of other recent books.

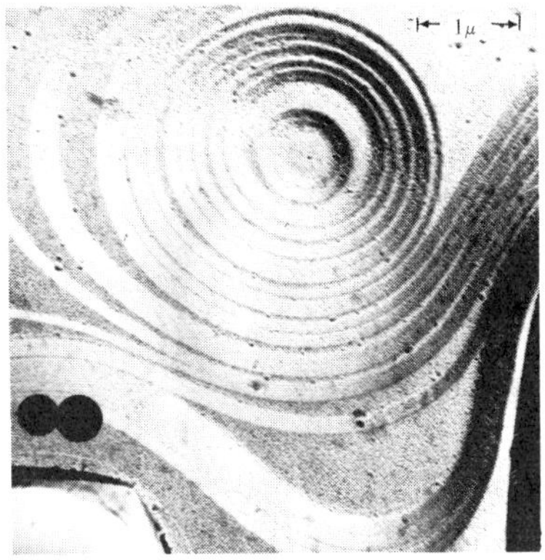

FIG. 7-2. Electron photomicrograph (×10,000) of a magnetic ceramic (CoZ). This surface structure was formed by vapor deposition. (Courtesy W. C. Bigelow and J. Drobek.)

Since, generally speaking, electron microscopy uses only grain morphology for phase identification, it is necessary, when using the electron microscope, to depend even more extensively on the supplemental means of initial phase identification than when using reflected light techniques. However, such information is commonly available to the ceramist working with a specific ceramic product. Selected electron photomicrographs of ceramic materials are presented at the end of the chapter.

7-3 Grain size

A *grain* can be considered an individual crystal within a material.* Generally a grain has an irregular boundary which is a consequence of its juxtaposition with other grains. However, it is not uncommon to find euhedral grains; i.e., grains which have sides that are determined by their internal structure (Section 7-4).

Just as ceramic materials may have microstructure ranging from submicroscopic to coarsely macroscopic (Section 7-1), there may be variations in grain sizes within any one material. Also, the grain size may be varied through heat treatment (Chapter 14).

Grain size measurement. A common procedure is to measure the "diameter" of grains with a microscope. This is a useful and quick procedure and gives an index of the size. However, it must not be used as an absolute measure, unless appropriate provisions are made for statistical variations. For example, a reflected light microsection (or a thin-section examined by transmitted light), shows a two-dimensional cut through three-dimensional grains. Under such conditions, the *median* of the observed diameters would equal 0.87 of the true

---

* Sometimes a grain is considered to be a separable particle which may or may not be a single crystal, e.g., a particle of fireclay grog which contains many small crystals. This definition of a grain will not be used in this chapter, where we are concerned with the microstructural characteristics of ceramics; however, it does have significance in ceramic forming processes (see Chapter 13).

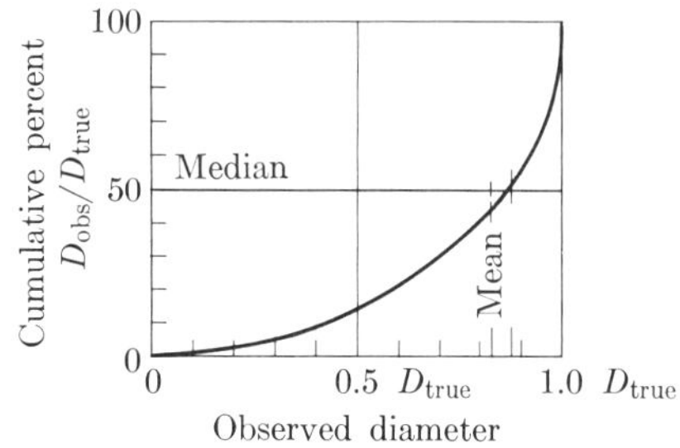

FIG. 7–3. Observed diameters: spheres with random sectioning.

diameter of spherical grains (Fig. 7–3). Furthermore, the *mean* diameter of a two-dimensional section is only 0.82 of the true diameter.

In practice, it is usually the relative size of the grains of two different structures that is of interest. Therefore the choice of measurement procedures is not critical, so long as we remember, when comparing data, that we must compare data gathered by the same procedure.

Grain growth. Isothermal grain growth proceeds under the driving force of interfacial energy (Section 5–4). A coarse-grained material has less interfacial area and therefore lower surface free energy. Figure 7–4 shows an example of grain growth in which solid wüstite (FeO) grains are growing within a $FeO\text{-}SiO_2$ liquid matrix. Atoms along the interface with a smaller radius of curvature (smaller grains) are less stable than are atoms at the boundary of a large grain. Therefore, the small grain is dissolved by the liquid, and the atoms precipitate onto the surface of the larger grains.

Single-phase ceramics also exhibit grain growth. When such a ceramic material is porefree, the growth proceeds by transfer of atoms across the grain

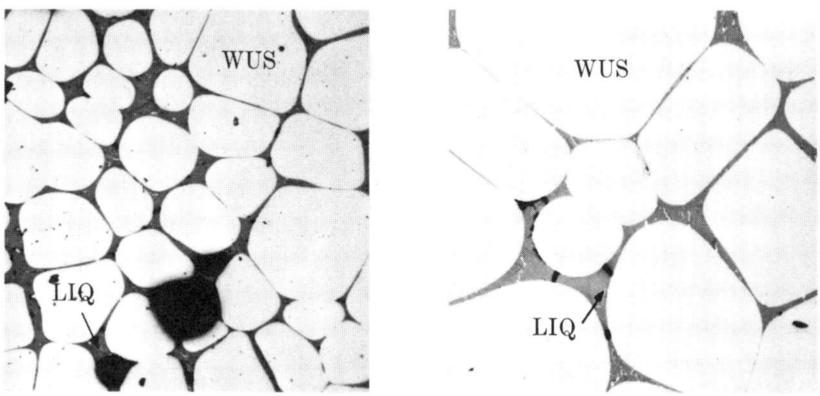

(a) (b)

FIG. 7–4. Grain growth (×150; reflected light). WÜStite (FeO) in an $FeO\text{-}SiO_2$ LIQuid matrix: (a) four hours at 1200°C (2190°F), (b) 64 hours at 1200°C. The black areas are holes.

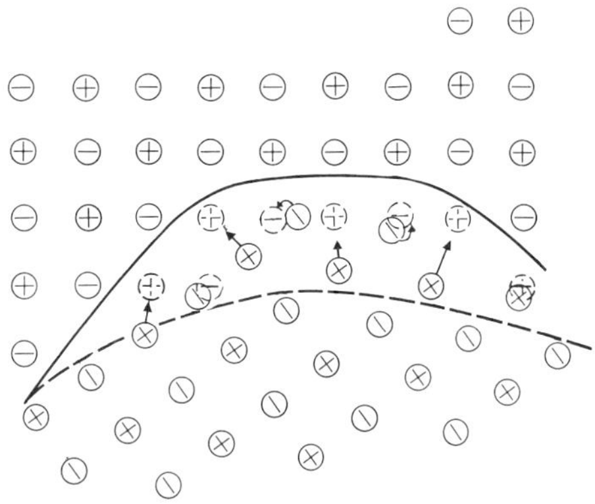

FIG. 7-5. Grain boundary movement. The interface area is reduced and the radius of curvature of the boundary is increased, resulting in a reduction of free energy.

boundary from the higher-energy convex surface to the lower-energy concave surface (Fig. 7-5).

Grain growth in porous ceramics is more complicated. Atoms must move from the small to the large grains by surface migration or by solid diffusion. Both procedures are possible, but both are slow in comparison with grain growth in the presence of a liquid phase. Often pores effectively arrest grain growth, because the curvature is reversed if the boundary tries to move past the pores. In such a situation, growth can proceed only as fast as the pores can be moved or eliminated.

Growth rates. In single-phase nonporous materials, the diameter D of a grain varies with time t, according to the following relationship:

$$dD/dt = k/D, \tag{7-1}$$

and

$$D_2^2 - D_1^2 = 2k(t_2 - t_1). \tag{7-2a}$$

Or, if the initial diameter is small,

$$D = (2kt)^{1/2}. \tag{7-2b}$$

The value of k is constant for any particular temperature or composition; its value increases with increasing temperature.

In general, the rate constant for isothermal grain growth is related to the rate of self-diffusion within a solid. This can be anticipated because the atom movements are related and require similar activation energies.

7-4 Grain shape and distribution

Grains may vary in shape from euhedral to anhedral. The external shape of the euhedral (or idiomorphic) grain is controlled by its internal structure, and certain planar faces and angles are observed. But the anhedral grain has no regularity of outline other than that imposed on it by its neighbors. The term

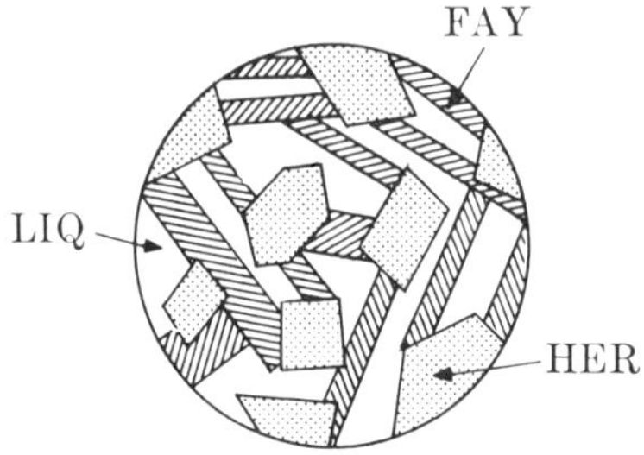

FIG. 7–6. Crystallization sequence (FeO-Al_2O_3-SiO_2 system). HERcynite ($FeAl_2O_4$) was the primary phase, and FAYalite (Fe_2SiO_4) is the secondary phase. The remaining glassy matrix is a supercooled eutectic LIQuid.

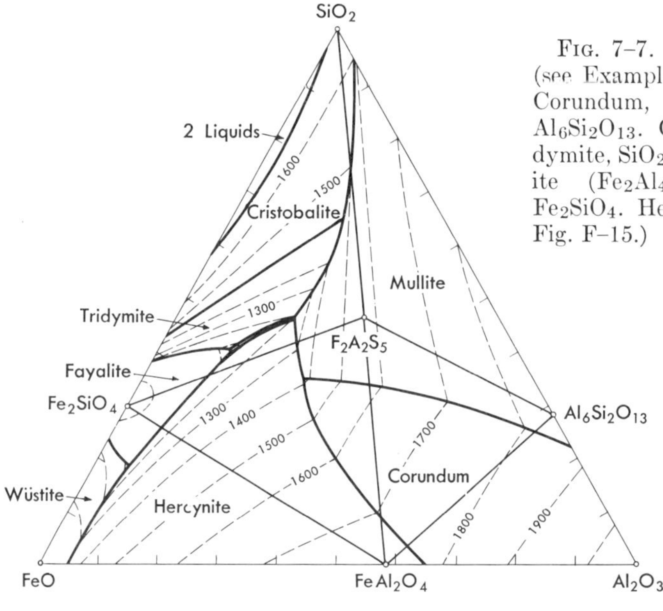

FIG. 7–7. Refractory alteration (see Example 7–1). Wüstite, FeO. Corundum, Al_2O_3. Mullite, $Al_6Si_2O_{13}$. Cristobalite, SiO_2. Tridymite, SiO_2. $F_2A_2S_5$, iron cordierite ($Fe_2Al_4Si_5O_{18}$). Fayalite, Fe_2SiO_4. Hercynite, $FeAl_2O_4$. (Cf. Fig. F–15.)

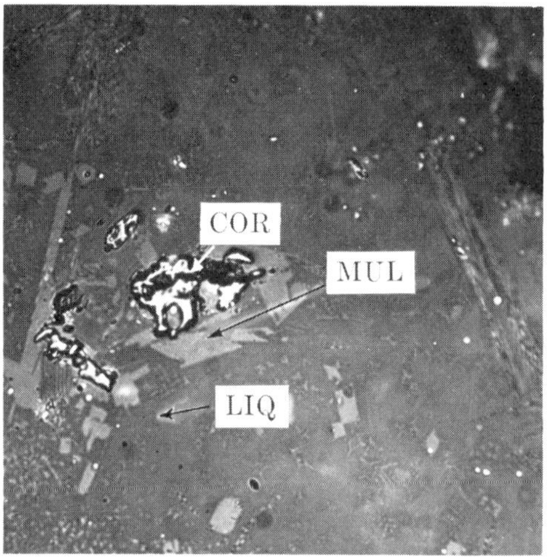

FIG. 7–8. Phase reactions (×500; reflected light). CORundum (Al_2O_3) reacting with a siliceous glass (LIQuid) to give MULlite ($Al_6Si_2O_{13}$).

"subhedral" is sometimes applied to a grain which has been subjected to solution, or surface tension, so that the euhedral shape is partially destroyed.

Phase reactions. The grain shape can often be used to help identify the sequence of crystallization. For the composition shown schematically in Fig. 7-6, the hercynite ($FeAl_2O_4$) had the freedom to develop an euhedral outline within the liquid. When the fayalite (Fe_2SiO_2) started to crystallize as a secondary solid phase at lower temperatures, it had its growth and shape limited by the existing hercynite crystals. (This particular composition may be cooled sufficiently fast that the silica-containing eutectic does not crystallize.) Microstructures such as these, when they are related to the appropriate phase diagram, can be of help in an analytical procedure. In this case, the quantities of the three component oxides can be estimated to within a few percent of their correct values (Section 6-4).

EXAMPLE 7-1. If the material in Fig. 7-6 contains 45, 35, and 20 weight percent (w/o) respectively of hercynite ($FeAl_2O_4$), fayalite (Fe_2SiO_4), and liquid at 1100°C (2010°F), what is the overall composition (w/o) of the structure?

Answer: From Fig. 7-7, the liquid would contain 49 w/o FeO, 12 w/o Al_2O_3, and 39 w/o SiO_2. A graphical solution involves making a triangular plot, with the three phases (hercynite, fayalite, and liquid) as the apexes (Fig. 6-9), and determining the composition of the center of gravity. Thus we obtain

53 w/o FeO, 29 w/o Al_2O_3, and 18 w/o SiO_2.

This may be determined mathematically as well as graphically.

The microstructures may also reveal alteration of a solid. Figure 7-8 shows corundum (Al_2O_3) which is being altered to mullite ($Al_6Si_2O_{13}$) by a siliceous glass. Originally present in the form of euhedral grains, the corundum was partly resorbed, while mullite rims were formed around corundum grains.

Phase distribution. A comparison of Figs. 7-4 and 7-9 reveals a difference in the distribution of the phases. The cristobalite (SiO_2) grains of Fig. 7-9 maintain a solid-to-solid contact. The wüstite (FeO) grains of Fig. 7-4 have been almost completely separated by a liquid film at the heat-treating temperatures. The differences in microstructure arise from the differences in interfacial energies (Fig. 5-4c). Since two of the three boundaries are equivalent, Eq. (5-6) can be rearranged to

$$\gamma_{S/S} = 2\gamma_{S/L} \cos \theta/2. \tag{7-3}$$

In this equation, $\gamma_{S/S}$ is the energy of the solid/solid grain boundary, $\gamma_{S/L}$ is the energy of the solid/liquid interface, and θ is the equilibrium dihedral angle which the liquid makes between the grains. Complete grain boundary wetting occurs when $\gamma_{S/S} > 2\gamma_{S/L}$.

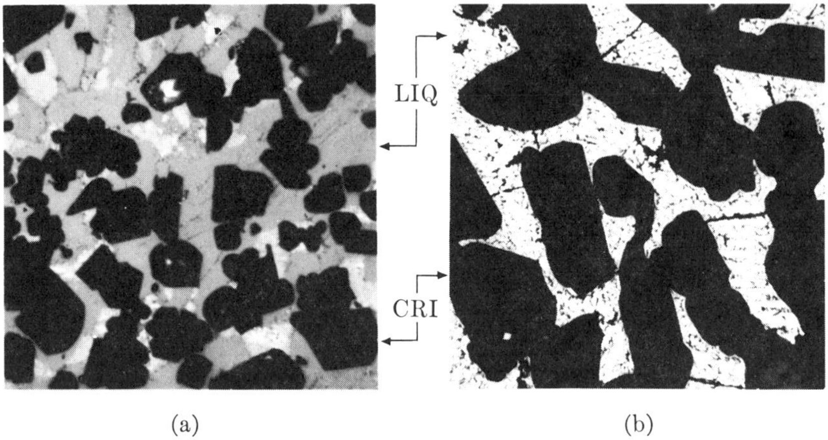

FIG. 7-9. Anhedralization (×300; reflected light). CRIstobalite (SiO_2) grains in an iron oxide-silica LIQuid at 1480°C (2700°F): (a) two hours, (b) 18 hours. The former euhedral grains have been replaced by larger subhedral and anhedral grains. (During the cooling, the liquid crystallized to magnetite and iron silicate.)

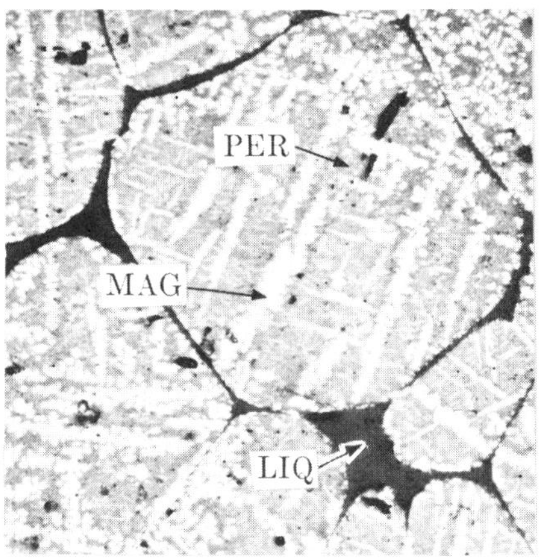

FIG. 7-10. Exsolution (×400; reflected light). Precipitation of MAGnesio-ferrite ($MgFe_2O_4$) from PERiclase (MgO). The grain boundary matrix is a glass resulting from the cooling of a eutectic MgO-Fe_2O_3-SiO_2 LIQuid. At high temperatures, only a MgO-Fe_2O_3 solid solution and a liquid were present. The magnesio-ferrite precipitated from the solid solution during cooling, leaving a purer MgO periclase. Precipitation was nucleated along the periclase crystal planes. (Courtesy R. G. Wells.)

Exsolution. A decrease in solubility in a solid permits supersaturation and a solid-from-solid exsolution (i.e., precipitation). Figure 7-10 shows such precipitation of magnesio-ferrite ($MgFe_2O_4$) from periclase (MgO). At high temperatures, the Mg^{2+} ions of a periclase structure can be replaced by Fe^{3+} ions (plus vacancies). As the entropy decreases at lower temperatures, most of the Fe^{3+} ions precipitate within a spinel ($R^{2+}R_2^{3+}O_4$) with an appropriate number of Mg^{2+} and O^{2-} ions, so that the remaining periclase is purer MgO. Solid precipitation is nucleated most readily along selected crystal planes. Hence the microstructure exhibits a Widmanstätten-like pattern.

7-5 Phase quantities

It is only natural to assume that the properties of ceramic materials will be influenced by the relative quantities of the various phases which are present. Quantitative measurements are therefore important.

There are several methods of determining the quantity of any selected phase within a microstructure. All provide volumetric answers; so, of course, subsequent calculations are necessary if we wish to find weight percentages. Point and linear analyses (Fig. 7-11) are the most widely recommended.

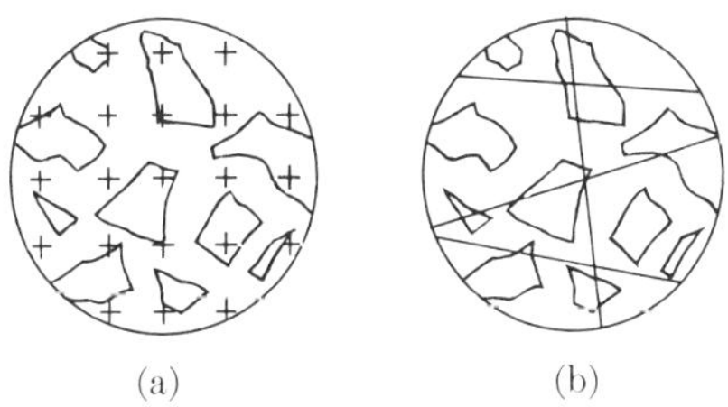

FIG. 7-11. Point and linear analyses (30 v/o of minor phase). (a) Point analysis; 30% of random points fall within the minor phase. (b) Linear analysis; 30% of the total length of random lines fall within the minor phase.

Point analysis is based on the fact that the probability that any random point within a material will fall within one phase is equal to the volume percentage of the phase present in the material. Linear analysis is based on a similar probability; viz., a random line will lie within any phase in proportion to the amount of that phase. With a given amount of effort on the part of the micrographer, point analyses can be slightly more accurate than linear ones. In each case, special effort must be made to avoid any nonstatistical selection of points or lines, and to prevent one phase being lost during sample preparation; e.g., the holes in Fig. 7-4. Assuming that the microscopic section is representative of the whole structure, and that statistical randomness has been maintained, the accuracy of data is given by the following relationship:

$$\sigma = \sqrt{F(1.0 - F)/N}, \tag{7-4a}$$

or

$$F = \left(1 \pm \sqrt{1 - 4N\sigma^2}\right)/2. \tag{7-4b}$$

In these equations, σ is the standard deviation of successive samplings, F is the true fraction, and N is the number of randomly selected observation points.

Example 7–2. Refer to Fig. 7–4(b). (a) Using either point or linear analysis, calculate the volume percentages of FeO and of liquid in this microstructure. (b) What is the composition of the material? (Specific gravity of wüstite is 5.7, and of the liquid, approximately 4.7.)

Answer: (a) By linear analysis: 8 v/o liquid.

(b) The composition is

$$\text{w/o liquid} = \frac{(8)(4.7)}{(8)(4.7) + (92)(5.7)} = 6.7 \text{ w/o.}$$

From Fig. 7–7, the liquid contains 75 w/o FeO and 25 w/o SiO_2.

$$\text{Total FeO} = (100 - 6.7) + (6.7)(0.78) = 98.5 \text{ w/o.}$$

$$\text{Total } SiO_2 = (6.7)(0.22) = 1.5 \text{ w/o.}$$

7–6 Colloidal structures

The microstructures of ceramic slips (suspensions) and pastes are important in ceramic technology. However, these microstructures cannot be examined directly because of (1) the liquid surrounding them, and (2) the submicron size of the colloidal particles (Section 5–2). Therefore it becomes necessary to infer the nature of the colloidal structures indirectly from studies of rheology, adsorption, and electrodialysis.

Slips and pastes differ primarily in their solid/liquid ratios. Slips have sufficient liquid content to flow; pastes have yield strengths which are high enough to maintain their shape for a significant length of time. Obviously there can be gradations between these two conditions. Furthermore, for given proportions of solid and liquid, there can be variations in structure, with a resultant effect on properties. It is the structural variation which we shall consider in this section. The flow properties will be discussed in the next chapter.

Deflocculation. The solid/liquid interface in a two-phase suspension develops a charge as a result of bond fracture and/or surface adsorption (Fig. 5–1). When the surface is markedly negative, colloidal particles have the characteristics of very large anions. Adjacent particles are repelled by each other; moreover, the particles are attracted to the anode within an electrolytic cell. When there is mutual repulsion between adjacent particles, a suspension can be deflocculated (i.e., dispersed). There are several consequences of deflocculation that will be discussed in the following chapter, among which is the fact that settling will occur more slowly in a deflocculated than in a flocculated suspension.

Because there is a repelling surface charge, the effective volume of a small particle is greater than its true volume. It may be shown that the repulsive

TABLE 7-1

EFFECTIVE VOLUMES OF COLLOIDAL PARTICLES

Particle diameter	Effective particle diameter*	Effective volume fraction**
1.0μ	1.04μ	0.112
0.5	0.54	0.125
0.1	0.14	0.27
0.05 } 500 A }	0.09	0.59†

* Adsorbed layer, approximately 200 A.
** Based on 0.10 true volume fraction (spheres).
† The packing factor for the closest packing of spheres is 0.74.

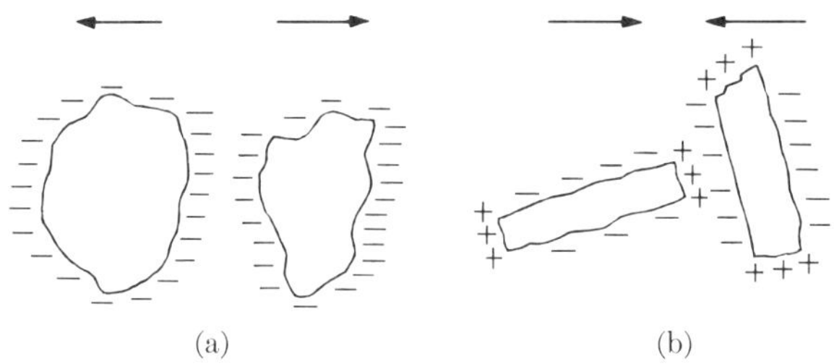

FIG. 7-12. Colloidal interactions: (a) like surface charges—repulsion and deflocculation; (b) mixed surface charges—attraction and edge-to-face flocculation.

"shell" in the presence of ionic surface charges is about 200 A. Table 7-1 shows the effective volumes of these colloidal particles; note that, as the particle size decreases, the effective volume becomes a large fraction of the total volume.

Flocculation. The surface structure of anisotropic particles varies with the crystal structure, and it is possible for the surface adsorption to vary (Section 5-2). In fact, parts of the surface may be negative and other parts positive in character. An extreme example is kaolinite particles, which have positive edges and negative faces (Fig. 7-12b) when they are suspended in a weak acid solution (pH = 4-6). Such a configuration develops coulombic attraction, and flocculation (aggregation) is a result of edge-to-face attraction. Large aggregates of particles will settle faster and otherwise behave differently from dispersed particles (Section 8-8).

If the pH is increased so that the positive edge charges disappear, the edge-to-face attractions also disappear, and deflocculation results. Similarly, addition of even a few polyanions can disperse a flocculated suspension if the anions are preferentially adsorbed on the edges of the colloidal particles.

In the absence of the extensive adsorption shown in Fig. 7–12, there will be opportunity for face-to-face flocculation as a consequence of relatively weak van der Waals forces. Such flocculation provides a much denser aggregate than that resulting from edge-to-face flocculation. Sometimes we refer to the two forms of flocculation by the descriptive terms of "card-house" and "card-deck" (Fig. 7–13).

(a) (b)

FIG. 7–13. Flocculation of colloidal plates: (a) edge-to-face (card-house) flocculation; (b) face-to-face (card-deck) flocculation. (See Reference 7–8.)

Colloidal deposits. The solid content of a colloidal suspension may be increased by physically removing the liquid, or by gravity settling and the formation of a sediment. In ceramic processing, the more usual procedure is removing the liquid. Dewatering is achieved either by filtration or by a process of water absorption into a porous mold wall (Chapter 13).

The structures of colloidal deposits depend on the same surface forces cited above. If there is a strong tendency toward coulombic edge-to-face attractions, the sediment can build up quite rapidly. In this case, the liquid content is still quite large, the density is low, and the permeability is high. When the deposit is subsequently removed and dried, it has a high shrinkage factor. With face-to-face van der Waals flocculation, the process is slower. However, a denser, more perfectly oriented deposit is formed, which contains less liquid and shrinks less during drying.

References for Further Reading

7–1. BLOSS, F. D., *Methods of Optical Crystallography.* New York: Holt, Rinehart, and Winston, 1961. Crystal optics, with detailed attention to optical anisotropy. A knowledge of refraction laws is necessary.

7–2. COBLE, R. L., "Effect of Microstructure on the Mechanical Properties of Ceramic Materials," *Ceramic Fabrication Processes* (W. D. Kingery, editor). New York: John Wiley & Sons, 1958, pp. 213–228. Attention directed to properties rather than micrographic procedures.

7–3. COMER, J. J., "The Electron Microscope in the Study of Minerals and Ceramics," *Microscopy*, STP 257, Philadelphia: American Society of Testing and Materials, 1959, pp. 99–120. Specific reference to ceramic materials; a rather complete bibliography.

7–4. ECONOMOS, G., "Effect of Microstructure on the Electrical and Magnetic Properties of Ceramics," *Ceramic Fabrication Processes* (W. D. Kingery, editor). New York: John Wiley & Sons, 1958, pp. 201–213. Attention directed to properties; assumes a knowledge of electrical parameters.

7–5. FISCHER, R. B., *Applied Electron Microscopy*. Bloomington, Ind.: Indiana University Press, 1953. Although this book is older than others, it presents the principles of electron microscopy in an excellent manner.

7–6. INSLEY, H., and V. D. FRECHETTE, *Microscopy of Ceramics and Cements*. New York: Academic Press, 1955. Techniques are presented briefly but amply; main attention given to the use of the microscope to examine raw materials, whitewares, refractories, glass, cements, enamels, structural clay products, foundry sands, slags and abrasives.

7–7. KINGERY, W. D., *Introduction to Ceramics*. New York: John Wiley & Sons, 1960. Gives special attention (Chapter 13) to the microstructure of ceramics. Presents topological requirements as well as chemical relationships.

7–8. MICHAELS, A. S., "Rheological Properties of Aqueous Clay Systems," *Ceramic Fabrication Processes* (W. D. Kingery, editor). New York: John Wiley & Sons, 1958, pp. 23–31. Gives attention to the aggregation of colloidal particles.

7–9. SMITH, C. S., "Grains, Phases, Interfaces: An Interpretation of Microstructure," *Trans. A.I.M.E.*, **175**, 15, 1948. The classical paper on geometric equilibration in microstructures. Some familiarity with metallurgy required.

7–10. *Microstructures of Ceramics*, National Bureau of Standards Miscellaneous Publication No. 257, Washington, D.C.: U.S. Government Printing Office, 1964. A series of six lectures on ceramic microstructures and their effect on properties of ceramic materials.

Problems

7–1. Refer to Fig. 7–4. Predict the WUStite grain size for a 36-hr heat treatment under the conditions indicated.

7–2. Wollastonite ($CaSiO_3$) and kaolinite (Table 3–5) are mixed in equal proportions. Indicate the probable microstructural changes during their firing.

7–3. State what phases would be expected in an electrical insulator made from 85 w/o talc (Table 3–5) and 15 w/o Al_2O_3, assuming (a) complete equilibrium, (b) only partial equilibrium.

7–4. The microstructure of Fig. 7–1 is a consequence of FeO and Al_2O_3 additions to a fireclay (kaolinite) refractory, and heating for an extended period of time at 1600°C (2910°F), followed by rapid cooling. (a) What is the total composition? (b) What was the FeO/Al_2O_3 ratio in the addition? (c) What fraction of the total was fireclay (calcined)? (d) What is the composition of the glass (LIQuid)?

7–5. The surface of an iron casting is oxidized during solidification in a sand (quartz) mold at 1500°C (2730°F). Predict the reactions and subsequent microstructures. [*Note:* silica phase transformations will occur but are sluggish.]

7–6. Figure 7–14 shows a microstructure of quartz grains altered by liquid FeO. Account for the various phases present.

7–7. With 20 v/o solids in suspension, calculate the mean interparticle distance if the particle size is (a) 1μ, (b) 0.1μ, (c) 0.01μ. Assume an equiaxed shape.

7–8. A uniform suspension contains 30 v/o solids. The particles are platelike, with a thickness of 0.01μ and a larger dimension of 0.1μ. What is the average distance between particles?

Photomicrographs of Ceramic Microstructures

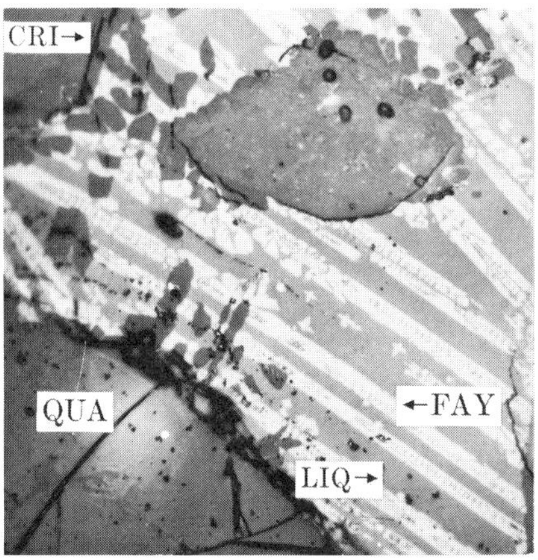

FIG. 7–14. Silica sand altered by an FeO liquid (×450; reflected light). See Problem 7–6. QUArtz (SiO_2), FAYalite (Fe_2SiO_4), CRIstobalite (SiO_2), LIQuid (FeO-SiO_2) eutectic. (See Fig. F–10.)

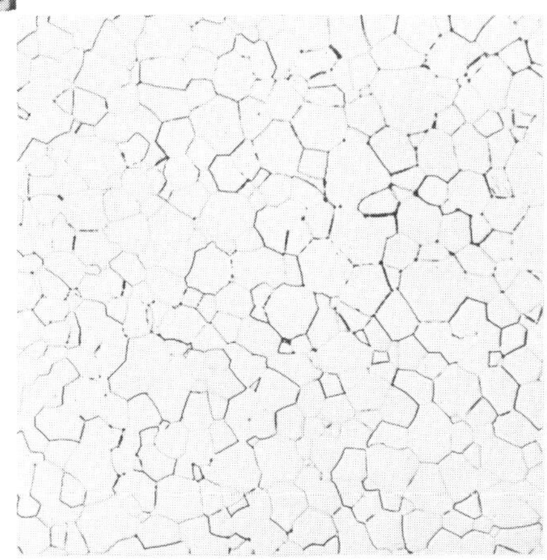

FIG. 7–15. Microstructure showing single-phase and equiaxed grains (×225; reflected light). The grain boundaries of high-density periclase, MgO, were revealed by an H_3PO_4 etch. (Gardner, R. E., and G. W. Robinson, Jr., "Improved Method for Polishing Ultra-High Density MgO," *Journ. Amer. Ceram. Soc.*, **45**, 46, 1962.)

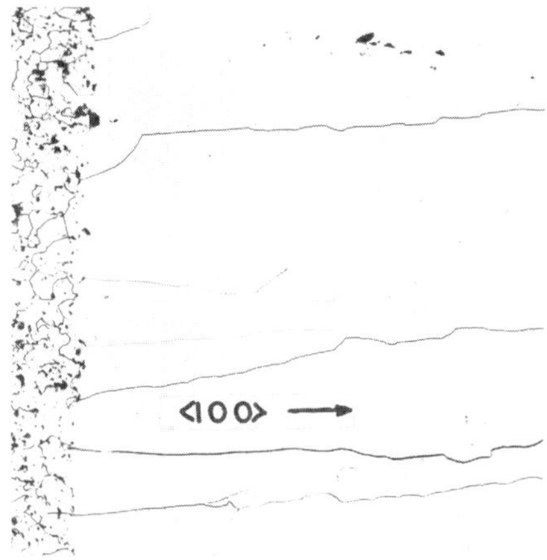

FIG. 7–16. Microstructure, showing columnar grains (×70; reflected light). The UO_2 had been melted and then solidified within a temperature gradient. (MacEwan, J. R., and V. B. Lawson, "Grain Growth in Sintered Uranium Dioxide: II, Columnar Grain Growth," *Journ. Amer. Ceram. Soc.*, **45**, 42–46, 1962.)

FIG. 7–17. Widmanstätten microstructure (×450; reflected light): UC_2 (bright) and UC have related orientations. Pressed at 1000°C and 35 kilobars for 20 minutes; etched in nitric acid, acetic acid, and water. (Chang, R., and C. G. Rhodes, "High-Pressure Hot Pressing of Uranium Carbide Powders, and Mechanism of Sintering of Refractory Bodies," *Journ. Amer. Ceram. Soc.*, **45**, 379–82, 1962.)

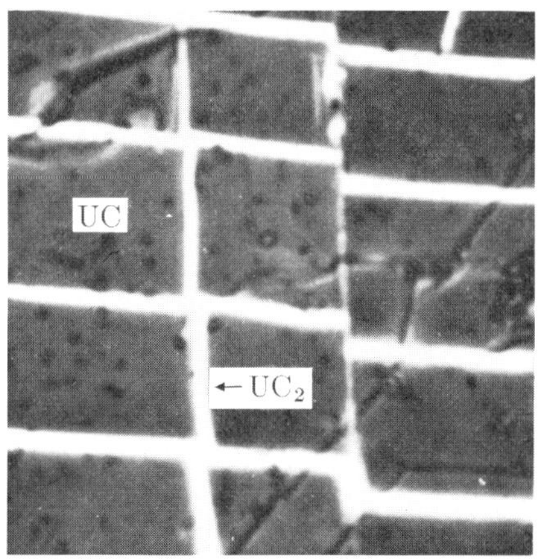

FIG. 7–18. Microstructure, showing fracture. UC_2 (bright) in UC, as in Fig. 7–17 (×900; reflected light). The fracture is revealed as a displacement on the Widmanstätten structure. Source: same as Fig. 7–17.

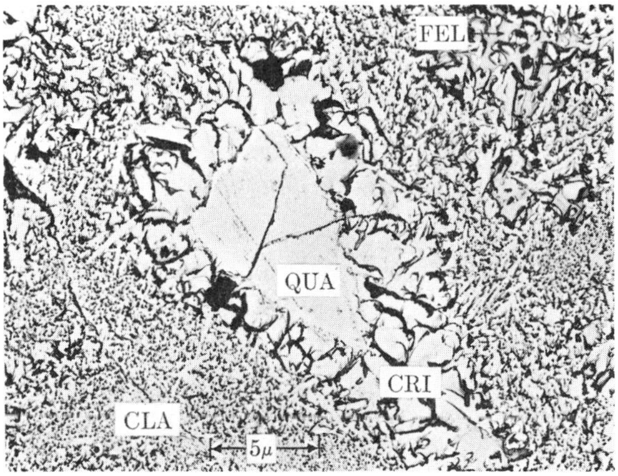

FIG. 7–19. Microstructure of a fired product: porcelain made from clay, feldspar and quartz. CLA, former clay, now fine mullite and glass. QUA, unaltered quartz, but cracked by the high-to-low transformation (Section 3–6). CRI, cristobalite, formed by high-temperature surface alteration of the quartz grain. FEL, former feldspar, now containing a significant amount of glass and coarse mullite. (Courtesy of S. Lundin, Royal Technological Institute, Stockholm. Electron micrograph from a replica; ×2500.)

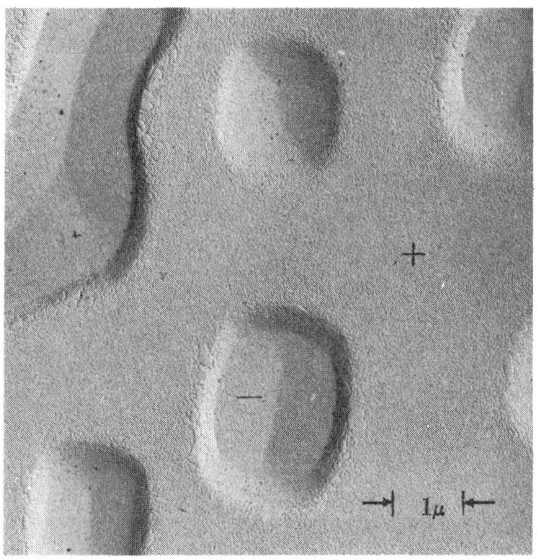

FIG. 7–20. Microstructure, showing domains (×8500; electron photomicrograph from a replica). $BaTiO_3$ aligned with the *c*-axes perpendicular to the picture, etched by H_3PO_4. The positive ends of the domains etched more rapidly than the negative ends. (Belitz, R. K., "Differential Etching of $BaTiO_3$ by H_3PO_4," *Journ. Amer. Ceram. Soc.*, **45**, 617–618, 1962.)

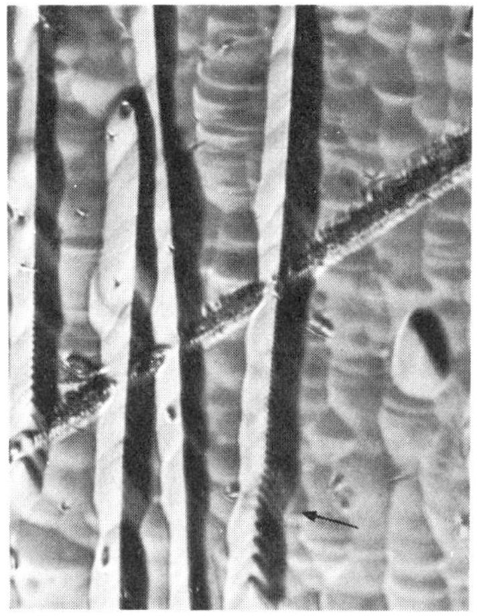

FIG. 7–21. Microstructure, showing dislocations (×850; reflected light). Rutile (TiO_2) after deformation and annealing at 1000°C for 25 hours. (Hirthe, W. M., and J. O. Britton, "Dislocations in Rutile as Revealed by the Etch-Pit Technique," *Journ. Amer. Ceram. Soc.*, **45,** 546–554, 1962.)

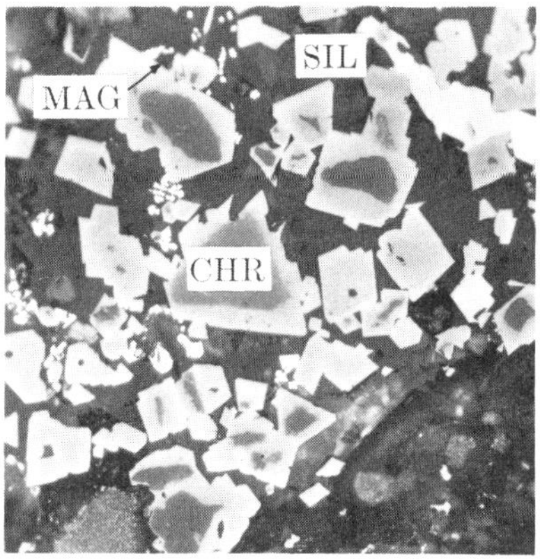

FIG. 7–22. Microstructure, showing alteration (×200; reflected light). CHRomite ($FeCr_2O_4$) grains undergoing MAGnetite ($FeFe_2O_4$) overgrowth and absorption. The two compositions are isomorphous. The higher index of refraction of the magnetite overgrowth gives it higher reflectivity than the chromite cores, while the low-index SILicate matrix has essentially no reflectivity. (Treffner, W.S., "Changes in Microstructure of Chromite Spinel During Service," *Journ. Amer. Ceram. Soc.*, **45,** 455–63, 1962.)

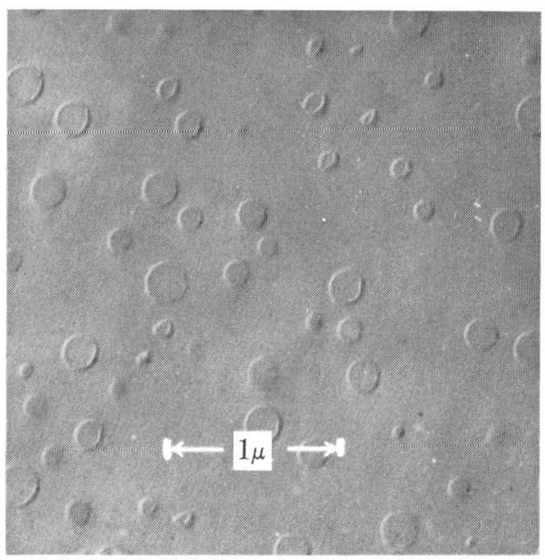

FIG. 7–23. Microstructure, showing glass immiscibility (×18,000; electron photomicrograph from a replica). Fractured surface of a Li_2O-B_2O_3-SiO_2 glass replicated after exposure in humid atmosphere. (Ohlberg, S. M., H. Golob, C. M. Hollabaugh, "Fractography of Glasses Evidencing Liquid-in-Liquid Colloidal Immiscibility," *Journ. Amer. Ceram. Soc.*, **45,** 1–4, 1962.)

Part 2

PROPERTIES AND BEHAVIOR OF CERAMIC MATERIALS

Part 1 of this text was devoted to the Nature of Ceramic Materials, with particular emphasis on structures (phases) which exist in the absence of external forces or fields.

Part 2 will focus on the response of these structures to imposed forces or fields. Chapter 8 will discuss *mechanical properties:* when forces originate externally to the phase or grain, and the response affects the structure of the phase, e.g., through deformation. Chapter 9 considers *thermal properties:* the energies of atoms within the structure are subjected to modification by heat. Chapter 10 deals with *electrical and magnetic properties* which are a consequence of intra-atomic (chiefly electronic) response to external energy fields. The subjects of these three chapters differ somewhat, in the sizes of the responding structural units: (a) phase, (b) atom, and (c) electron. However, there is some overlap; e.g., the mobility of electrons is affected by deformation of the phases.

Ceramic materials may be affected by radiation in both the electromagnetic and the particulate forms. *Optical* response to electromagnetic radiation is of major importance in Chapter 11. Thus Chapter 11 is related not only to the electrical properties of the preceding chapter but also to the radiation of higher intensity encountered in *nuclear* applications (Chapter 12).

These chapters on properties and behavior of ceramics provide the background for Part 3: The Processing of Ceramic Products.

Chapter 8

MECHANICAL PROPERTIES AND APPLICATIONS

8–1 Introduction

Although under applied mechanical stresses ceramic materials behave differently from metals (Section 1–2), they are used in many engineering structural applications. This chapter will review the response of ceramic materials to stresses, as revealed by their behavior in various elastic, fracture, and deformation situations. We shall also give attention to ceramic materials used in structural and abrasive applications. In general, we shall limit our discussion to the behavior of ceramics at ambient temperatures. The effects of temperature on mechanical behavior will be considered in greater detail in the following chapter.

8–2 Elastic properties

Any material is deformed slightly when stresses are applied. If the stress is small enough, or the period of time is short enough, the strain (deformation) disappears when the stress is removed. Strain which is thus recovered is elastic strain and the stress is elastic stress. Excessive stresses produce permanent atomic displacements, or plastic deformation.

Moduli of elasticity. With elastic strains, the atoms maintain their original neighbors. The strain simply alters the interatomic spacings. The modulus of elasticity E is a measure of the amount of stress σ required to produce unit elastic strain ϵ:

$$E = \sigma/\epsilon. \tag{8–1}$$

There are numerous empirical relationships which might be anticipated on the basis of our initial knowledge about interatomic spacings (Section 2–3). Since these relationships are generally substantiated by experiments, they will be reviewed here.

As a class, ceramic materials have higher moduli of elasticity than other materials (10^7 to 6×10^7 psi), which is not surprising, since the modulus of elasticity is an index of the strength of the interatomic bonds. Other indices

TABLE 8-1

MODULI OF ELASTICITY VERSUS MELTING TEMPERATURES

Compound	Average modulus of elasticity (10^6psi)	Melting temperature, °C
Titanium carbide	45	3180
Tungsten	60	2996
Silicon carbide	50	>2800
Periclase (MgO)	30	2800
Beryllia (BeO)	45	2585
Spinel ($MgAl_2O_4$)	35	2160
Corundum (Al_2O_3)	53	2050
Iron	30	1539
Copper	16	1083
Halite (NaCl)	5	801
Aluminum	10	660
Magnesium	6	650
Polystyrene	0.4	<300
Nylon	0.4	<300
Rubber	0.01	<300

of interatomic bond strength are melting temperature and hardness. Table 8-1 compares elasticities with melting temperatures, and although the correlation is not perfect, it is significant.

Graphically, the modulus of elasticity is the differential of the sum curve of Fig. 8-1. Therefore any change in the interatomic distances will affect this modulus. Three examples can be cited: (1) A compression stress will increase the modulus. This fact has been verified by compressibility experiments. (2) A tension stress will decrease the modulus.* This can also be measured in certain materials; however, because fracture occurs with relatively low tension stresses in ceramic materials (Section 8-3), there are no large increases in interatomic distance. (3) The modulus decreases as the temperature increases because thermal expansion increases the interatomic distance. This accounts for the gradual decrease in the modulus data in the 0-800°C range of Fig. 8-2. (The sharper drop at higher temperatures is a consequence of anelastic behavior, discussed in Section 9-6.)

There are variations of the modulus of elasticity with the crystallographic orientation that may be demonstrated with periclase (MgO). A tensile force can produce a larger elastic deformation per unit stress in the ⟨110⟩ directions than in the ⟨100⟩ directions. In the latter directions, any strain produces a direct separation of the positive-negative, coulombic bonds. In the ⟨110⟩ directions, strain can occur through a distortion of the 90° lattice angle. Although

* An exception occurs with linear polymers.

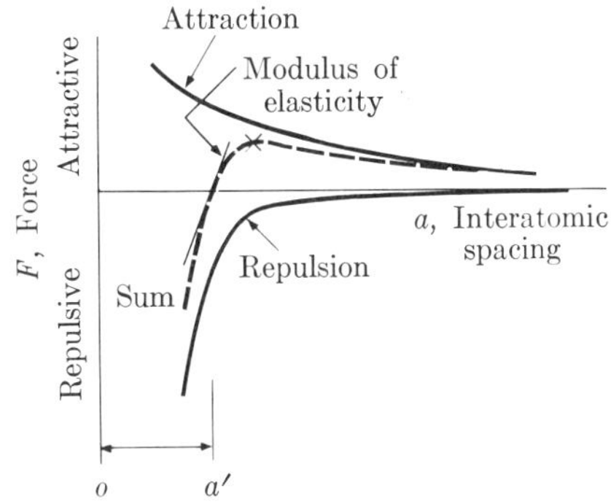

FIG. 8–1. Modulus of elasticity. A force is required to change the equilibrium interatomic distance o–a'. The slope of the sum curve is related to the modulus of elasticity. With small strains ($<1\%$), the force and the strain are proportional. (Cf. Fig. 2–7a.)

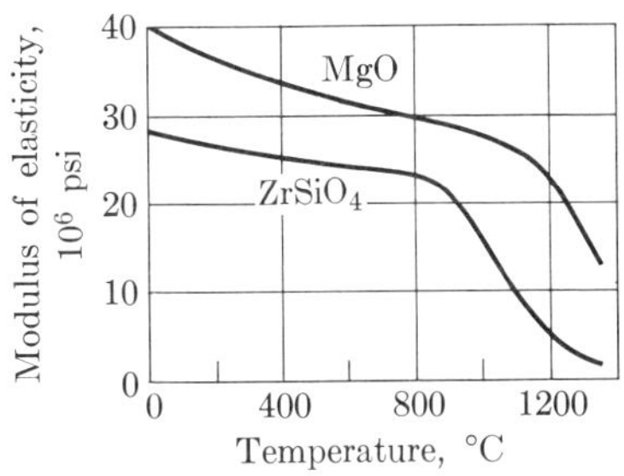

FIG. 8–2. Modulus of elasticity versus temperature: periclase (MgO) and zircon ($ZrSiO_4$). (After Jaffee, R. I., and D. J. Maykuth, "Refractory Materials," Defense Metals Information Center Memorandum 44, Battelle Memorial Institute, Columbus 1, O.

the average modulus of elasticity (Table 8–1) is generally appropriate, because most ceramic materials are polycrystalline with random grain orientations, the variation of modulus from grain to grain does produce a nonuniform stress pattern. Those grains which are so oriented that they are strained in the direction of a lower modulus will have lower stresses than adjacent grains aligned in a direction of the higher modulus. Since grains of both orientations will be deformed equally, some grains will be subject to plastic deformation or fracture before other grains are.

Shear modulus. The paragraphs above pertain to tensile or compressive stresses, which are unidirectional. *Shear stresses* will also distort the structure of a solid material. The shear modulus G is relatively constant over small strain ranges, and is the ratio between shear stress τ and shear strain γ:

$$G = \tau/\gamma. \tag{8–2}$$

The shear modulus G and the elastic modulus E are related through Poisson's ratio μ, which is the ratio of transverse unit strain, $\Delta d/d$, to axial unit strain, $\Delta l/l$:

$$E = 2G(1 + \mu), \tag{8–3}$$

and

$$\mu = -\frac{\Delta d/d}{\Delta l/l}. \tag{8–4}$$

Since Poisson's ratio is always less than 0.5, the shear modulus is always between $\frac{1}{3}$ and $\frac{1}{2}$ of the elastic modulus. It must be recognized, however, that any anisotropy in one modulus is reflected in the other.

EXAMPLE 8–1. Al_2O_3 has a Young's modulus of 53,000,000 psi and a Poisson's ratio of 0.4 Predict the shear modulus.

Answer:

$$G = \frac{E}{2(1 + \mu)} = \frac{53{,}000{,}000}{(2)(1.4)} = 19{,}000{,}000 \text{ psi.}$$

The value of Poisson's ratio tends to be lower in those materials which do not have close-packed structures. In such cases, strain can occur in the prime direction without equivalent lateral adjustments. It is not surprising, therefore, that many open-structured ceramic phases have lower Poisson's ratios than many metals.

Second phases (and pores). The elastic modulus of a two-phase material is intermediate between the moduli of the two components. Assuming that the two components have the same Poisson's ratios and have interfacial coherence, Eq. (8–5) permits an estimation of the modulus:

$$E = E_a V_a + E_b V_b, \tag{8–5}$$

where V_a and V_b are the volume fractions of the two phases. The assumptions involved are probably good enough for approximate calculations of moduli for glass-reinforced plastics, cermets, and also for semivitreous materials, which contain crystals in a glassy matrix.

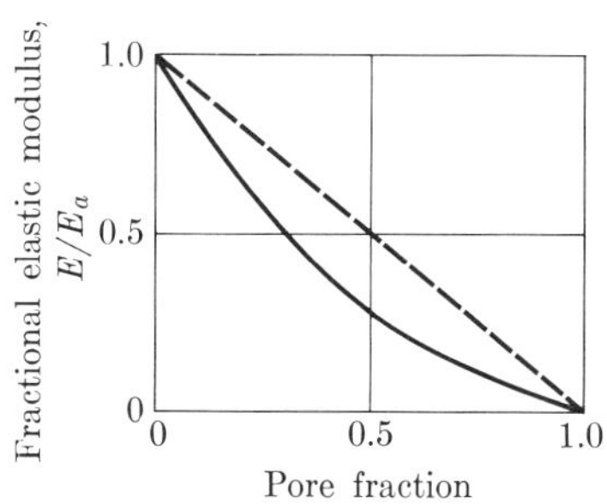

FIG. 8–3. Elastic moduli of porous materials [see Eq. 8–6)]. After Mackenzie, J. K., "The Elastic Constants of a Solid Containing Spherical Holes," *Proc. Phys. Soc.* (London), **B63,** 2 (1950).

Pores may be considered as a second phase without an elastic modulus. Hence, for pores, the above assumptions do not hold and the modulus of elasticity is lower than would be calculated from Eq. (8–5.) The empirical relationship

$$1 - E/E_a = 1.9P - 0.9P^2 \tag{8–6}$$

appears satisfactory for making approximations. The volume fraction of pores is P; E/E_a is the fractional modulus as related to a nonporous solid of the same material (Fig. 8–3).

EXAMPLE 8–2. Compare the Young's modulus of an Al_2O_3 insulator containing (a) 95 v/o corundum and 5 v/o glass (modulus of elasticity $= 12 \times 10^6$) with (b) 95 v/o corundum and 5 v/o porosity.

Answer:

(a) $E = (53{,}000{,}000)(0.95) + (12{,}000{,}000)(0.05) = 51{,}000{,}000$ psi.

(b) $E = (53{,}000{,}000)\,[1 - (1.9)(0.05) + (0.9)(0.05)^2] = 48{,}000{,}000$ psi.

8–3 Strength of ceramics

The strength of a material may be limited by its susceptibility to fracture or deformation. Since ceramics are essentially nonductile and brittle, fracture is the more common mechanism of failure in many ceramic materials.

Fracture stresses. The stress required for fracture is indicated schematically in Fig. 8–1 as the force at the maximum of the sum curve (point ×). Tension is required to separate the atoms to this point. Beyond this point, separation can continue with lower stresses. As shown in Fig. 8–1, materials cannot fail by true compression, because no critical stress value is encountered as the interatomic spacing is reduced.

Fracture-producing stress varies markedly within a crystal and is called the critical normal fracture stress (cnfs). Mica provides a vivid example because it has pronounced cleavage on one plane across which there are but few primary bonds. The cnfs for the other two mutually perpendicular planes are markedly higher because many covalent bonds must be broken.

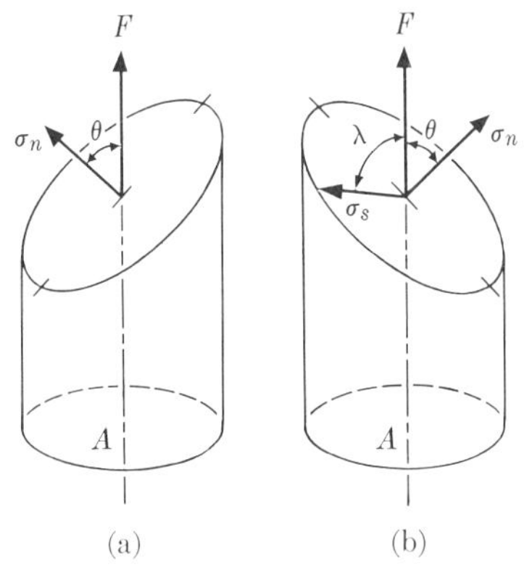

FIG. 8–4. Critical stresses: (a) critical normal fracture stress (cnfs), (b) critical shear stress (css).

If the principal stress is at some angle to the normal direction for a plane, only the vectorial component will be applicable (Fig. 8–4a). Thus if F is the force, A is the cross-sectional area, and θ is the angle of inclination from the normal,

$$\sigma_n = (F/A)\cos^2\theta. \qquad (8\text{–}7)$$

This means that the plane with a low cnfs may be subjected to fracture even though it is at an angle to the stress. It also means that, as a material is deformed and reoriented, new planes may become aligned for easy fracture.

A noncrystalline substance such as a glass has no specific plane of low normal fracture stress. Therefore, in the absence of residual stresses, a glass will tend to fracture on a plane that is normal to the tensile stress.

Strength. Since ceramics are usually nonductile, their strength characteristics are appreciably different from most metals. A nonductile ceramic material is strong in compression because it does not fail by slip, and therefore the compressive behavior of a ceramic is directly related to its interatomic forces. On the other hand, the tensile strength is usually much lower than would be initially predicted from interatomic forces. A calculation based solely on interatomic forces would indicate that the strength should be about 10% of the elastic modulus (that is, $E/10$). In practice, commercial ceramic products have tensile strengths of only 1% of this calculated value (that is, 0.01 of $E/10$, or $E/1000$). Any rationalization of the strength of ceramic materials must account for this discrepancy.

In the 1920's, Griffith postulated that fracture could be a consequence of stress concentrations at the tip of existing cracks or other structural flaws.* The stress concentration σ_c at the tip of the crack should exceed the normal stress σ_n in the following manner:

$$\sigma_c/\sigma_n = 2\sqrt{f/r}. \tag{8-8}$$

The length of the flaw is f, and the radius of the flaw tip is r. The tip radius would be approximated by atomic dimensions (<2 angstroms) for nonductile materials. Thus a crack of one micron depth would provide a stress concentration, σ_c/σ_n, of approximately 100.

Although the theory of Griffith's flaws is not wholly satisfactory, there are several observations which support his postulate for brittle materials. In the first place, glass, mica, and other brittle materials may have high tensile strengths when their surfaces are fresh. For example, freshly prepared glass rods or fibers have strengths in excess of 10^6 psi, but these high strengths disappear if the surface contacts abrasive dust, however fine. Also, many glass surfaces possess cracks with micron dimensions. As a result, the strength of glass is more commonly about 10^4 psi.

Statistical strengths. The microstructures of most ceramic materials, including glass, are such that there are points where stresses are concentrated. These points may be pores, grain corners, or true microcracks; thus there are many places for cracks to be initiated. Once initiated, the crack readily continues through the brittle phase because the f/r ratio of Eq. (8-8) increases in the absence of ductility at the tip of the crack. The tensile strength of a brittle

---

* Griffith, A. A., "The Phenomena of Rupture and Flow in Solids," *Phil. Trans. Royal Soc.* (London), **A221,** 163–198, 1921.

material is thus governed by the probability that a defect will be encountered which will produce a stress concentration in excess of the critical normal fracture stress. Several experimental consequences are observed: (1) Considerable scatter is found in the tensile test data. (2) In many cases, the average strength for large test specimens is lower than for small test specimens because there is a greater probability that there will be points of high stress concentration. (3) A porous, brittle material loses strength more rapidly than density (Fig. 8–5) because pores concentrate the stresses unevenly across the true cross-sectional areas.

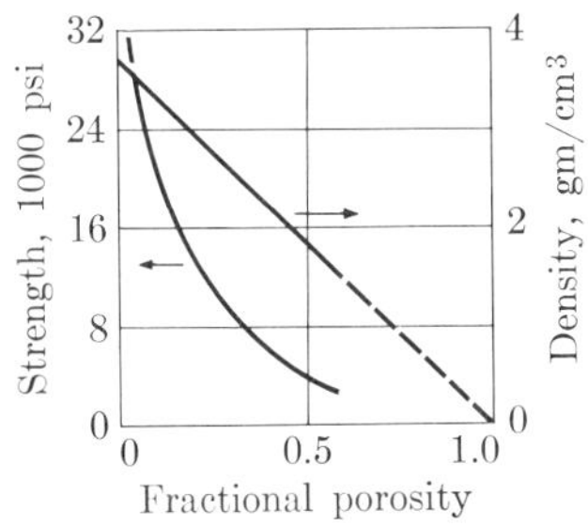

FIG. 8–5. Strength versus porosity in polycrystalline alumina. [Adapted from Coble, R. L., "Effect of Microstructure on Mechanical Properties," *Ceramic Fabrication Processes*. New York: John Wiley Sons (1958), page 223.]

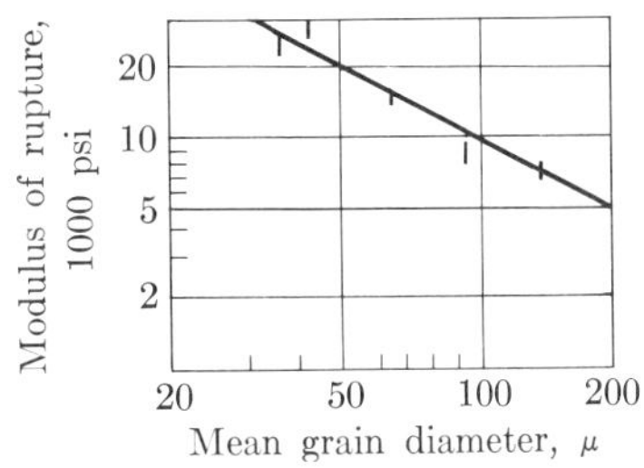

FIG. 8–6. Strength (modulus of rupture) versus grain size, in BeO. [After Coble, R. L., "Effect of Microstructure on the Mechanical Properties of Ceramic Materials," *Ceramic Fabrication Processes*. New York: John Wiley & Sons (1958), page 217.]

The strength of a nonporous, brittle material varies with grain size (Fig. 8-6) In the case of polycrystalline solids, this is because the variation of elastic modulus with orientation produces different stresses in adjacent grains (Section 8–2). There are more severe stress concentrations in solids composed of coarse crystals than in those composed of fine ones. Therefore, it is more probable that critical normal fracture stress will be exceeded as the grain size is increased.

EXAMPLE 8–3. The modulus of rupture expresses the maximum stress in a nonductile material as the outer "fiber" stress in three-point loading:

$$\text{Modulus of rupture} = 3FL/2bd^2, \tag{8–9}$$

where F is the breaking force, L is the span, and b and d are the cross-sectional dimensions. Five samples of brick (2 in. × 4 in. × 8 in.) are broken with loads of 2100, 2550, 2600, 2950, and 3100 lb. (The bricks were laid flat on a 6-in. span.) What is the mean modulus of rupture?

Answer: From Eq. (8–9), the five moduli are 1180, 1430, 1460, 1660 and 1740 psi respectively.

$$\text{Modulus of rupture (mean)} = 1490 \text{ psi}.$$

8-4 Structural ceramics

Static loading is the most common service requirement for nonmetallic, inorganic (i.e., ceramic) construction materials. Other requirements vary with individual engineering design, but may include (a) intermittent dynamic loading, e.g., concrete roadways; (b) thermal insulation, as in fiber products; (c) inertness to environments; (d) resistance to freezing and thawing; and even (e) light transmission, e.g., window glass. Usually cost is also a significant factor; and, of course, installation and maintenance costs must be incorporated with material costs and pro-rated over the period of useful life.

TABLE 8-2

PHYSICAL SPECIFICATIONS OF BUILDING BRICK (ASTM C 62-58)

Grade	Minimum compression strength		Maximum H_2O adsorption	
	Average of 5 bricks, psi	Of one brick, psi	Average of 5 bricks, %	Of one brick, %
SW*	3000	2500	17	20
MW†	2500	2200	22	25
NW**	1500	1250	No limit	No limit

* SW: Brick intended for use where a high degree of resistance to frost action is desired, and the exposure is such that the brick may be frozen when permeated with water.

† MW: Brick intended for use where exposed to temperatures below freezing but unlikely to be permeated with water, or where a moderate and somewhat uniform degree of resistance to frost action is needed.

** NW: Brick intended for use as backup or interior masonry.

Vitrified products. This category of ceramic products includes a wide variety of clay-base materials such as brick, tile, and whitewares. The processing methods presented in Chapters 13 and 14 involve the formation of a glassy or vitrified bond within the product. In these cases, the firing temperatures and times are such that liquefaction starts and a silicate glass is formed which wets the remaining solids. During subsequent cooling, the glass becomes a rigid (but still noncrystalline) solid that gives coherency to the product. In addition, vitrification reduces the porosity and permeability, which gives the product better resistance to environmentally caused deterioration. These consequences of porosity changes are reflected in Table 8-2. Ideally, any vitrified product should contain a significant amount of glass to ensure high strengths and other mechanical properties. However, excessive glass content usually increases the brittleness and thereby reduces the strength. It also permits viscous flow during firing (Chapter 14), making dimensional specifications more difficult to maintain.

Cements and concretes. In contrast to vitrified ceramic products which gain their mechanical strengths through the presence of glass, concretes receive their strengths through the formation of hydrated phases, or related compositions such as carbonates, commonly colloidal in character. The chemistry of their formation will be discussed in Chapter 13. Let it suffice here to note that greatest strength is attained with maximum packing factors for the aggregated components, and with minimum void space within the cement addition.

Composite materials. Many ceramic materials are used in conjunction with other materials. Examples include (1) reinforcement of plastic with glass fiber, (2) reinforcement of concrete with metal, (3) vitreous coatings (enamels) on sheet metal, and (4) cermets, which have an intimate mixture of ceramic and metallic phases. The reasons for these combinations may be several-fold. When we observe the mechanical properties illustrated by the data in Table 8-3, it is immediately apparent that the amount and geometry of the components are important. Less obvious is the fact that optimum mechanical properties are achieved only when there is coherency between the materials. This coherency is achieved, in the case of reinforced plastics, by silicone coatings which are compatible with both the glass and the plastic. In contrast, the interlock between steel and concrete results from surface irregularities rolled into the steel, and the bonding of enamels to metal depends on the primary chemical bonds described in Chapter 2. These chemical bonds are obtained by saturating the enamel with the oxide of the underlying metal.

TABLE 8-3

GLASS-REINFORCED PLASTIC OF POLYESTER RESIN AND GLASS LAMINATES*

Fiber pattern	Glass content, w/o	Stress direction	Tensile strength, psi
Chopped strand	0	—	5,000
	25	—	20,000
	33	—	21,000
	50	—	23,000
Continuous strand:			
unidirectional	48	With strand	75,000
	48	Across strand	2,500
cross-laminated	48	With width	40,000
	48	Across width	40,000
	48	45° to width	10,000
random	48	—	25,000

* Data from Morgan, P., *Glass Reinforced Plastics*, London: Iliffe Books, Ltd. (1954), pp. 19, 23, 147.

Test procedures for structural ceramics. Since many construction applications are sufficiently well established for service performance to be anticipated from laboratory results, standardized tests for the majority of ceramic materials used in construction are described in the *ASTM Standards**. However, one must use caution in evaluating laboratory data for use in service conditions, since laboratory testing usually either restricts parameters to a measurable few, or requires a "scaling-up" of certain parameters to compensate for laboratory time limitations. This statement is not intended to detract from the demonstrated value of standardized tests, but only to emphasize that all test data must be interpreted in the light of sound technical judgment, augmented by an understanding of the nature and behavior of the materials in question.

8-5 Abrasive materials

When we look at the upper listings of Table 8-4, we see that many of the hardest materials fall into the category of ceramic compounds. Among these, alumina and silicon carbide are frequently used commercially in abrasive applications. Although hardness, *per se*, is an important requirement for abrasives,

TABLE 8-4

HARDNESS OF ABRASIVE MATERIALS

Material	Composition	Mohs hardness number	Approximate Knoop hardness	Melting temperature, °C
Diamond	C	10	8000	>3500
Boron carbide	B_4C	—	3500	2450
Silicon carbide	SiC	—	3000	>2700
Titanium carbide	TiC	—	2800	3190
Tungsten carbide	WC	—	2100	2770
Corundum	Al_2O_3	9	2000	2050
Topaz	$SiAl_2F_2O_4$	8	1500	Dissociates
Quartz	SiO_2	7	1000	Transforms
Orthoclase	$KAlSi_3O_8$	6	600	Dissociates
Apatite	$Ca_5P_2O_{12}F$	5	500	~1400
Fluorite	CaF_2	4	200	1330
Calcite	$CaCO_3$	3	150	Dissociates
Gypsum	$CaSO_2(OH)_4$	2	50	Dissociates
Talc	$Mg_3Si_4O_{10}(OH)_2$	1	20	Dissociates

* The 1964 edition has standards for cements, limes and plasters in Part 9, concrete in Part 10, mortar, brick, and building products in Part 12, and refractories and glass in Part 13.

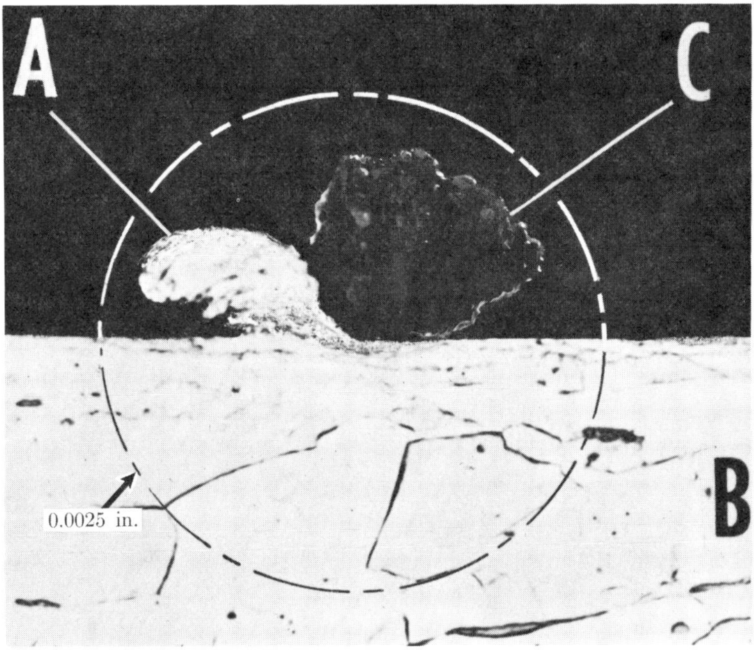

FIG. 8-7. Photograph (×1000) of grinding action. The steel chip (A) is shown being cut from the workpiece (B) by grinding grit (C). Photograph courtesy Cincinnati Milling Machine Company.

toughness is equally important, especially when the abrasive is used for grinding high-strength ductile materials. Also, since a grinding action is essentially a cutting action (Fig. 8-7), high local temperatures develop, and refractoriness is required. Fortunately this does not present a major problem, because the strong interatomic bonds provide both hardness and refractoriness (Table 8-4).

Hardness. The hardness of a material is most commonly indexed by one of two scales. The *Mohs scale* concerns dynamic hardness, or hardness by cutting, and rates materials according to results when a softer material is scratched by a harder phase. The *Knoop scale* is based on the degree of penetration of a soft material by a harder one. Thus it is a static test. The two tests, though closely related, are not perfectly correlatable. The Mohs scale is a simple and widely used index of hardness, but not satisfactory for measuring the hardness of ceramic abrasives, because these all have ratings of 9 or above, as we can see in Table 8-4. The Knoop scale is somewhat more quantitative, since figures are available for gradations of hardness.

Toughness. The relative toughness or friability of different abrasive materials may depend on a number of variables. Let us take Al_2O_3 as an example. When we have a pure corundum, the abrasive grains (white) are extremely friable and

Vitrified alumina example

27* A 120 G 3 V 7†

Abrasive type	Grain size	Grade	Structure	Bond
A, Alumina	Coarse	"Soft"	Dense	V, Vitrified
C, Silicon	4	A	0	S, Silicate
carbide	6	B	1	R, Rubber
	8	C	2	B, Resinoid
	10 70	D	3	E, Shellac
	12 80	E N	4	O, Oxychloride
	14 90	F O	5	
	16 100	G P	6	
	20 120	H Q	7	
	24 150	I R	8	
	30 180	J S	9	
	36 220	K T	10	
	46 240	L U	11	
	50 280	M V	12	
	60 320	W	13	
	400	X	14	
	500	Y	15	
	600	Z	16	
	Fine	"Hard"	Open	

C 36 R 10 R

Silicon carbide–rubber example

*The manufacturer's code for abrasive mix is optional with each manufacturer.

†Optional code for bond.

FIG. 8–8. Designations for bonded abrasives.

fracture readily under the stresses of grinding. In contrast, an Al_2O_3 which contains Fe_2O_3 and TiO_2 in solid solution (red or brown) is appreciably tougher and more resistant to wear and fracture (for reasons which experiment has not yet been able to explain). The commercial significance of this difference between the pure Al_2O_3 and its variations necessitates careful choice of the right abrasive for the right job. For grinding hard materials, the friable grain is best, since it maintains its angular cutting edges, even though this implies faster wear of both abrasive and work. As with hardness, there can be no fundamental quantitative test of the toughness or friability of an abrasive, because so many factors are involved: strength, ductility, geometry, cutting speed, etc.

Abrasive products. The engineer gives more attention to bonded abrasives in the form of grinding wheels and related products than he does to other products such as abrasive grains, polishing compounds, and cutting tools. The coherency within these bonded abrasives is accomplished either by organic or by glassy cements. The many varieties of bonded-abrasive grinding wheels has led to standardization of identification codes, as in the two examples in Fig. 8–8. The terms "abrasive type," "grain size," and "bond" are self-explanatory, but "grade" and "structure" require explanation here.

The *grade* of the wheel refers to the friability of the bonded product. Grade may be controlled in manufacture by (1) the nature of the grain, (2) the nature of the bond, or (3) the relative amounts of bond and grain. A "soft" grinding wheel implies that the grain is readily torn from the wheel. A "hard" grinding wheel resists attrition during use.

The term *structure* refers to the pore space among the atoms within the grinding wheel or other bonded abrasive product. Surface grinding, for example, requires an open structure to allow ample clearance of chips. On the other hand, in precision grinding of bearing surfaces to prescribed dimensions and contours, a dense structure is required; hence grinding is slower.

Ceramic cutting tools. Cutting tools made of ceramics offer an excellent example of (1) the property gradations between metals and ceramics (Section 1–2), and (2) the interrelationship between mechanical and thermal properties. If we consider the series: (1) tempered martensitic steels (with Fe_3C), (2) high-speed steels (with VC, Cr_4C, and WC), (3) cemented carbides (with WC), (4) cermets (with TiC), and (5) oxide tools (with Al_2O_3), we see the spectrum of abrasives used for cutting tools, from the metals to the ceramics. At first, it is hard to understand why the VC is considered to be a metallic phase, and the TiC of a cermet a ceramic phase, when they are both intermediate compounds. However, a look at the extreme cases of the above series provides a partial answer, on the basis of solubility. In the course of processing, the Fe_3C (cementite) can be dissolved into the austenitic (fcc iron) phase. The solubility of Al_2O_3 in a solid-metal phase is so restricted that such a consideration does not enter in problems of production or application. The making of high-speed steels always includes a solution step to dissolve a portion of the VC (or WC or Cr_4C). On the other hand, cermets are made entirely by sintering an aggregate (of TiC particles) without any metallurgical heat-treating. Generally, when a carbide can be dissolved in a metal, it is considered to be metallic.

Ceramic cutting tools are usually composed of Al_2O_3, which is chosen because of its high hardness, and because it becomes dense during sintering. Also it may be toughened to a certain extent by adding Fe_2O_3 and TiO_2 in solid solution. It has two main advantages which permit oxide cutting tools to cut up to three times faster than relatively harder cemented carbides (Table 8–5): (1) It doesn't depend on a metal bond, and (2) it will not weld to the chip (Table 8–6). The latter fact, and the insoluble behavior of Al_2O_3 described above, are related. There are disadvantages as well as advantages to oxide cutting tools, the most significant being their brittleness, which has made it necessary to develop new types of mounts, machines, and techniques in order to fully utilize ceramic tools. But cemented carbides also required modifications in machines and techniques when they were first introduced, because they were more brittle than previously used tool steels.

Polishing powders. The basis for good polishing is not satisfactorily understood. It involves not only a removal of surface irregularities but also the flow of surface materials from projecting to recessive areas. There is some evidence to indicate that a degree of coherency must be established between the powder and the surface. Therefore, polishing powders are not always just finer versions

TABLE 8-5

CUTTING SPEEDS*

Tool	Cutting speed, sfpm		
	After 1 minute	After 5 minutes	After 50 minutes
Oxide (0-30)	7000	3500	800
Carbide (831)	1600	1100	500
High-speed steel	—	200	140

* Adapted from Metcalfe, A. G., "Why Oxide Tools Cut Faster," *American Machinist,* Dec. 3, 1956.

TABLE 8-6

COMPOUNDS USED IN CERMETS*

Material	Hardness, Vickers hardness number	Melting point, °F	Welding-on temperature, °F
Al_2O_3	3000	3700	2800
AlB_{12}	3800	—	2600
TiC	3000	5700	2500
WC	—	5100	2400
ZrB_2	3600	5400	2250
TiB_2	4200	5300	2250

* Adapted from Metcalfe, A. G., "Why Oxide Tools Cut Faster," *American Machinist,* Dec. 3, 1956.

of abrasive grains. For example, an extremely fine SiC powder does not provide a fully satisfactory polishing compound. Further, the best Al_2O_3 structure for polishing is γ-Al_2O_3, which is cubic and not the normal hexagonal corundum. Rare-earth oxides, such as Ce_2O_3, also make good polishing compounds. In general, the best polishing agents are refractory oxides, such as Fe_2O_3, which also have hydrated forms.

8-6 Plastic deformation in crystals

Permanent deformation can occur in ceramic materials by means of several mechanisms. In this section, we shall discuss plastic deformation of ceramic crystals. Subsequent sections will cover viscous deformation of glasses and the rheology of colloidal ceramics.

Although plastic deformation of crystals is common in metals, it occurs relatively infrequently in ceramic materials. Thus it is not practical to fabricate ceramic products by deformation of the crystals themselves. However, the

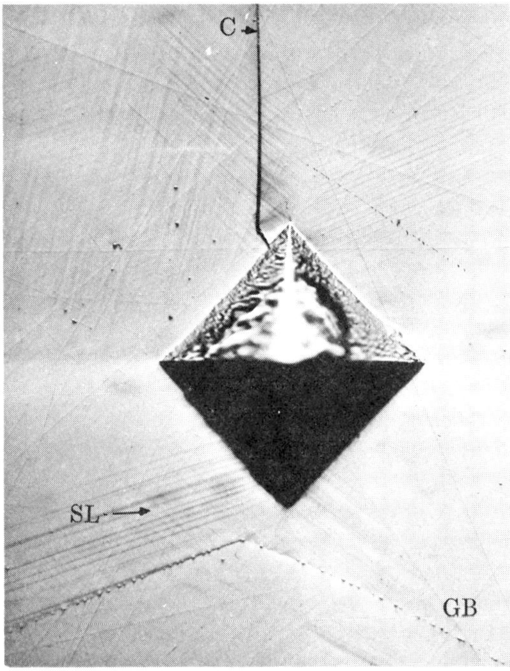

FIG. 8-9. Ductile deformation of alabandite (MnS). Although MnS is nominally nonductile, it is possible to plastically deform this phase by the compressive stresses under a diamond indentor (pyramid). SL, slip lines (parallel); GB, grain boundaries (impurities present); C, crack. Random lines are polishing scratches. Photomicrograph (×500) by Robert J. Warrick.

ceramist has an interest in crystalline deformation, because even the slightest yielding of crystals may have significant effects on the strength of the product. For example, if deformation permits any adjustment in the radius of the tip of a flaw (Eq. (8-8)), the stress concentrations are greatly reduced.

Plastic slip. Under certain conditions, a material that is nominally nonductile can be plastically deformed. This is illustrated in Fig. 8-9 for MnS, by the slip lines adjacent to an indentation made by a diamond pyramid. In this situation, the stresses are primarily compressive and are resolved into shear stresses. In contrast, when tensile stresses are applied to MnS, fracture occurs. Limited slip can also be observed in other ceramic phases, such as corundum (Al_2O_3), periclase (MgO), and calcite ($CaCO_3$), plus almost all of the phases with a layered structure (Section 3-4).

Critical shear stress. Before one crystal plane can slide over another, a minimum shear stress is required. This is called the critical shear stress (css), and may be calculated as shown in Fig. 8-4(b), and by Eq. (8-10):

$$\sigma_s = (F/A)(\cos\theta \cos\lambda). \tag{8-10}$$

In this equation, σ_s is the shear stress; F, A, and θ have the same significance as in Eq. (8-7). The angle λ is the angle between the force axis and the direction of slip. It should be noted that θ and λ are complementary only in the special case when the slip direction is in the plane defined by the force direction and the normal to the slip plane.

The critical shear stress, like the fracture stress, is much lower than would be predicted solely on the basis of interatomic forces. It has been well established that this apparent discrepancy is a consequence of dislocation movements; hence the whole plane does not shear at once and high stresses are not required.

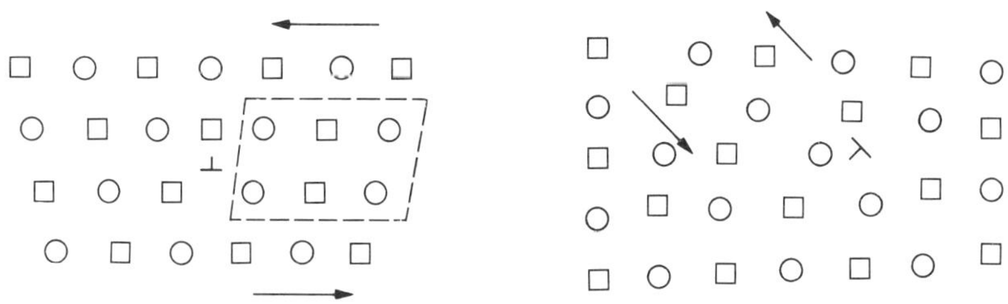

(a) (b)

FIG. 8-10. Deformation of MgO. (a) Slip in the [100] direction. (b) Slip in the [110] direction. In the [100] direction, adjacent like charges prevent dislocation movement. Symbols: □, Mg^{2+}; ○, O^{2-}.

The critical shear stress values vary markedly with the type of crystallographic plane, as illustrated in Fig. 8-10(a). Slip in the [100] direction is resisted because ions of like charges are brought into close contact. This does not occur in the [110] direction (Fig. 8-10b). Herein lies the difference between deformation of ceramics and deformation of metals. In metals, each atom is essentially equivalent, so that there are more directions with low critical shear values. The three-dimensional deformation in periclase (MgO) and similar highly ionic crystals involves $\langle 110 \rangle$ slip on $\{110\}$ planes. The same crystals at very high temperatures, or similar crystals which are less strongly ionic (such as AgCl) will also allow $\langle 110 \rangle$ slip on $\{100\}$ planes.

The chief consequence of the restrictions in slip planes and slip directions is reflected in the strength of the crystal. With fewer slip possibilities, there is greater probability that the critical normal fracture stress will be exceeded before the critical shear stress is met. This situation naturally causes considerable attention to be paid to current research on the shear mechanism, since a slight increase in critical shear stress could make a significant reduction in the probability of a fracture.

EXAMPLE 8-4. A periclase (MgO) crystal has a tensile stress applied in the [100] direction. It deforms by slip along the (110) plane. What is the maximum ratio of critical shear stress (110) to critical fracture stress (100)? Slip is in the $[1\bar{1}0]$ direction.

Answer:

$$\begin{aligned}
\theta_{(100)} &= 0°; \\
\theta_{(110)} &= 45°, \lambda_{(110)} = 45°; \\
\text{css}_{(110)} &= (F/A)\cos\theta\cos\lambda = (F/A)\cos 45°\cos 45°; \\
\text{cnfs}_{(100)} &= (F/A)\cos^2\theta = (F/A)\cos^2 90°; \\
\text{Maximum ratio} &= \cos 45°\cos 45° \\
&= 0.5.
\end{aligned}$$

(Of course, the ratio may be less than this since slip occurred in preference to fracture.)

8–7 Viscous deformation

Noncrystalline materials (Chapter 4) do not deform by dislocation movements; nor do they have a critical shear stress, because they have no crystal planes. For the same reason, there is no critical normal fracture stress. However, fracture and deformation can occur in glasses and related noncrystalline materials. Fracture can be propagated at stress concentrations in directions perpendicular to the resolved tensile stresses, and viscous deformation can occur in the direction of maximum resolved shear stress, if there is sufficient force to break the more highly stressed bonds.

In an amorphous material such as that indicated schematically by a glass in Fig. 4–1(a), the interatomic distances are not precisely equal. In fact, some distances may vary markedly from that shown as the equilibrium distance in Figs. 2–7(b) and 8–1, to such a point that they are near the bond fracture stress point (point × of Fig. 8–1). When this is so, an external load does not have to be high to actually fracture bonds. Once a bond has been fractured, all of the neighboring atoms adjust their interatomic spacings, and subsequent bonds of higher-than-average interatomic distances are subject to fracture. Any persistent stress will therefore provoke continuous deformation; hence viscous flow occurs.

The relationship between external stresses and the resulting viscous flow can be expressed by

$$\eta = \sigma_s/(dv/dx), \tag{8–11a}$$

where η is the ratio of shear stress σ_s, or F/A, to the flow-velocity gradient, dv/dx. This ratio is called the viscosity coefficient, and is expressed in poises or dyne·sec/cm^2. An atomistic explanation of the above relationship recognizes that higher stresses of external origin bring a proportionally greater number of bonds beyond their breaking point (point × of Fig. 8–1) within any given period of time, thus increasing the velocity gradient, dv/dx. Also, since bond strengths vary markedly from material to material, we find viscosity coefficients which vary over many orders of magnitude, ranging from 10^{-2} poise (1 centipoise) for water to 10^{20} poises for silica glasses.

The effect of temperature on viscosity will be discussed in the following chapter. However, it can be noted here that higher temperatures induce greater thermal vibrations among the atoms, and therefore there is a greater probability that any particular atom pair may have its bond broken by externally applied forces. Thus increased temperatures produce additional viscous flow for a given stress σ_s, and reduce the viscosity coefficient (Fig. 9–11).

Depolymerization from a network structure (Fig. 4-1a) to a more ionic structure (Fig. 4–2) reduces the number of covalent bonds between specific pairs of atoms, with the result that flow occurs more readily. Thus additions of small amounts of network modifiers (Section 4–3) to glasses become important in glass-forming operations (Chapter 13).

Whether viscous flow or fracture results is directly related to the rate of deformation. If stress is applied slowly, there is time for viscous flow, as described above. After each bond is broken, the neighboring atoms can adjust before additional bonds are broken. However, if the rate of deformation is rapid, after one bond is broken, stresses are concentrated on the next or succeeding bonds, in order to break them. For example, glass at elevated temperatures normally exhibits viscous flow. Under slow loading, the glass deforms as expected; but under impact loading at the same temperature, the glass will shatter just as it would under stresses at room temperature.

8–8 Rheology of colloidal structures

Viscous deformation was presented in Eq. (8–11a), which may be modified as follows:

$$\sigma_s = \eta\, dv/dx. \tag{8–11b}$$

Again, σ_s is the shear force per unit area, η is the viscosity coefficient, and dv/dx is the velocity gradient.

Suspensions. A suspension has a higher viscosity than the suspending liquid. In dilute suspensions the effect is relatively small and may be considered simply a consequence of an increased dv/dx between particles in the liquid, as contrasted to the average dv/dx through the bulk of the suspension. In other words, because of the presence of the solids, the value of σ_s of Eq. (8–11b) must be increased, even though the viscosity of the liquid proper remains the same. Thus,

$$\eta_b/\eta_l = (dv/dx)/(dv/dx)_{\text{ave}}. \tag{8–12}$$

The subscripts, b, l, and ave, refer to bulk, liquid, and average, respectively.

In more concentrated suspensions there is mutual interference of the suspended particles, so that the bulk viscosity increases quite rapidly. Norton*

---

*Norton, F. H., *Elements of Ceramics*, Reading, Mass.: Addison-Wesley (1952), p. 74.

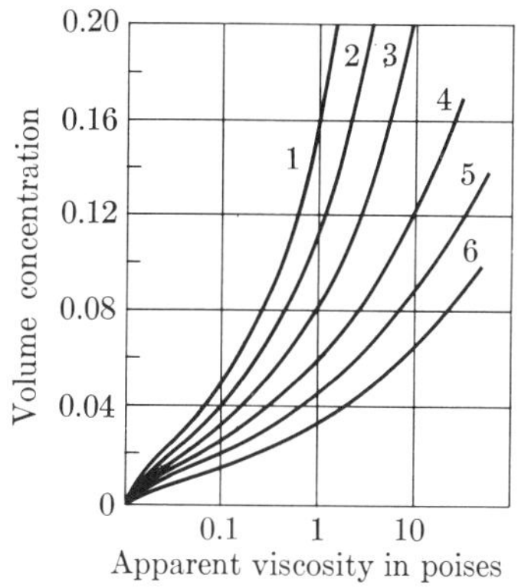

FIG. 8-11. The bulk or apparent viscosity of several suspensions, with decreasing sizes of particles which have been flocced from monodisperses. The equivalent spherical diameters of the flocced particles are: (1) 6.4 to 12.8μ; (2) 3.2 to 6.4μ; (3) 1.6 to 3.2μ; (4) 0.8 to 1.6μ; (5) 0.4 to 0.8μ; (6) 0.2 to 0.4μ. [Norton, F. H., *Elements of Ceramics*. Reading, Mass.: Addison-Wesley (1952), page 76.]

gives the empirical relationship for deflocculated suspensions as

$$\eta_b = \eta_l(1-C) + 0.08C + 7.5C^3, \tag{8-13}$$

where C is the volume fraction of solids present. Michaels* presents

$$\eta_b/\eta_l = 1 + [1.25C/(1-C/0.74)]^2. \tag{8-14}$$

Each equation indicates a rapid increase in viscosity as the solid content is increased to produce a concentrated suspension. The choice between these two empirical relationships is less important than the definition of that fraction of the volume which acts as a solid in suspension. It must be an effective volume and include not only the solid proper, but also the adsorbed surface layer, which significantly increases the volume at colloidal dimensions (Section 5-2). Thus the degree and nature of flocculation are important variables which are not incorporated into Eqs. (8-13) and (8-14). These variables are shown graphically in Fig. 8-11 for flocculated particles of the "card-house" type (Fig. 7-13a). On the other hand, if flocculation is face-to-face (Fig. 7-13b), the effective volume of the solids is much smaller, and in fact is less than in a deflocculated suspension.

The effect of flocculation on viscosity is dramatically demonstrated by some work of Norton, shown in Fig. 8-12. At pH values below 6, there is edge-to-face flocculation of the flat colloidal clay particles. As base is added and the pH increased, deflocculation occurs, and the viscosity decreases to about 1% of the flocculated value. Such changes are of major significance in the forming processes of ceramic materials (Chapter 13).

The bulk viscosity of a suspension is also sensitive to the rate of shear. This is shown in Fig. 8-13, which plots the bulk viscosity versus the reciprocal of the shear rate $(dx/d\dot{v})$. In the more concentrated suspensions, the viscosity decreases as the shearing rate increases (and the reciprocal of the shear rate

---

* Michaels, A. S., "Rheological Properties of Aqueous Clay Systems," *Ceramic Fabrication Processes* (W. D. Kingery, editor). New York: John Wiley & Sons; and Cambridge, Mass.: The Technology Press (1958) p. 24.

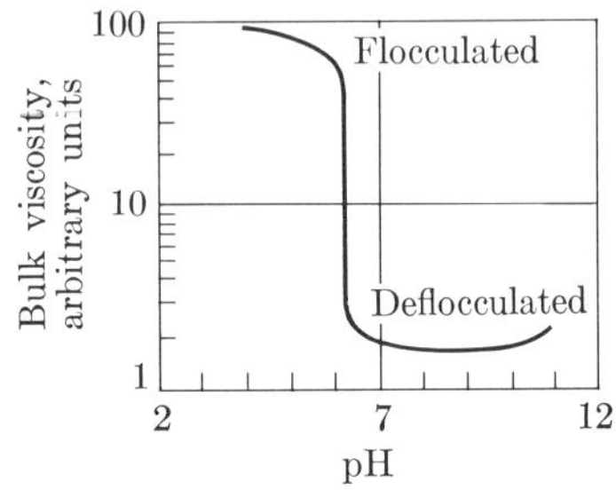

FIG. 8-12. Viscosity versus pH in colloidal kaolinite particles. Flocculation occurs by edge-to-face attractions (see Fig. 7-12). Deflocculation occurs when the change in pH removes the positive edge charges. [Adapted from Norton, F. H., *Elements of Ceramics*. Reading, Mass.: Addison-Wesley (1952), page 87.]

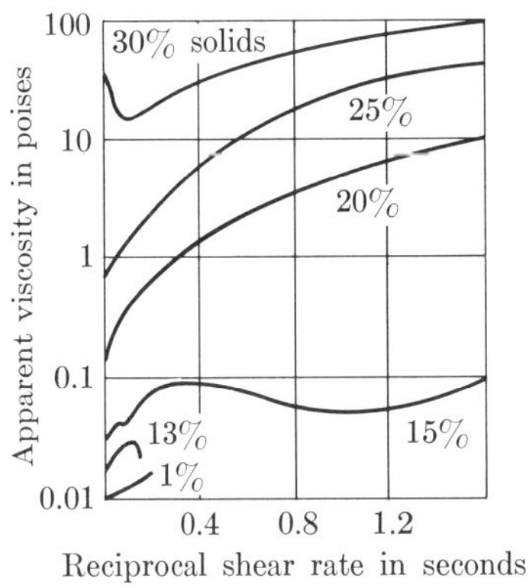

FIG. 8-13. Bulk viscosity versus the reciprocal of the shear rate (dx/dv). [Norton, F. H., *Elements of Ceramics*. Reading Mass.: Addison-Wesley (1952), page 75.]

decreases). This nonideal rheological property is called *thixotropy*, and is common to both slips and pastes.

Our attention has been directed toward colloids with two-dimensional particles because their structural variation has a noticeable effect on flocculation and flow behavior; however, many other materials exhibit colloidal behavior of note. For example, in suspensions with low pH values, colloidal alumina develops a positive surface. Amphoterically, in high-pH suspensions, it develops a negative surface. In each case, dispersion or deflocculation occurs as a result of mutual particle repulsion. In neutral suspensions, there is an intermediate pH range where the particles can be flocced because both positive and negative adsorption occurs.

Pastes. Pastes can be defined as solid–liquid colloidal structures that are plastic and behave as semisolids. As such, they have a volume fraction of solids large enough for the fluid characteristics to be lost. For example, gravitational forces do not immediately produce a level surface. In fact, the paste always has a finite yield strength under reasonably fast rates of strain.

In general, an increase in liquid content of a paste decreases both the yield strength and the shear modulus, and increases the rupture strain (Table 8-7). However, the overall picture is not always simple; a paste with an edge-to-face structure may be partially rigid, while one with a face-to-face structure would permit slip with small shear stresses, although the liquid content of each is the same. Furthermore, plasticity may be influenced by any factor which affects

TABLE 8-7

PASTE BEHAVIOR VERSUS LIQUID CONTENT OF CLAY WITH AN INTERMEDIATE DEFORMATION RATE*

Percent H_2O	Shear yield stress, psi	Elastic shear modulus, psi	Rupture strain, %
35	15	220	11
40	8	120	42
50	2	60	65

* Adapted from Norton, F. H. *Elements of Ceramics*, Reading, Mass.: Addison-Wesley (1952), page 73.

structure. For example, assume that the "card-house" structure of Fig. 7-13 is agitated. The superimposed external forces can temporarily destroy the edge-to-face structure and thus effectively decrease the rigidity, producing a thixotropic behavior. On the other hand, a low-liquid, face-to-face colloidal structure may lose some of its plasticity if vibration destroys the highly oriented slip planes. The resulting decrease in plasticity and "apparent decrease" in liquid content is termed dilatancy or rheopexy.

References for Further Reading

8–1. Charles, R. J., "A Review of Glass Strength," *Progress in Ceramic Science,* **1** (J. E. Burke, editor). New York: Pergamon Press, 1961, pp. 1–38. A review of the strength of noncrystalline materials which fail by fracture.

8–2. Coble, R. L., "Effect of Microstructure on the Mechanical Properties of Ceramic Materials," *Ceramic Fabrication Processes* (W. D. Kingery, editor). New York: John Wiley & Sons, 1958, pp. 213–228. An excellent general review of mechanical properties of polycrystalline ceramic products.

8–3. Gilman, J. J., "Mechanical Behavior of Ionic Crystals," *Progress in Ceramic Science,* **1** (J. E. Burke, editor). New York: Pergamon Press, 1961, pp. 146–199. A thorough presentation of dislocation movements, with prime attention to ionic single crystals. Advanced level.

8–4. Hibbard, W. R., "Introductory Review of Strengthening Mechanisms," *Strengthening Mechanisms in Solids.* Metals Park, O.: Amer. Soc. for Metals, 1962, pp. 1–16. An outline of strengthening mechanisms, with emphasis on plastic behavior.

8–5. Johnston, T. J., C. H. Li, and R. J. Stokes, "The Strength of Ionic Solids," *Strengthening Mechanisms in Solids.* Metals Park, O.: Amer. Soc. for Metals, 1962, pp. 341–374. Comparable to Gilman's presentation (8–3); however; more attention is given to polycrystalline solids.

8–6. Kingery, W. D., *Introduction to Ceramics.* New York: John Wiley & Sons, 1960. Chapters 16 and 17 pertain specifically to mechanical behavior. Advanced level.

8–7. Marin, J., *Mechanical Behavior of Engineering Materials.* Englewood Cliffs, N. J.: Prentice-Hall, 1962. General treatment, from a nonatomistic approach, of mechanical properties of materials.

8–8. Norton, F. H., *Elements of Ceramics.* Reading, Mass.: Addison-Wesley, 1952. An introductory presentation of ceramic materials. Chapter 8 discusses flow properties of ceramic slips and pastes.

8–9. Pask, J. A., "Mechanical Properties of Ceramic Materials," *Mechanical Behavior of Materials at Elevated Temperatures* (J. E. Dorn, editor). New York: McGraw-Hill, 1961, pp. 288–318. Nonmathematical presentation, with emphasis on plastic deformation.

8–10. *Mechanical Behavior of Crystalline Solids,* National Bureau of Standards Monograph 59, Washington, D.C.: U.S. Government Printing Office, 1963. A series of six papers on the flow and fracture of crystalline ceramics. Introductory approach.

Problems

8–1. How much glass is required to give a glass-reinforced plastic a modulus of elasticity of 7×10^6 psi? The glass and plastic have moduli of 9×10^6 psi and 2.5×10^5 psi, respectively.

8–2. A glass plate with a $\frac{1}{2}$-in. hole is subjected to a normal stress of 5000 psi. What is the stress concentration?

8–3. Compare the atom density in the (100) and (111) planes of diamond. (a) Along which plane would cleavage be most probable? (b) On what other planes would cleavage be expected?

8–4. An Al_2O_3 crystal can be plastically deformed under appropriate conditions. (a) Calculate the force needed to produce slip in a single crystal rod (0.1 in. diameter) if the css is 180,000 psi. The normal to the slip plane and the slip direction make angles of 55° and 60°, respectively, with the tensile direction. (b) What is the value of the normal stress on this same plane?

8–5. A prism of MgO (0.1 in. × 0.1 in. × 1.0 in.) is bent as a small beam. Predict the mechanism of slip if the prism surfaces are {100} planes.

8–6. If slip in MgO occurs only in the ⟨110⟩ directions on the {110} planes, compare the force required in the [001] direction to produce slip (a) with the force required in the [101] direction; (b) with the force required in the [111] direction.

8–7. A BB shot goes through a windowpane. Account for the shape of the hole.

8–8. Provide a rationale for the specifications cited in Table 8–2.

8–9. A selected glass has a Knoop hardness number which is the same as that of a specified steel. Compare their behavior (a) during grinding; (b) during polishing.

8–10. Review the polishing procedures for the preparation of micrographic samples. Explain the purpose of each step.

Chapter 9

THERMAL PROPERTIES AND HIGH-TEMPERATURE BEHAVIOR

9–1 Introduction

When the temperature of a material changes, its internal energy changes also, giving rise to several properties of interest to the engineer. Among these are: (1) thermal expansions, (2) heat capacities, and (3) thermal conductivities. In addition, various thermal behaviors are direct consequences of the atomic energies within a material. The more important of these include: (1) thermal stresses, with attendant spalling and relaxation of stresses, and (2) high-temperature deformation that involves anelastic behaviors, creep, and viscous flow. In this chapter we shall discuss all of the above, and give examples of refractory products in which behavior is related to the nature of the ceramic. For a discussion of phase stability, the reader is referred to Chapter 6.

9–2 Thermal expansion

Thermal expansion coefficients of a number of materials are given in Table 9–1. The volume changes which accompany temperature changes may be explained directly from Fig. 9–1 (see also Fig. 2–7b), which indicates the basis for structural dilatation. Three conclusions may be drawn.

(1) Expansion occurs because the energy trough for interatomic spacings is asymmetrical (Eq. (2–9)). (2) The thermal expansion coefficient is greater at higher temperatures. (3) The energy trough is deeper and more symmetrical for materials which melt at high temperatures. As a result, there is a relationship between melting temperature and thermal expansion.

There are two other facts concerning thermal expansion that are not as evident from previous considerations: (4) Densely packed phases, such as ionic crystals, have greater thermal expansion than phases with more open structures, such as glasses. (5) The thermal expansion varies with the symmetry of the crystalline phase.

The asymmetry of the energy trough for interatomic spacings derives directly from the fact that the attractive and repulsive forces, expressed as functions of interatomic distance, show different variations. As discussed in Section 2–5, the net coulombic attractive energy in ionic crystals increases as a function of

TABLE 9–1

THERMAL EXPANSION

Material: polycrystals	Thermal expansion coefficient: 20 to 1000°C—mean temperature,* cm/cm/°C
Cordierite ($Mg_2Al_4Si_5O_{18}$)	2×10^{-6}
Corundum (Al_2O_3)	9
Periclase (MgO)	14
Quartz (SiO_2)(20–400°)	17
Spinel ($MgAl_2O_4$)	9
Zircon ($ZrSiO_4$)	4
Fused silica	0.6

* The coefficient at 20°C is somewhat less than the above figures. Linear expansion generally increases with temperature.

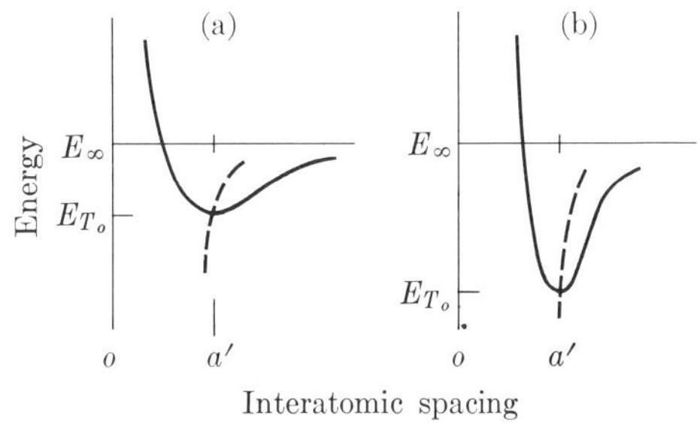

FIG. 9–1. Schematic drawing showing the origin of thermal expansion (see also Fig. 2–7b). (a) Weakly bonded structures. (b) Strongly bonded structures (e.g., materials with a high melting point). Addition of thermal energy increases the mean interatomic spacing more in a weakly than in a strongly bonded structure.

$1/a$, while the electronic repulsive energy decreases as a function of $1/a^n$, where n is approximately 9 (Eq. (2–8)). Thus the mean interatomic spacing increases with increasing thermal energy.

The linear coefficient of thermal expansion α_l may be expressed as

$$\alpha_l = (dl/l)(1/dT), \tag{9-1}$$

where dl/l has dimensionless units such as in./in. or cm/cm. As shown in Fig. 9–1, the slope of the curve of the mean interatomic spacing changes at higher energies. Hence, the linear coefficient of thermal expansion is a function of temperature (Fig. 9–2a). The relationship

$$\alpha_l = a + bT + cT^2 \tag{9-2}$$

is generally an empirical formulation of experimental data. Theoretical derivations are possible, but require numerous assumptions.

The correlation between thermal expansion and melting temperature is indicated in Table 9–2 and Fig. 9–2(b). This correlation could be predicted,

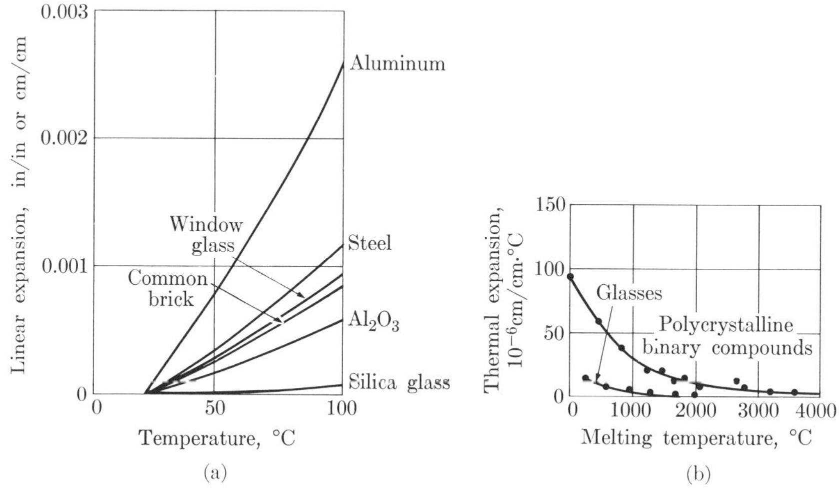

FIG. 9-2. (a) Thermal expansion of common structural materials. (b) Thermal expansion coefficients of close-packed binary compounds and glasses (see Table 9-2).

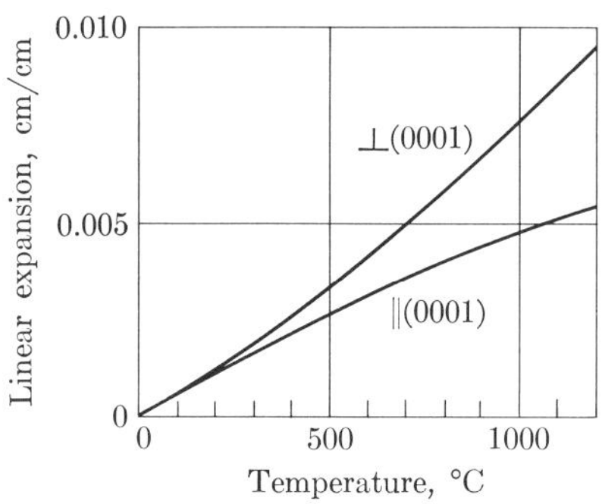

FIG. 9-3. Thermal expansion of ZnO. The anisotropic expansion derives from the hexagonal structure. [Adapted from Beals, R. J., and R. L. Cook, "Directional Dilatation of Crystal Lattices at Elevated Temperatures," *Journ. Amer. Ceram. Soc.*, **40,** 279 (1957).]

since materials with high melting temperatures require more thermal energy to overcome the greater attractive forces present within these materials. But the correlation in Table 9-2 is not perfect, which indicates that factors other than those just cited are also present. One such factor is the structure of the phases: If the structure has a low atomic packing factor providing internal "voids," interatomic expansion can occur with small overall dimensional changes. Examples are the glasses and those crystals which have low atomic coordination numbers and therefore lower thermal expansions (Table 9-2).

Anisotropic structures have an asymmetrical distribution of interatomic bonds. Therefore, phases which do not have cubic (or amorphous) structures (Table 3-2) have anisotropic thermal expansions. This is illustrated in Fig. 9-3 for zincite (ZnO). Zincite, in common with wurtzite (ZnS), has a hexagonal structure

TABLE 9–2

THERMAL EXPANSION VERSUS MELTING TEMPERATURE*

Material	Thermal expansion, cm/cm/°C, at 0°C†	Melting temperature, °C
I. Polycrystalline binary compounds		
H_2O	90×10^{-6}	0
CsCl	54	646
NaCl	40	800
PbS	20	1120
CaF_2	20	1330
Fe_2O_3	9	1560
SiO_2	12	1710
Al_2O_3	8	2050
ZrO_2	10	2700
MgO	9	2800
TiC	7	3190
Graphite	5	>3500
II. Glasses and amorphous materials		
Bismuth	11	271
Antimony	9	630
Arsenic	5	>1000
Pyrex	3	~1400
Vycor (96% SiO_2)	0.5	~1550
Fused silica	0.3	1710

* Adapted from Zwikker, C., *Physical Properties of Solid Materials*, London: Pergamon Press (1954), and others.

† Anisotropic mean.

(see also Fig. 3–9d). The variation in thermal expansion of zincite derives from the fact that, as temperature increases, thermal vibrations produce greater dilatation perpendicular to the stacking planes than parallel to the planes. Also, let us note that the interatomic forces between these planes are more easily ruptured by normal and shear stresses; thus mechanical weakness can arise in these areas.

Table 9–3 shows the variation of thermal expansion for a few selected hexagonal and tetragonal crystals. The negative value for calcite ($CaCO_3$) is not unique among ceramic phases; however, it does require explanation. Calcite's characteristic of high thermal expansion in the direction of weaker interatomic forces produces a contraction in perpendicular directions. Thus, although there is a negative linear coefficient, it should be noted that the volume expansion α_v of

TABLE 9-3

THERMAL EXPANSION OF SELECTED UNIAXIAL CRYSTALS

Phase	Thermal expansion, cm/cm/°C, at 0°C	
	$\parallel$ (0001) or (001)	$\perp$ (0001) or (001)
Zincite (ZnO)	6×10^{-6}	5×10^{-6}
Quartz (SiO_2)	14	9
Calcite ($CaCO_3$)	−6	25
Zircon ($ZrSiO_4$)	4	6
Rutile (TiO_2)	7	8

calcite, for small temperature changes, is still positive:

$$\alpha_v = (1 + \alpha_x)(1 + \alpha_y)(1 + \alpha_z) - 1, \tag{9-3a}$$

or

$$\alpha_v \approx \alpha_x + \alpha_y + \alpha_z. \tag{9-3b}$$

EXAMPLE 9-1. Determine the volume coefficient of thermal expansion for calcite ($CaCO_3$), using the data of Table 9-3.

Answer:

$$1 + \alpha_v = (1 + \alpha_\perp)(1 + \alpha_{||})(1 + \alpha_{||}).$$

Therefore

$$\alpha_v \approx \alpha_\perp + 2\alpha_{||} \approx 13 \times 10^{-6} \text{ cm/cm/°C}. \tag{9-3c}$$

There are rare cases where the volume thermal expansion can be negative. But in each case, these apparent anomalies result from a change in atomic coordination from open to close packing. The change of ferrite (body-centered-cubic iron) to austenite (face-centered-cubic iron) at 910°C (1670°F) is such a case: Here, the change in coordination number from 8 to 12 actually produces a phase change. Another example, in which the changes in internal structure are more subtle than those in iron, is the contraction of water between 0°C and 4°C. The densification of molecular packing that occurs when ice changes to water is continued to give this small contraction. Other examples with small negative expansions are some of the lithium alumino-silicates that are used for electrical insulating purposes. Here experiment has not yet defined the structural changes responsible, but it is known that these changes are related to atomic packing.

Cycling of temperature produces thermal growth (or thermal ratcheting) in materials with anisotropic expansions if the differential changes in dimensions introduce cracking or permanent deformation. The microcracks thus formed produce a slight porosity which is not removable on the return cycle. Because of a gradual increase in bulk volume, succeeding cycles aggravate the situation.

9–3 Heat capacity

The heat capacities of ceramics are important because: (1) they are a basis for processing and fuel requirements, (2) they can provide information regarding thermodynamic equilibria, (3) a change in heat content can serve as a basis for analytical procedures (i.e., differential thermal analysis), and (4) heat capacity is a factor in thermal spalling (Section 9–5).

The molar heat capacity C_p is the differential of the heat content H of a material with respect to temperature T. Hence

$$C_p = dH/dT, \tag{9–4}$$

or, as shown in Eq. (6–2),

$$H_2 - H_1 = \int_{T_1}^{T_2} C_p \, dT.$$

Figure 9–4 plots both the molar heat capacity and the heat content (enthalpy) of quartz and other SiO_2 phases from 298°K (25°C or 77°F) to 1200°K (927°C or 1700°F). Two facts may be noted: (1) the heat capacity increases with temperature but approaches a maximum, and (2) there is a discontinuity in the enthalpy curve at 573°C.

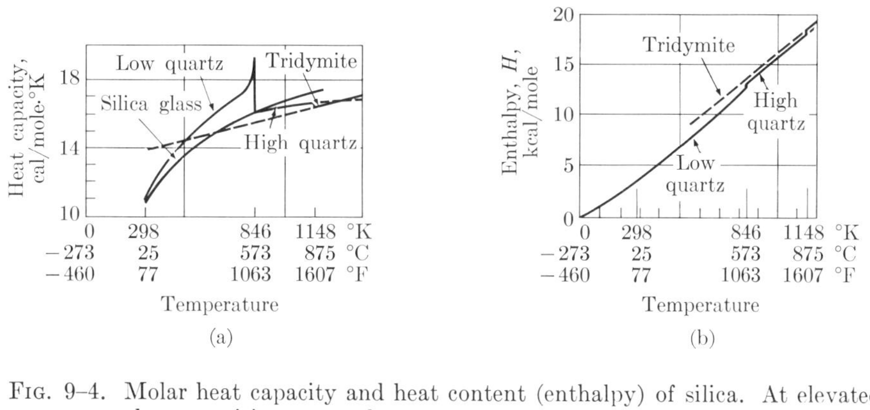

FIG. 9–4. Molar heat capacity and heat content (enthalpy) of silica. At elevated temperatures, heat capcities approach a constant value.

The shape of this heat-capacity curve is typical of those for most other materials. Empirical relationships are established from experimental data where

$$C_p = a + 2bT - cT^{-2}. \tag{9–5}$$

Temperature must be in absolute units (°K). As shown in Table 9–4, the applicability of such heat-capacity data is limited to a definite temperature range. Utmost caution should be used in extrapolation below this range because the plots for all heat capacities reverse their curvature in a manner not covered in Eq. (9–5). Some extrapolation beyond the upper end of the range is tolerable if phase changes do not occur. In any case, the molar heat capacity approximates $6M$ at high temperatures, where M is the number of atoms per mole (Example 9–2).

The discontinuity in the enthalpy data at 573°C (1063°F) corresponds to the displacive transformation (Fig. 3–22) from the low to high temperature

TABLE 9–4

MOLAR HEAT CAPACITIES OF SELECTED CERAMICS

Material	Molar heat capacity C_p, cal/gm-mole·°K	Temperature range, °K
Low quartz (SiO_2)	$11.22 + 8.20 \times 10^{-3}T - 2.70 \times 10^5/T^2$	298– 848
High quartz (SiO_2)	$14.41 + 1.94 \times 10^{-3}T$	848–2000
Glass (SiO_2)	$13.38 + 3.68 \times 10^{-3}T - 3.45 \times 10^5/T^2$	298–2000
Corundum (Al_2O_3)	$27.43 + 3.06 \times 10^{-3}T - 8.47 \times 10^5/T^2$	298–1800
Periclase (MgO)	$10.18 + 1.74 \times 10^{-3}T - 1.48 \times 10^5/T^2$	298–2100

modifications of quartz. A heat of transformation (+290 cal/gm) is required, which represents an increase in entropy that accompanies the bond straightening.

Equilibrium conditions require that quartz change to tridymite at 875°C (1610°F) as shown in Fig. 6–3. Since this is a sluggish reaction, heat contents for quartz can be measured considerably above this temperature. The quartz-to-tridymite transformation also requires a heat of transformation and a further increase in entropy.

EXAMPLE 9–2. Check the approximation cited above that the molar heat capacity is $6M$ for the compounds of Table 9–4 at 500 and 1000°C. [*Note:* The approximation is better suited for the constant-volume heat capacity C_v than for constant-pressure heat capacity, that is, C_p of Table 9–4. However, in solids, C_v and C_p are almost identical.]

Answer (calculated from data in Table 9–4):

Material	$6M$	500°C (773°K)	1000°C (1273°K)
Low quartz (SiO_2)	18	17.11	—
High quartz (SiO_2)	18	—	16.88
Glass (SiO_2)	18	15.65	17.85
Corundum (Al_2O_3)	30	28.32	30.81
Periclase (MgO)	12	11.27	12.31

9–4 Thermal conductivity

The transfer of thermal energy (heat) in ceramics is noticeably different from that in metals. Two related factors contribute to this difference: (1) ceramics have very few free electrons, and (2) ceramic phases are transparent to radiant energy. The lack of free electrons reduces the thermal conductivity, and the transparency increases the transfer of radiant energy, particularly at high temperatures.

Thermal conductivity k can be expressed mathematically as the coefficient which relates the rate of heat transfer per unit area, $(dQ/dt)(1/A)$, with the thermal potential or temperature gradient $(-dT/dx)$:

$$(dQ/dt)(1/A) = -k(dT/dx). \tag{9–6}$$

Of these, only the thermal conductivity k is a property of the material itself; therefore, we shall give our attention to it. The remaining independent variables, t, T, A, and x depend on geometry and service conditions.

Conductivity mechanisms. The value of k may be determined from kinetic-energy considerations, where

$$k = sc\lambda/3. \tag{9–7}$$

Again k is the thermal conductivity in cal/(sec·°C·cm), s is the volumetric specific heat of the energy carrier in cal/(cm^3·°C), c is the carrier velocity in cm/sec, and λ is the mean free path of the carrier (in cm). In the absence of gaseous molecules, there are two types of energy carriers in a solid: (1) electrons, and (2) phonons, or quantized elastic lattice vibrations.

The rate of transfer of electronic energy in metals is high because there are many free electrons which have a considerable amount of energy per unit volume, high velocity, and a reasonably long mean free path. Therefore they contribute substantially to the transfer of thermal energy. In ceramics, the number of free electrons is low, except in the case of semiconducting materials. Even semiconductors do not have enough free electrons to conduct significant amounts of thermal energy until they reach high temperatures (Section 10–2).

Phonons (lattice vibrations) are a major mechanism of thermal conductivity in ceramics. The vibrations are wavelike and may therefore be quantized in a way similar to that of photons of electromagnetic radiation (Section 2–2). Hence, Eq. (9–7) is applicable because: (1) the lattice vibrations have an energy s that is the volumetric heat capacity of the solid; (2) there is a definite wave velocity c which is the velocity of sound; and (3) there is a mean free path λ by which the phonon will move before it is diffracted by irregularities in the lattice. Any change in structure, environment, or composition which alters these three factors will affect the thermal conductivity of the materials.

Effect of phase structures. Crystalline phases have higher conductivities than glasses of the same composition (Table 9–5) because the regularity of their crystalline structure provides less interference for phonon movements than the glass structure does, and therefore permits longer mean free paths.

TABLE 9–5

THERMAL CONDUCTIVITY OF SILICA*

Temperature, °K	Thermal conductivity, cal·cm^{-1}·sec^{-1}·°C^{-1}	
	Quartz†	Glass
25	1.0	—
100	0.10	0.002
200	0.05	0.003
300	0.03	0.0035

* Adapted from Kittel, C., *Introduction to Solid State Physics*, New York: John Wiley & Sons (1953).
† Parallel to the *c*-axis.

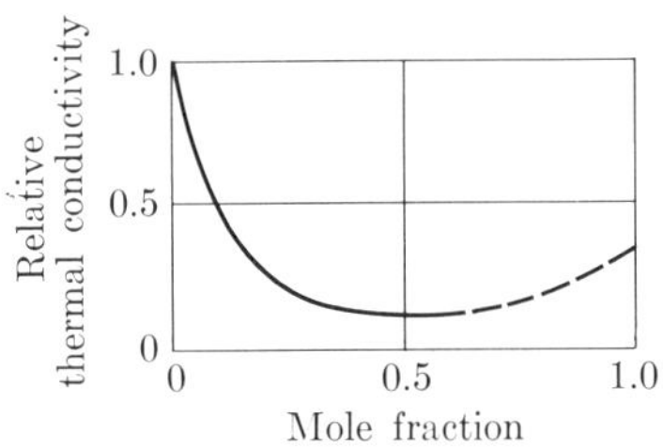

FIG. 9–5. Thermal conductivity of a MgO-NiO solid solution. [Adapted from Kingery, W. D., *Property Measurements at High Temperatures*. New York: John Wiley & Sons (1959), page 96.]

When crystals are solid solutions, the regularity of the crystal structure is altered and the phonons are diffracted more, with a consequent decrease in conductivity. There are relatively few data on the thermal conductivity of ceramic solid solutions; however, there is always a rapid decrease of thermal conductivity with small concentrations of solute. The effect, as shown in Fig. 9–5, is entirely compatible with metallic phenomena.

Thermal conductivity is essentially isotropic in cubic crystals such as periclase (MgO), or in amorphous materials such as glass. Anisotropic crystals have anisotropic conductivity. For example, the thermal conductivity of quartz is about 50% greater parallel to the *c*-axis than perpendicular to it, and graphite has thermal conductivity values of 0.006 and 0.6 (cal·cm)/(°C·cm^2·sec), respectively, in directions perpendicular and parallel to the main crystal plane. Almost certainly other layered structures, such as talc, kaolinite, and mica (see Table 3–5), have similar anisotropies in thermal conductivities; however, accurate data are still lacking. Actually, most ceramic products contain crystals of sufficient randomness so that mean values are appropriate.

Effect of temperature. Above ambient conditions, temperature changes do not make significant changes in the velocity of sound which enters into Eq. (9–7) as *c*, the carrier velocity. Further, the heat capacity approaches a constant value as the temperature is raised (see Fig. 9–4 and Table 9–4). However, the

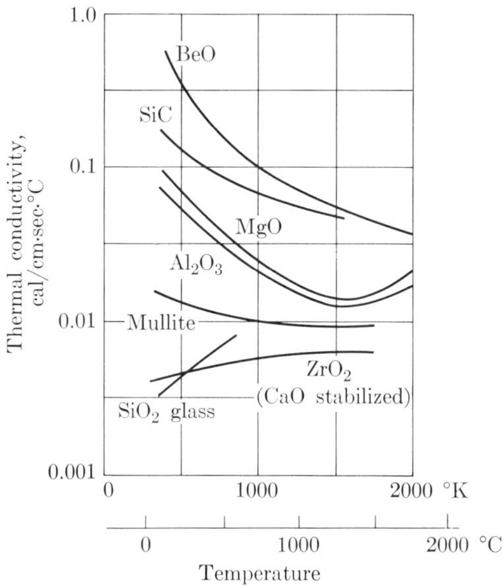

FIG. 9-6. Thermal transfer. [Adapted from Kingery, W. D., "Thermal Conductivity: XII, Temperature Dependence of Conductivity for Single-Phase Ceramics," *Journ. Amer. Ceram. Soc.*, **38,** 251–255 (1955).]

mean free path of phonon movement decreases as the temperature is raised,

$$\lambda \approx 1/T, \tag{9-8}$$

because the more intense the thermal vibrations the less regular the lattice structure. Thus thermal conductivity normally decreases as the temperature is raised to higher levels.

Apparent exceptions to the above generalization may be rationalized satisfactorily. As shown in Table 9–5, increased temperature gives glass a slightly increased conductivity (beyond that imparted by radiation). The explanation lies in the fact that glass has no lattice regularity; therefore the mean free path of phonons is already short, and this path does not decrease significantly as the temperature is raised. The only contributing factor for the k-versus-T variation is the increase in the heat capacity of silica that occurs at lower temperatures (Fig. 9–4).

A second exception applies to all transparent ceramics at high temperatures, where radiation transfer supplements the normal conductivity. Radiation varies as the fourth power of the temperature. This significantly reverses the downward trend in thermal conductivity at the very high temperatures to which many ceramics are exposed (Figs. 9–6 and 9–7), and accentuates the upward trend of thermal transfer in glass.

Effect of microstructures (and pores). There are several possible variations in microstructure which can affect thermal conductivity: (1) grain shape, (2) phase

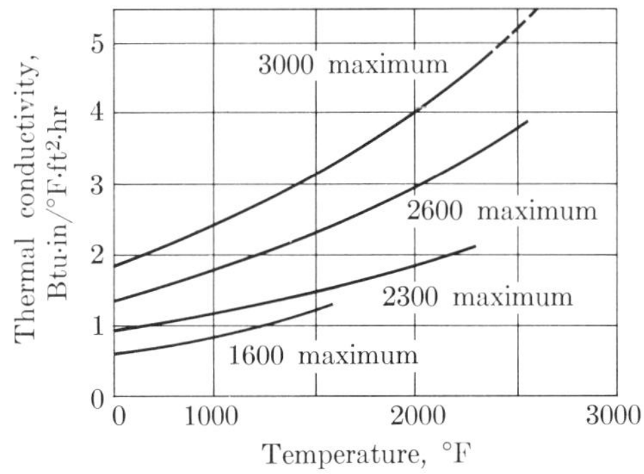

FIG. 9-7. Thermal conductivity of aluminosilicate insulating brick. Those bricks for higher service temperatures must be less porous.

quantity, (3) grain orientation, and (4) porosity (Section 7-1). Grain size has no significant effect so long as the other factors are constant.

The conductivities of phase mixtures are intermediate to those of the component phases. However, the change in conductivity does not always vary proportionally to the phase quantities, because the heat flows through those microstructural paths of higher conductivity and is restricted by the lower-conductivity phase. Two extreme cases may be cited by assuming a structure which is composed of alternate layers of high- and low-conductivity phases. When the conduction is parallel to the layers, the bulk conductivity coefficient k_B is

$$k_B = V_L k_L + (1-V_L)k_H, \tag{9-9}$$

where V_L is the volume of phase L, and k_L and k_H are the conductivities of the two phases. In this atypical case, the conductivity changes porportionally to the quantity of the two phases. The other extreme involves conduction perpendicular to the layers. In series, the total conductivity is related to the component conductivities by:

$$1/k_B = V_L/k_L + (1 - V_L)/k_H. \tag{9-10}$$

Under these conditions, the lower-conductivity phase markedly restricts the heat flow to low figures. If $k_L \ll k_H$, k_B approaches k_L/V_L but never decreases below k_L. Thermal conductivities as shown by Eqs. (9-9) and (9-10) are directly comparable to electrical conductivities in parallel and in series.

The previously described extremes, though not typical of ceramic microstructures, do place limits on bulk conductivities. A more realistic situation arises when one phase is in the form of dispersed particles within a matrix of another phase. This has been handled mathematically by Kingery as follows:

$$k_B = k_c \left[\frac{1 + 2V_d(1 - k_c/k_d)/(2k_c/k_d + 1)}{1 - V_d(1 - k_c/k_d)/(2k_c/k_d + 1)} \right], \tag{9-11}$$

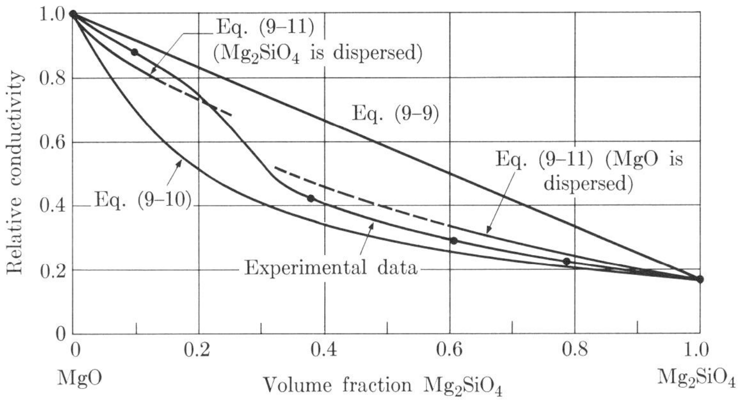

FIG. 9–8. Thermal conductivities of two-phase microstructures [periclase (MgO) and forsterite (Mg_2SiO_4)]. [After Kingery, W. D., *Introduction to Ceramics*. New York: John Wiley & Sons (1960), page 502.]

where the subscripts d and c refer to the dispersed and continuous phases. He supports his calculations with data (Fig. 9–8) for microstructures of periclase (MgO) and forsterite (Mg_2SiO_4). The agreement is satisfactory at the compositional extremes, where the minor phase is truly dispersed. In intermediate compositions, each phase maintains continuity and the necessary transition occurs.

A special case of importance arises when the second "phase" is an isolated pore. Francl and Kingery (*Journ. Amer. Ceram. Soc.*, **37,** 99–107 (1954)) found that the bulk conductivity could be expressed by

$$k_B = k_s(1 - P_c), \tag{9–12}$$

where k_s is the conductivity of the solid material and P_c is the cross-sectional pore fraction. There are limitations on the above relationship if the solid components are not continuous, or if the temperature exceeds 500°C and the emissivity is high. Thus Eq. (9–12) does not hold for compacted but unsintered ceramic raw materials. The temperature limitation is necessary for materials of high emissivity, because radiation transfer, as well as conductive transfer, become significant.

9–5 Thermal stresses

If the dimensional changes caused by temperature changes are not uniform, the resulting differential strains cause stresses within a material. These stresses are relatively more significant in ceramics than in metals because there is a lack of ductility to relieve the stresses; consequently, cracking and spalling are more common.

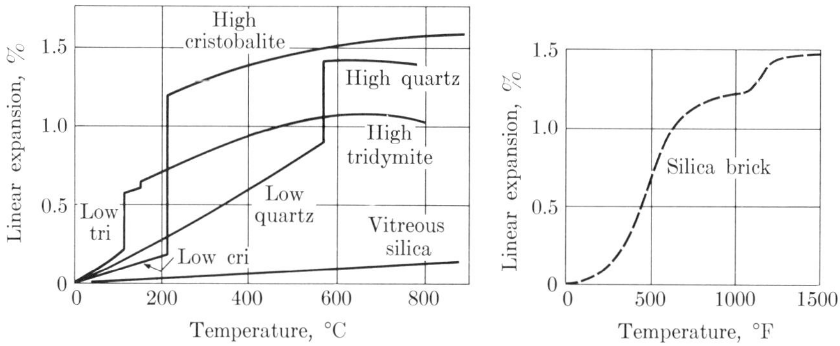

FIG. 9-9. Expansion of silica: (a) individual phases; (b) refractory containing a mixture of silica phases. The smoothing off of curve (b) is partially a consequence of nonuniformities of temperature during heating.

Nonuniform dimensional changes may have several different origins. (1) Phase transformations. These may be exemplified either by the reconstructive transformation of quartz (SiO_2; $\rho = 2.65$ gm/cc) to cristobalite (SiO_2; $\rho = 2.32$ gm/cc), or by the displacive transformation (Section 3-6) of low cristobalite to high cristobalite (Fig. 9-9). Either change involves heat-induced expansion, which produces tension stresses in the unaltered part of the material. (2) Differential expansions. These are typified by mullite ($Al_6Si_2O_{13}$) porcelains, which exhibit differential expansion in polyphase materials, because mullite has a linear coefficient of 5×10^{-6} cm/cm/°C, while the surrounding glassy matrix has an expansion of 7×10^{-6} cm/cm/°C. (3) Anisotropic expansions within crystals. These provide similar differentials in single-phase, polycrystalline materials. As an example, an alumina insulator can have adjacent grains which expand 8.3 and 9.0×10^{-6} cm/cm/°C if their *c*-axes are perpendicular to each other. (4) Temperature differentials and their effect on dimensional changes. Here the simplest example is glass, which can be cooled so that the outside contracts while the center is still hot and semielastic. Further cooling places the center under tension (and the surface under compression) because the rigid surface cannot continue to contract as the center contracts.

EXAMPLE 9-3. Stress analyses show that if thermal expansion is prevented during heating of a material, there develops a compressive stress which has the value of

$$\sigma = E\alpha_l(T_2 - T_1)/(1 - \mu). \qquad (9\text{-}13a)$$

Given that the various silica phases of Fig. 9-9(a) have an elastic modulus E of 10^7 psi and a Poisson's ratio μ of 0.25, determine the stress σ which would be developed during heating cristobalite (SiO_2) from 25°C to 325°C, if thermal expansion is prevented.

Answer: In Eq. (9-13a), $\alpha_l(T_2 - T_1)$ is the unit expansion; in Fig. 9-9(a), the unit expansion is $(1.3 - 0.05)$ percent, or 0.0125 in the 25-325°C temperature interval:

$$\sigma = 10^7(0.0125)/(1 - 0.25) = 167{,}000 \text{ psi}.$$

Thermal cracking (*spalling*). The four kinds of dimensional changes just cited provide stresses that lead to cracking in brittle materials. If the fracture stress σ_f is the strength of the material, Eq. (9–13a) can be used to determine the temperature differential ΔT_f, which will produce fracture

$$\Delta T_f = \sigma_f(1 - \mu)/E\alpha, \tag{9–13b}$$

or

$$\Delta T_f = f\left(\frac{S}{E\alpha}\right), \tag{9–13c}$$

where S is the strength of the material. The value of T_f is also an inverse function of thermal diffusivity h, which is equal to $k/c_p\rho$:

$$\Delta T_f = f\left(\frac{1}{h}\right) = f\left(\frac{c_p\rho}{k}\right). \tag{9–13d}$$

In this relationship, k is the thermal conductivity, c_p is the specific heat capacity, and ρ is the density. Equation (9–13d) is significant because materials with high thermal conductivities or low-volume heat capacities ($c_p\rho$) permit rapid temperature equalization and reduced thermal stresses.

It is convenient to utilize spalling resistance indices, SRI, as

$$\text{SRI} = hS/\alpha E. \tag{9–14a}$$

As such, the SRI is the ratio of $(S/\alpha E)$ and $(1/h)$ of Eqs. (9–13c) and (9–13d). Equation (9–14a) could also be written:

$$\text{SRI} = kS/\alpha E c_p\rho. \tag{9–14b}$$

Furthermore, since the above is only an index and not a fundamental property, variants are commonly used, such as $\text{SRI} = f(kS/\alpha E)$. Table 9–6 shows how

TABLE 9–6

SPALLING INDICES OF BeO*

Temperature, °C	Relative spalling resistance, $hS/\alpha E$	
	Hot pressed, 2% porosity	Slip-cast, 13% porosity
300	10	4.5
500	5	3
700	3	1.5

* Calculated from Eq. (9–14a). Adapted from Murray, P., D. T. Livey, and J. Williams, "The Hot Pressing of Ceramics," *Ceramic Fabrication Processes*, New York: John Wiley & Sons (1958), p. 159.

the spalling resistance is affected by variations in production and service conditions. These effects result primarily from differences in thermal conductivity and strength of the bulk material as the porosity and temperature are altered.

EXAMPLE 9–4. Determine the thermal diffusivity at 100°C of (a) quartz crystals, and (b) a bed of fine quartz sand (bulk density = 103 lb/ft³). (c) Compare the units of thermal diffusivity with those of mass diffusion. The "conductivity" of air is 6×10^{-4} (cal·cm)/(cm²·°C·sec) under these conditions.

$$h = k/c_p\rho = \frac{(\text{cal·cm})/(\text{cm}^2\text{·°C·sec})}{(\text{cal/gm·°C})(\text{gm/cm}^3)} = \frac{\text{cm}^2}{\text{sec}}.$$

Answer: (a) From Table 9–5:

$$k = 0.02 \text{ cal/(cm·°C·sec)}.$$

From Table 9–4:

$$c_p = [11.22 + 0.0082\,(373) - 270{,}000/(373)^2]/60.1 = 0.2 \text{ cal/(gm·°C)}.$$

From Appendix E:

$$\rho = 2.65 \text{ gm/cm}^3;$$
$$h = (0.02)/(0.2)(2.65) = 0.037 \text{ cm}^2\text{/sec}.$$

(b) First we find that

$$\text{Porosity} = 1 - 103/(2.05)(62.4) = 0.38.$$

From Eq. (9–11):

$$k = (6)(10)^{-4}[1 + 2(0.62)(1 - \text{nil})/(\text{nil} + 1)]/[1 - 0.62(1 - \text{nil})/(\text{nil} + 1)]$$
$$= 3.5 \times 10^{-3}\text{(cal·cm)/(cm}^2\text{·°C·sec)};$$
$$h = 3.5 \times 10^{-3}/(0.2)(2.65)(0.62) = 10^{-3}\text{cm}^2\text{/sec}.$$

(c) Thermal diffusivity and mass diffusion have units of cm²/sec (Section 6–5).

Stress relaxation. Stress-induced atom movements can relieve strains that exist within materials. This is particularly true in glasses above their fictive temperatures, where the atoms have sufficient energy to move to new neighbors. We recall from Section 4–4 that the fictive temperature is a function of the cooling rate, and that slow reheating permits atom movements at lower temperatures (Example 4–3 and Fig. 4–5). Thus it is to be expected that residual stresses in glass may relieve themselves with time, and that the higher the temperature, the more rapid the relaxation.

Stress relaxation in crystalline ceramics occurs differently from the way it occurs in glass, and usually with higher activation energies than in glass, because dislocation movements must take place. As a result, if stress relief is to occur in reasonable periods of time, higher temperatures are required. Stresses in polycrystalline ceramics may also be relaxed by grain-boundary movements because the boundary atoms have low order and high energy, and therefore creep readily.

9–6 High-temperature deformation

Deformation of ceramic materials can be accentuated by high temperatures because atom movements occur more readily. We shall briefly consider three effects: (1) anelastic behavior, (2) creep, and (3) viscous flow.

Anelasticity. Figure 8–2 showed a gradual decrease in the modulus of elasticity with increasing temperature. In addition, at high temperatures, there is a *marked* decrease. This major decrease occurs because the stress at high temperatures induces movements of atoms from one site to another. In this respect, the strain is not truly elastic, because some atoms gain new neighbors, but is partially reversible, in that a release or reversal of the stress permits a return of many of the atoms to their original positions. The term "anelasticity" has been applied to describe this behavior, and a hysteresis loop occurs when stress and strain are cycled (Fig. 9–10). The anelastic displacement of atoms is more extensive at higher temperatures, or with more time or lower loading frequencies.

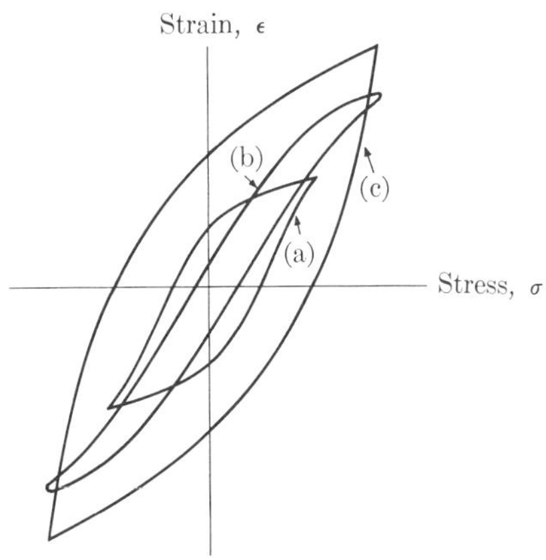

FIG. 9–10. Anelastic behavior under cyclic loading (schematic): (a) low stress, high temperature; (b) high stress, low temperature; (c) high stress, high temperature.

Damping capacity is closely related to anelastic behavior. Specifically, the energy absorbed per stress cycle is equal to the area within the loop of Fig. 9–10, and the total energy absorbed by damping is obtained by integrating over the total number of cycles. The greatest energy absorption occurs in temperature ranges in which the value of $-dE/dT$ is largest (see Fig. 8–2).

Creep. The slow deformation which occurs in materials under static loads at high temperatures has been called creep. Predictably enough, the rate of creep is a function of both stress and temperature. Mechanisms involved include

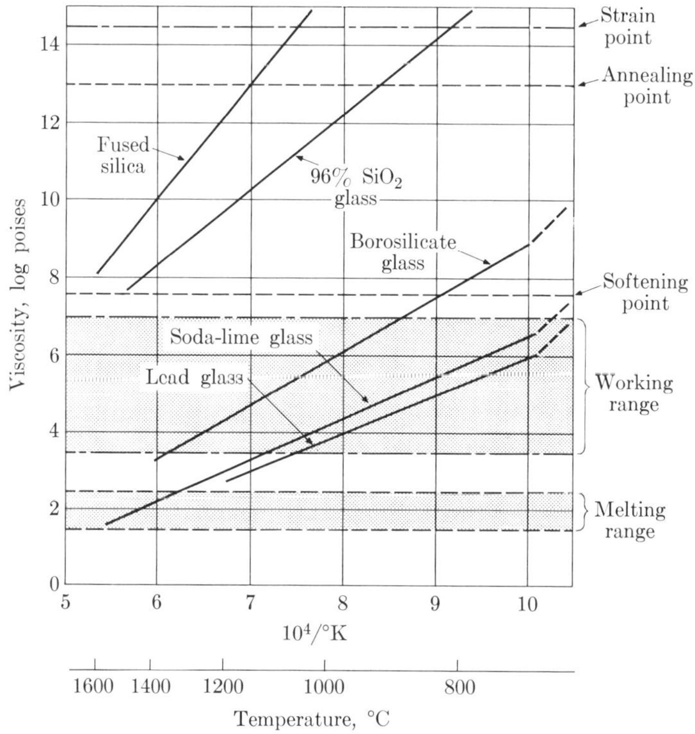

FIG. 9-11. Viscosity versus temperature. The temperature at which the atomic mobility is sufficient to permit stress relief in 15 minutes is called the *annealing temperature*. This corresponds to a viscosity of 10^{13} poises.

grain-boundary deformation, dislocation movements, and diffusion. Therefore, there is no simple relationship by which we may anticipate creep rates.

Viscous deformation. The high-temperature deformation of glass is viscous: That is, the flow-velocity gradient is proportional to the applied stress (Eq. (8-11a)). Furthermore, the proportionality constant or viscosity coefficient, η, represents an activation process in which

$$\eta = Ae^{-E/RT}. \tag{9-15}$$

As before, A is a constant, and E, R, and T have the usual connotations. Figure 9-11 shows the variation of viscosity with temperature for several typical glasses. When we recall the foregoing discussions, the significance of the a nnealing and softening points becomes evident. The annealing point is the temperature at which the viscosity is such that atom movements occur rapidly enough for residual thermal stresses to be relieved in 15 minutes (a period that is compatible with production schedules). This viscosity is 10^{13} poises.* The softening point, with its lower viscosity ($10^{7.6}$), is the maximum temperature at which glass can be handled without serious dimensional changes. Still lower viscosities meet the requirements of production practices as to working and melting range (Chapter 13).

9–7 Refractory products

Many materials can be used at high temperatures; we do not need to categorize all of them here. However, since the behavior of a refractory material is closely associated with its nature and properties, we shall list the general categories, giving specific attention to the origins of service-limiting behaviors.

Silica refractories. The widespread occurrence of relatively pure silica, plus the fact that its melting temperature is in excess of 3000°F (>1650°C), has made it a widely used refractory. Next to fireclay, discussed below, silica is probably the cheapest refractory; therefore it is considered whenever its characteristics are appropriate.

In addition to ready availability and low cost, silica refractories have the advantage of low bulk densities (Table 9–7) and high softening temperature. The latter characteristic is of great significance (see Fig. 9–12). It is this property which is responsible for the long-continued use of silica refractories in metallurgical furnaces, where it would seem that they would otherwise be an illogical choice.

The strength of silica refractories at high temperatures is unique enough to warrant explanation. There are two contributing factors: (a) phase equilibria,

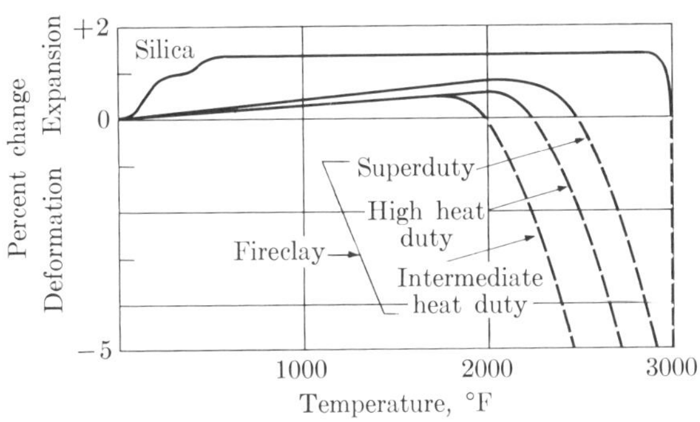

Fig. 9–12. High-temperature deformation: 50 psi load; $\Delta T/\Delta t$ = 100°C per hour above 1100°C (2000°F).

* As formerly defined, the annealing point was the temperature at which the viscosity is $10^{13.4}$ poises. Another reference temperature for a glass is the strain point of $10^{14.5}$ poises. This is usually about 40°C lower than the annealing point, so that four hours are required for stress relief.

TABLE 9-7

SELECTED PROPERTIES OF COMMON (UNALTERED) REFRACTORY MATERIALS*

Brick	Composition	PCE † or Softening point	True specific gravity	Approx. weight, 9″ brick	Porosity	Linear coefficient of expansion
High-heat-duty firebrick	Al_2O_3: 35–42% SiO_2: 52–60%	31–33 (3060–3170° F)	2.60–2.70	7.5–8 lb	15–24%	(21–1300° C) 0.0000054
Superduty	Al_2O_3: 43–44% SiO_2: 51–53%	33–34	2.60–2.70	8–8.3 lb	12–16%	
High alumina	Al_2O_3: 50–60–70 or 80%	34–39 (3200–3390° F)	2.60–2.70	8–8.3 lb	20–31%	
Kaolin	Al_2O_3: 44–45% SiO_2: 51–53%	31–34 (3060–3200° F)	2.60–2.70	8–8.3 lb	18–28%	(21–1610° C) 0.0000043
Bonded mullite	Al_2O_3: 60% SiO_2: 36%	38 (3340° F)	3.00	8.4 lb	16–24%	(200–1600° C) 0.0000045
Molten cast mullite	Al_2O_3: 71–72%	38 (3340° F)	3.3	10–10.25 lb	0.5–1%	(21–900° C) 0.0000060
Zircon	$ZrSiO_4$	Above 2015° C 3660° F	4.6	12 lb	20–26%	(21–1550° C) 0.0000042
Silica	SiO_2: 95–96%	31–33 (3060–3170° F)	2.30–2.40	6 lb	22–30%	(21–1550° C) 0.0000083
Recrystallized silicon carbide	SiC: 98–99%	Dissociates at 2250° C 4080° F	3.17	9 lb	20%	(21–1000° C) 0.0000045
Fused alumina	Al_2O_3: 88–90%	2050° C 3720° F	3.90	11 lb	21%	(21–800° C) 0.0000081
Chrome	Cr_2O_3: 30–45% Al_2O_3: 15–33% SiO_2: 11–17% FeO: 3–6%	1950–2200° C 3540–3990° F	3.80–4.10	11 lb	20–28%	(21–1540° C) 0.0000104
Magnesite (burned)	MgO: 83–93% Fe_2O_3: 2–7%	2200° C 3990° F	3.40–3.60	9.5 lb	20–28%	(21–1700° C) 0.0000147
Magnesite (unburned)	MgO: 83–93% Fe_2O_3: 2.7%	2200° C 3990° F	3.40–3.60	10.5 lb	10–22%	(21–1700° C) 0.0000147
Forsterite	$2\ MgO{\cdot}SiO_2$	1910° C 3470° F	3.30–3.40	9 lb	24–27%	(21–1500° C) 0.0000125

* Courtesy of Charles Taylor and Sons, a subsidiary of National Lead Company, Cincinnati, Ohio.
† Pyrometric cone equivalent.

TABLE 9–7 *(continued)*

Thermal conductivity, CGS units	Specific heat	(SiC–100) Spalling index	Deformation under load, 25 lb/in²	Constancy of volume	Slag resistance
(200–1000° C) 0.00339	(25–1000° C) 0.254	60–90	3–6% at 2460° F	0.5 to 2% shrinkage at 2550° F	Rapidly attacked by basic slags, particularly those high in Fe_2O_3; moderate resistance to acid slags
Approximately the same as high-heat-duty firebrick		90–95	1–2% at 2460° F	1% max. shrinkage at 2900° F	
Approximately the same as high-heat-duty firebrick		60–80	Somewhat less than high-heat-duty firebrick		More resistant to most alkalies than firebrick; readily attacked by Fe_2O_3
(200–1000° C) 0.0045	(250–1000° C) 0.254	75–85	1.0% at 2800° F	5% shrinkage at 3000° F	Approximately same as high-heat-duty firebrick; slightly better resistance to alkalies
(600–900° C) 0.0044	(20–1000° C) 0.20	90	0.0% at 2460° F, 1.0% at 2900° F	No shrinkage at 3000° F; 2.5% at 3200° F	Relatively insoluble in most slags and glasses, particularly those high in lime, alkalies, or fluorides; attacked by strongly basic slags or those high in Fe_2O_3
0.06 (No temperature range given)	(20–980° C) 0.25	50–60	No deformation at 1350° C	No appreciable change at 1590–1650° C	Chemical characteristics of mullite: low porosity and high density; offers maximum resistance to corrosion
(200–1000° C) 0.0046	(20–51° C) 0.132	70–90	Failed at 1550–1600° C	No appreciable change at 1550° C	Slightly attacked by acid slags; strongly attacked by basic slags and fluorides
(200–1000° C) 0.0045	(25–1000° C) 0.265	10–30	0.0% at 1500° C, 50 lb/in²	Reversible expansion to 3170° F; no shrinkage	Resists acid slags and dust; attacked by basic slags, fluorides and Fe_2O_3
(200–1350° C) 0.0330	(20–1000° C) 0.186	100	0.0% at 1500° C, 50 lb/in²	No shrinkage at 1500° C; up to 10% permanent expansion due to oxidation	Attacked by basic slags, particularly those high in iron oxide
(200–1000° C) 0.0095	(20–1000° C) 0.20	60–70	2% at 1500° C, 50 lb/in²	No shrinkage at 1500° C	Properties similar to sillimanite
(200–1000° C) 0.0040	(20–1000° C) 0.217	25–70	Improved chrome brick show no deformation at 1425° C	No shrinkage at 1450° C; 1.3% at 1550° C	Neutral properties, resistant to both acid and basic slags
(200–1000° C) 0.0087	(20–1000° C) 0.278	20–40	Not recommended where high hot strength required	Shrinkage at temperatures above 1600° C	Highly resistant to basic slags; readily attacked by acid slags
(200–1000° C) 0.0087	(20–1000° C) 0.278	30–50	3.5% deformation at 1480° C	Some shrinkage at 1500° C; 2.6% at 1550° C	
Slightly lower than magnesite	(20–100° C) 0.22	20–40	Similar to silica	Negligible shrinkage on reheat to 1650° C	Similar to magnesite, but more readily attacked by either acid or basic slags

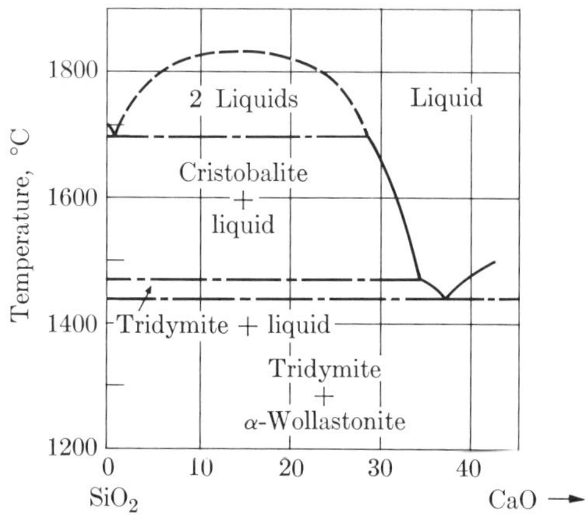

FIG. 9–13. CaO-SiO$_2$ phase diagram (partial). Compare with Fig. 6–12.

and (b) microstructure. A common procedure is to use about 2% Ca(OH)$_2$ as a catalyzer for the transformation of quartz to tridymite and cristobalite (Section 3–6). When this is fired, a CaO-SiO$_2$ glass results. The monotectic character of the phase relationships (Fig. 9–13) produces very little change in liquid content as the temperature approaches and even exceeds 3000°F (1650°C). Further, any liquid which does form does not penetrate along the grain boundaries. As a result, considerable solid-to-solid contact is maintained, even when the liquid content is high [Fig. 7–9 and Eq. (7–3)].

Silica refractories have a major disadvantage arising from their polymorphic changes. The displacive transformations are more severe than the reconstructive transformations because they occur much more rapidly and produce significant volume changes. Severe stresses accompany these changes, and spalling (i.e., cracking) will result unless extreme care is taken in heating. Since a fired silica refractory contains a large quantity of cristobalite and lesser quantities of quartz and tridymite, a volume expansion curve such as shown in Fig. 9–9(b) is common. The dimensional increases indicate the low-to-high displacive transformations of quartz and cristobalite. Because of these sudden changes, silica refractories require extremely slow heating and cooling rates.

Alumina and aluminosilicate refractories. Fireclay refractories are the commonest of this group; for that matter, they are the commonest of all refractories. There is a whole spectrum of commercial refractories, ranging from the fireclay composition to that of pure alumina compositions. When clays with very high kaolinite content are fired, the products are Al$_2$O$_3$ and SiO$_2$:

$$Al_2Si_2O_5(OH)_4 \rightarrow Al_2O_3 + 2SiO_2 + 2H_2O. \qquad (9\text{–}16)$$

In turn, the Al$_2$O$_3$ and SiO$_2$ of those proportions should form mullite and cristobalite (Fig. 9–14). However, crystallization is slow and a metastable glass is formed first. This permits viscous deformation to occur at subeutectic temperatures, as indicated in Fig. 9–12. The presence of impurities lowers the viscosity of the glass and increases the deformation rate, thus reducing the high-temperature load-bearing ability. Various "duty" refractories have been established by ASTM standardized tests (Fig. 9–12 and Table 9–7).

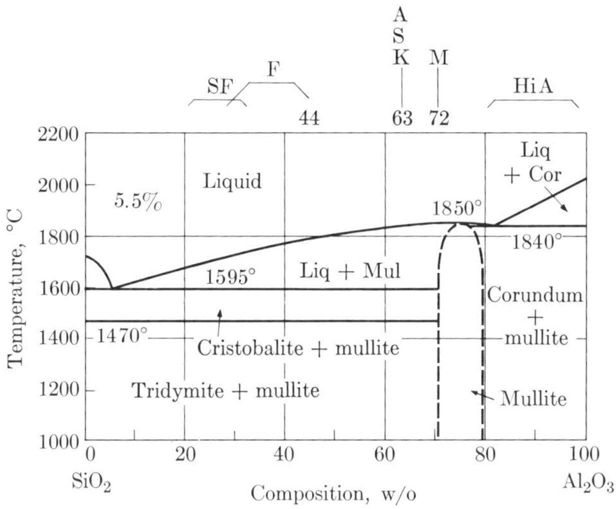

FIG. 9–14. Aluminosilicate refractories: F, fireclay; SF, siliceous fireclay; ASK, andalusite, sillimanite, kyanite; M, mullite; Hi A, high alumina. (This phase diagram shows mullite as a congruently melting phase. Recent evidence favors this relationship (cf. Fig. 6–12).

Refractories with Al_2O_3 at mullite levels or higher have a major property advantage. In the absence of other components, liquid is not a stable phase until 1840°C (3340°F). Further, since less silica is present, any liquid present will not be glassy, but will crystallize quite readily.

Basic refractories. The term "basic refractories" is applied to those ceramic products which contain relatively high levels of calcium or magnesium oxides. Common basic refractories include: (1) calcined dolomite ($\rightarrow$ CaO + MgO), (2) calcined magnesite ($\rightarrow$ MgO), and (3) forsterite (Mg_2SiO_4), which is an end member of an olivine series [$(Mg,Fe)_2SiO_4$]. Noticeably absent from this list is calcined calcite ($\rightarrow$ CaO). It has a high melting temperature and is widely available as relatively pure limestone; however, CaO hydrates so readily that it is difficult to maintain the oxide in a calcined form. This hydration not only evolves considerable heat, but also results in volume changes that create problems. In addition, the dehydration of $Ca(OH)_2$, if it were used at high temperatures, would have numerous serious consequences. To prevent hydration, calcined dolomite must be isolated from atmospheric moisture by closed containers or by tar coatings. Calcined magnesite can be stabilized as periclase (MgO) if there is enough grain growth to reduce the surface area available for hydration.

Periclase, as a pure refractory material, can withstand temperatures of 2800°C (5070°F), but it seldom reaches this temperature in practice, since SiO_2 in small amounts is nearly always present. This SiO_2 causes a liquid silicate film, similar to that in Fig. 7–4, to develop. Even when it is of slight volume, the liquid prevents solid-to-solid contact of the periclase grains. Therefore the strength of the material is greatly reduced. The increased use of basic refractories in recent years is a consequence, in part, of the effort to circumvent this microstructural characteristic, either by avoiding SiO_2 or by adding chromite [$(Fe,Mg)(Fe,Al,Cr)_2O_4$]. The latter, when it is combined with the periclase (MgO) of calcined magnesite, provides the chrome-magnesite brick of the metallurgical industry.

Carbon and carbides. Carbon, since it is not a compound of metallic and nonmetallic elements, does not conform to our previous definition of a ceramic material (Section 1–1). Of all the elements, carbon is most nearly in the intermediate zone between metals and nonmetals. Hence it is generally neglected by both metallurgists and organic chemists as being atypical. The extensive use of carbon and graphite as refractory materials has made them of utmost interest, however, to the ceramist.

Although both amorphous and graphitic carbon are used for refractory purposes, the latter is more important. Carbon has two characteristics which are significant for its application: First, although its stability temperature is high, the solid phase sublimes when it does dissociate. This starts to occur around 2200°C (4000°F) under reduced pressures. The second significant factor is its oxidation to gaseous CO. It is thus mandatory that carbon be used in low-oxygen environments. However, carbon oxidizes much more slowly than many metals do.

Graphitic carbon is very anisotropic in structure and properties. The atoms within any hexagonal basal plane are held together with extremely strong bonds, which have been described as intermediate between covalent and metallic. On the other hand, the bonds in the perpendicular direction are fewer in number and extremely weak. The effects of these characteristics are well known. Graphite is widely used as a dry lubricant, and the anisotropy of its electrical and thermal conductivities is currently being exploited through the use of pyrolytic graphite. When hydrocarbons come into contact with hot surfaces (>1700°C or 3100°F), they dissociate into smaller molecules and carbon. For example,

$$CH_4 \rightarrow 2H_2 + C_{(gr)}. \tag{9–17}$$

Such reactions are of interest because the graphite may be fabricated into the shapes of the surfaces of deposition. Because of the parallelism of its basal planes, graphite is highly oriented. However, there is relatively poor stacking (Section 3–4) of one plane over another. As a result, the average distance between planes is greater than in highly crystalline graphite. This only serves to accentuate the anisotropy which already exists. Therefore, the thermal conductivities are about 0.006 and 0.6 (cal·cm)/(°C·cm^2·sec), respectively, for the directions perpendicular to and parallel to the basal plane (Section 9-4). Electrical conductivity figures are 0.004×10^6 and 0.4×10^6 (ohm·cm)$^{-1}$, respectively. High-temperature service permits a reorientation which provides better stacking of successive layers. This lowers the anisotropy, and also gives graphite the unusual property of attaining increased mechanical strength as temperatures are increased (Fig. 9-15).

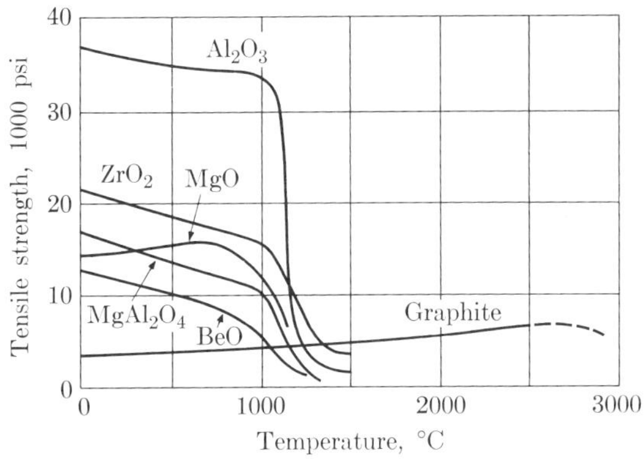

FIG. 9-15. Tensile strength versus temperature. (Adapted from Jaffee, R. I., and D. J. Maykuth, "Refractory Materials," Defense Metals Information Center Memorandum 44, Battelle Memorial Institute, Columbus 1, Ohio.)

The carbides of greatest refractory interest are those which form from semimetallic elements, such as SiC, TiC, ZrC, and B_4C. Silicon carbide is widely used, not only because it provides good thermal conductivity and thermal stability, but also because its electrical conductivity is such that it can be used as an electrical resistor. Like all carbide refractories, SiC is subject to oxidation, and the temperature at which it may be used is limited to 2800°F (1480°C). Above this temperature the refractory must be considered expendable, because gaseous SiO and CO form too rapidly. This particular refractory is protected at lower temperatures by the formation of a surface SiO_2 coating, which is a product of oxidation (Table 10-2).

Miscellaneous refractories. Figure 9-16 summarizes the melting temperatures of a number of specialized binary refractories. Although all of these exhibit high melting temperatures, many are subject to chemical alteration by common environments. A few warrant greater attention because of their general usefulness. Zirconia (ZrO_2) is one such example. It becomes useful only if its crystal structure is stabilized to avoid a monoclinic-to-tetragonal phase inversion at 1100°C (2000°F), since that inversion involves a 7.5% volume change. Zirconia is usually stabilized by the addition of CaO, thus engendering a solid solution and a stable cubic form.

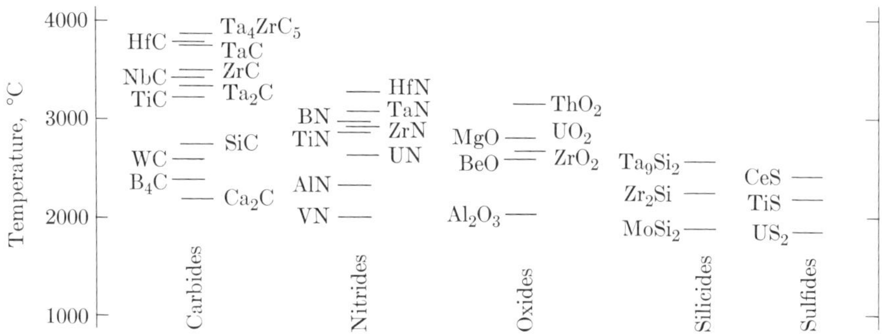

FIG. 9-16. Melting points of selected binary ceramic compounds. (Adapted from Jaffee, R. I., and D. J. Maykuth, "Refractory Materials," Defense Metals Information Center Memorandum 44, Battelle Memorial Institute, Columbus 1, Ohio.)

Other binary compounds which receive special attention include $MoSi_2$ and CeS, the former because it can be used to produce a protective SiO_2 coating on molybdenum metal, and the latter because it has one of the highest melting points of all of the sulfide compounds. However, CeS is so subject to oxidation that it is pyrophoric.

Some refractory ternary compounds also receive commercial attention, including zircon ($ZrSiO_4$) and spinel ($MgAl_2O_4$); both are used in cases which involve contact with molten metals.

References for Further Reading

9-1. AUSTIN, J. B., "Thermal Expansion of Nonmetallic Crystals," *Journ. Amer. Ceram. Soc.*, **35**, 243–253, 1952. A summary of thermal expansion versus structure. Discusses the anisotropy of thermal expansion, which produces thermal ratcheting.

9-2. BALDWIN, W. M., JR., *Residual Stresses*. Philadelphia: American Society of Testing and Materials, 1949. An introduction to residual stresses. Recommended because of its illustrative sketches which show the distribution of residual stresses.

9-3. CAMPBELL, I. E. (editor), *High-Temperature Technology*. New York: John Wiley & Sons, 1956. Three dozen papers on the materials, methods and measurements of high-temperature technology. For the specialist.

9-4. CHESTERS, J. H., *Steel Plant Refractories*. Sheffield, England: United Steel Companies, Ltd., 1944. Specific reference to refractory applications.

9-5. DORN, J. E., *Mechanical Behavior of Materials at Elevated Temperatures*. New York: McGraw-Hill, 1961. For the research worker interested in the interrelationship of strength and temperature.

9–6. Kelley, K. K., *High Temperature Heat-Content, Heat-Capacity, and Entropy Data for the Elements and Inorganic Compounds*, U. S. Bureau of Mines Bulletin 584. Washington: U. S. Government Printing Office, 1960. The most thorough compilation available of thermodynamic data on ceramic compounds.

9–7. Kingery, W. D., *Introduction to Ceramics*. New York: John Wiley & Sons, 1960. Thermal properties are discussed in Chapter 14, thermal stresses in Chapter 18. Advanced undergraduate and graduate level.

9–8. Kingery, W. D., *Property Measurements at High Temperatures*. New York: John Wiley & Sons, 1959. For the graduate and professional research worker. Very complete.

9–9. Kingery, W. D., "The Thermal Conductivity of Ceramic Dielectrics," *Progress in Ceramic Science* (J. E. Burke, editor). New York: Pergamon Press, 1962, pp. 181–235. A thorough discussion of phonon and photon conductivity in ceramics. For the advanced student.

9–10. M.I.T., "Thermal Conductivity," *Journ. Amer. Ceram. Soc.*, **37**, 67–110, 1954. A special issue, presenting ten papers on thermal conductivity. Graduate level.

9–11. Norton, F. H., *Refractories*, 3rd edition. New York: McGraw-Hill, 1949. For the manufacturer and user of tonnage refractories.

9–12. Rait, J. R., *Basic Refractories*. London: Iliffe & Sons, Ltd., 1950. Composition and phase relationship of basic refractories. For the specialist.

9–13. Ryshkewitch, E., *Oxide Ceramics*. New York: Academic Press, 1960. Unlike Norton (9–11), this book deals primarily with specialized, higher-purity refractory oxides. Detailed attention to alumina, spinel, magnesia, beryllia, zirconia, zircon, thoria and ceria.

9–14. Smoke, E. J., and J. H. Koenig, *Thermal Properties of Ceramics*. New Brunswick, N. J.: Rutgers University, 1958. Brief attention to the melting, thermal conductivity, thermal expansion, thermal endurance, and thermal conditioning of ceramics.

9–15. Stanworth, J. E., *Physical Properties of Glass*. Oxford: Clarendon Press, 1950. Annealing and stress relaxation within glasses at elevated temperatures.

Problems

9–1. Show that the volume thermal expansion α_v of a polycrystalline material is approximately $3\alpha_l$, where α_l is the linear expansion.

9–2. Calculate the volume thermal expansion for the phases given in Table 9–3.

9–3. Plot the heat capacities versus temperature of (a) corundum (Al_2O_3), and (b) periclase (MgO).

9–4. Thermal diffusivity h is equal to $k/c_p\rho$, where k is the thermal conductivity and c_p and ρ are specific heat and density respectively. (a) What are the dimensional units of the thermal diffusivity? (b) Calculate the thermal diffusivity of three of the phases in Table 9–4 at room temperature and at 500°C.

9–5. A silica brick contains 50 v/o cristobalite, 25 v/o tridymite, 15 v/o quartz, and the rest is glass. What is the linear expansion of the brick at 200°C, 300°C, 400°C, 500°C, and 600°C?

9-6. The following phases are heated from room temperature to 600°C with volume constrained so that only 1.0 v/o expansion can occur: (a) quartz, (b) tridymite, (c) cristobalite, (d) vitreous silica. What stresses are developed? Utilize the appropriate property data of Example 9-3 (page 152).

9-7. A large spherical grain (single crystal) of quartz is heated rapidly so that the center and surface have the following simultaneous temperatures during heating.

Center, °C	25	50	500	800	1300	1450
Surface, °C	25	400	900	1200	1425	1450

(a) Describe the stresses and the possibility of fracture within the grain. (b) Cite phase changes, if any.

9-8. An insulating brick of diatomaceous earth contains skeletal particles of amorphous SiO_2. Its bulk density is 1.2 gm/cm^3. Estimate its thermal conductivity at room temperature. The density of amorphous silica is 2.2 gm/cm^3.

9-9. Residual stresses in glasses are commonly relaxed exponentially. Polarized light measurements indicate that stresses in a glass decrease from 1000 psi to 800 psi when it is heated to 400°C for one day. (a) How long will it take to reduce the stresses to 200 psi at the same temperature? (b) To 100 psi? (c) To 50 psi?

9-10. When enameled steel is cooled to room temperature and restrained from warping, what are the stresses in the vitreous coating?

Characteristic	Enamel	Steel
Annealing point, °F	700	—
Mean thermal coefficient, in./in./°F	7.0×10^{-6}	8.0×10^{-6}
Modulus of elasticity, psi	10^7	3×10^7
Thickness, in.	0.01 (each side)	0.08

9-11. To what temperature must the soda-lime glass of Fig. 9-11 be raised in a tank 3 feet deep to permit bubbles 0.5 cm in diameter to escape within 1 hour? Bubbles 0.1 cm in diameter? See Eq. (13-3).

9-12. Cite factors which affect the life of SiC resistance elements.

9-13. Compare the resistances of the following refractories to attack at 1600°C by FeO: (a) silica, (b) fireclay, (c) mullite, (d) alumina, (e) zirconia, (f) periclase, (g) CaO, (h) carbon. [*Note:* Make use of Reference 6-2.]

9-14. Discuss the merits of a glaze (a) having a thermal expansion identical to that of the underlying body; (b) having a thermal expansion lower than that of the underlying body.

9-15. Discuss the thermal requirements of a ceramic nose cone.

Chapter 10

ELECTRICAL AND MAGNETIC PROPERTIES AND APPLICATIONS

10–1 Introduction

Electromagnetic fields and electrical charges in a material interact, with consequences that vary markedly with the freedom of the electrical charges to respond to the fields. If the structure permits the movement of electrons or of electronically unbalanced ions, electrical charges may be transferred along a potential gradient (see Sections 10–2 and 10–3). When ceramic materials have structures which do not permit charge transfer, there is still an internal charge displacement (Section 10–4) induced in the structure by an external electromagnetic field. Finally, because they are nonconducting, magnetic ceramics have numerous technological applications, and we shall discuss them in the last section of this chapter.

10–2 Electrical conductivity

The charge on an electron is 1.6×10^{-19} coulomb (or 4.803×10^{-10} esu). Hence this charge exists wherever an electron is present. In normal materials, there are equal numbers of protons and electrons, so that there is a neutral charge balance. However, if a sufficiently small volume is isolated, it is possible to find exceptions. For example, a sulfate ion $(SO_4)^{2-}$ has two more electrons than protons, and we indicate that such an anion has an electronic charge of -2 (or 9.6×10^{-10} esu, if we are making calculations which require those units). Similarly, an ammonia ion $(NH_4)^+$ has a positive charge of $+1$ because it has one less electron than it has protons. On a larger scale, the positive electrode of a capacitor has had some of its electrons removed. Therefore it has a charge that is a large multiple of 1.6×10^{-19} coulomb.

Charge carriers. An electrical charge is transferred from one point to another if an electric field (potential gradient) is applied, and if the charge is mobile. Since the charge can be carried by four different mechanisms, the amount of mobility of the charges varies markedly. These charge carriers include (1) anions, (2) cations, (3) electrons, and (4) electron holes. The transfer of charges by anions and cations will be discussed under ionic conduction; movements of electrons and electron holes will be presented under semiconduction.

Ionic conduction. The mechanism of ionic conduction in liquids is easily understood. There is a net drift of the vibrating ions in the electrical field; positive ions migrate toward the negative electrode and vice versa. The ionic conductivity σ_i is equal to the product of the charge density nq and the ionic mobility μ_i:

$$\sigma_i = nq\mu_i. \tag{10–1}$$

The charge density includes the number n of charge carriers per unit volume, while q is the charge per carrier and may be expressed as $(1.6 \times 10^{-19})\ Z$ coulomb, where Z is the valence of the ion. The ionic mobility μ_i is the net or drift velocity per voltage gradient [(cm/sec)/(v/cm), or cm^2/v/sec]. Large ions do not develop high mobilities because their relatively large masses prevent them from responding rapidly in an electric field. However, nonviscous liquids permit sizable charge transfers because the hindrances to movement are only minor.

EXAMPLE 10–1. Compare and contrast electrical conductivity (Eq. (10–1)) and thermal conductivity (Eq. (9–7)).

Answer: The equations are:

	ELECTRICAL	THERMAL
	$\sigma = nq\mu$	$k = sc\lambda/3$
Conductivity	σ, (coul/sec)/(v·cm)	k, (cal/sec)/(°K·cm)
Energy density	nq, coul/cm^3	s(°K)/3, cal/cm^3
Energy mobility	μ, (cm/sec)/(v/cm)	$c\lambda$/°K, (cm/sec)(cm)/°K or (cm/sec)/(°K/cm)

There is considerable interference with ion movements in crystalline solids or highly viscous glasses; therefore, the mobility and conductivity of ions in these substances are low. By the Einstein relationship, the mobility of an ion is directly related to its diffusion characteristics:

$$\mu_i = qD_i/kT, \tag{10–2}$$

where D_i is the ionic diffusion coefficient ((Eq. (6–9)). Combining Eqs. (10–1), (10–2), and (6–10), the following can be obtained:

$$\ln \sigma_i = \text{const} - \ln T - Q/kT. \tag{10–3a}$$

Of these terms, Q is the diffusion activation energy, T is the absolute temperature, and k is Boltzmann's constant. Since $\ln T \ll Q/kT$, Eq. (10–3a) approximates

$$\ln \sigma_i \approx \text{const} - Q/kT. \tag{10–3b}$$

Three generalizations are possible: (1) The higher the temperature the greater the diffusion coefficient (Eq. (6–10)). (2) Diffusion can occur more readily in structures that contain defects. This is illustrated schematically in Fig. 10–1 for interstitial (Frenkel) and vacancy (Schottky) defects as contrasted to ion interchanges. (3) The smaller the ion, the more readily diffusion takes place, since small ions require low activation energies. All of these relationships help to explain the characteristics of ionic conductivity in many ceramic materials.

Let us observe two examples of defect-controlled conduction for nonstoichiometric compounds. The FeO of Fig. 3–21, because of the presence of Fe^{3+} ions, has cation holes into which the adjacent cations will move, giving a net movement of cations toward the negative electrode, and of the holes toward the positive electrode. Thus the cation *hole* is a "negative carrier." On the other hand, if zirconia (ZrO_2) is treated with CaO, the charge balance is maintained by anion vacancies. (The Ca^{2+} substitutes directly for Zr^{4+}, and therefore one less O^{2-} is required.) An electrical gradient will thus move the anions toward the positive electrode, and the anion *hole* serves as a "positive carrier" as it moves toward the negative electrode.

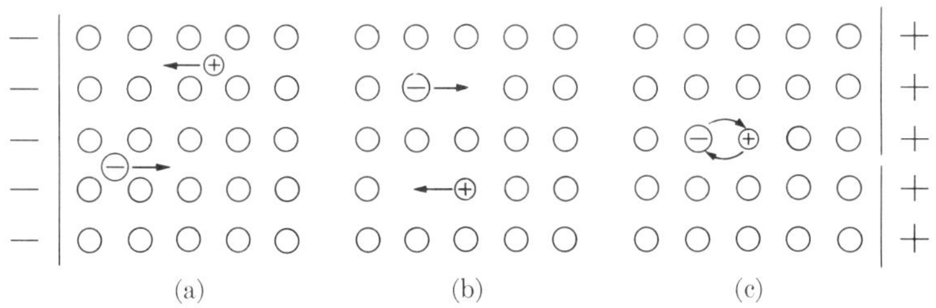

FIG. 10–1. Ion movements: (a) Frenkel; (b) Schottky; (c) interchange. In this drawing, the negative electrode is on the left.

EXAMPLE 10–2. When we add 7.5 w/o CaO to ZrO_2, how many ion holes providing positive carriers are there per cm^3? (ZrO_2 has the same structure as CaF_2, and the lattice constant is 5.45 A.)

Answer:

$$\begin{aligned}7.5 \text{ w/o CaO} &= \{7.5/(40+16)/[(7.5/(40+16)) + 92.5/(91.2+2(16))]\}\\ &= 15.1 \text{ m/o CaO}.\end{aligned}$$

Therefore

$$(Zr_{.849}\,Ca_{.151}O_{1.849})_4 \qquad \text{per unit cell,}$$

or

$$\begin{aligned}8 - (4)(1.849) &= 0.604\ O^{2-} \text{ vacancies/unit cell;}\\ 0.604/(5.45 \times 10^{-8})^3 &= 37 \times 10^{20} \text{ anion holes/cm}^3.\end{aligned}$$

EXAMPLE 10–3. At 800°C, the conductivity of the ZrO_2-CaO solid solution of Example 10–2 is 0.001/(ohm·cm). What is the mobility of the oxygen ions?

Answer:

$$\begin{aligned}\mu &= \sigma/nq \\ &= 0.001/(37 \times 10^{20})(2)(1.6 \times 10^{-19}) \\ &= 0.85 \times 10^{-6}\ (\text{cm/sec})/(\text{v/cm}).\end{aligned}$$

For comparable ionic conduction, stoichiometric compounds require higher temperatures than nonstoichiometric compounds because the defects and ion movements must be introduced by thermal activation. This is indicated schematically in Fig. 10–2 (line *a*) for sodium chloride. The straight-line relationship shows correspondence with Eq. (10–3b). Although the ion movements become excessively small at lower temperatures, there is still measurable conductivity because there are always impurity ions which can carry the charge. Thus, as shown in Fig. 10–2, there is commonly a break in the ($\ln \sigma$ versus $1/T$) curve. In effect, this figure presents the conductivities of two situations: At temperatures below the break, the high mobility of the relatively few impurity ions predominates; at higher temperatures, the mobility of the majority of the ions predominates. The location of change from one effect to the other depends on the purity of the substance.

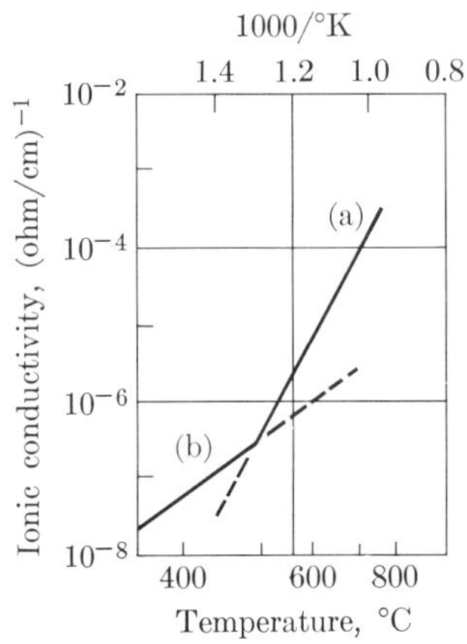

FIG. 10–2. Ionic conductivity of NaCl. Above 500°C, the Na^+ ions are the main charge carriers. Below that temperature, conduction is primarily through diffusion of impurities.

Ionic conductivity in glass depends on the number and kind of modifier oxides (Section 4–3). For example, Na_2O increases the conductivity more markedly than CaO, for several reasons: (1) There are more cations added with each oxygen. (2) The monovalent Na^+ ion is held in the network less strongly than the divalent Ca^{2+}. (3) The Na^+ has a slightly smaller radius than Ca^{2+} (0.98 versus 1.06 A). The fictive temperature is also important (Section 4–4). Above the transformation range of the fictive temperature, the network structure can adjust to cation movements and conductivity is increased.

Semiconductors. In contrast to ionic conduction, which is the movement of charged ions, semiconduction is the movement of the electrons themselves. However, the principles are the same. (1) Mobility varies with charge site. (2) The mobility of the charges is proportional to the potential gradient. The

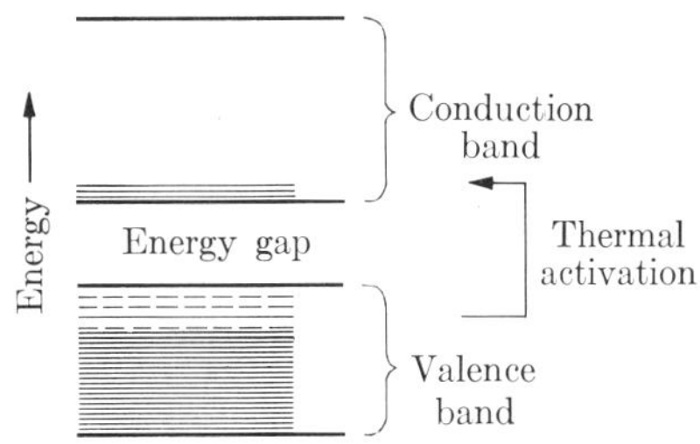

FIG. 10-3. Intrinsic semiconduction. Electrons have been raised to the conduction band by thermal activation. This leaves conducting levels in the valence band.

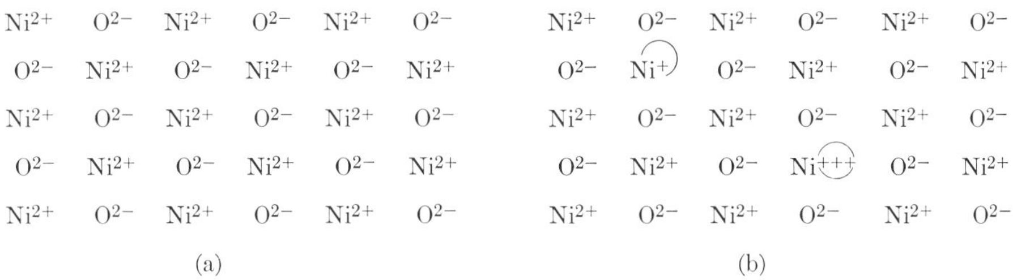

FIG. 10-4. Intrinsic valence semiconduction of NiO: (a) low temperatures; (b) high temperatures. The conductivity is increased by the mechansim shown in Fig. 10-3.

higher the temperature, the more charges are freed for movement by an activation energy relationship which can be expressed as $e^{-E/kT}$. An electron's freedom for movement depends on how readily it can be raised to a higher energy level. This activation takes place quite readily in metals (Fig. 2-6b) because the conductivity band is not filled. On the other hand, a stoichiometric ceramic compound has its conduction band filled; hence energy must be supplied to raise the electron beyond the energy gap (Fig. 2-6c) so that electron movement is possible. If the energy gap is small, it is possible for the electron to receive sufficient energy by thermal activation (Fig. 10-3). An atomistic explanation of the activation is given in Fig. 10-4 for bunsenite (NiO), which has the NaCl structure. At low temperatures there is a regular array of Ni^{2+} ions among the O^{2-} ions. With increasing temperatures, 3*d*-electrons of some atoms are transferred, leaving Ni^{3+} ions and forming Ni^{+} ions:

$$Ni^{2+} \rightleftharpoons Ni^{3+} + e^{-}, \tag{10-4a}$$

$$e^{-} + Ni^{2+} \rightleftharpoons Ni^{+}. \tag{10-4b}$$

In effect, this means that the Ni^{+} ion has an electron within the conducting band, and the Ni^{3+} ion has an electron vacancy (Fig. 10-3), thus producing intrinsic semiconduction. These electron vacancies are called *electron holes*.

A similar effect results from the presence of impurities, which gives extrinsic semiconduction, as demonstrated in Fig. 10-5 for SiC. Here we see impurity

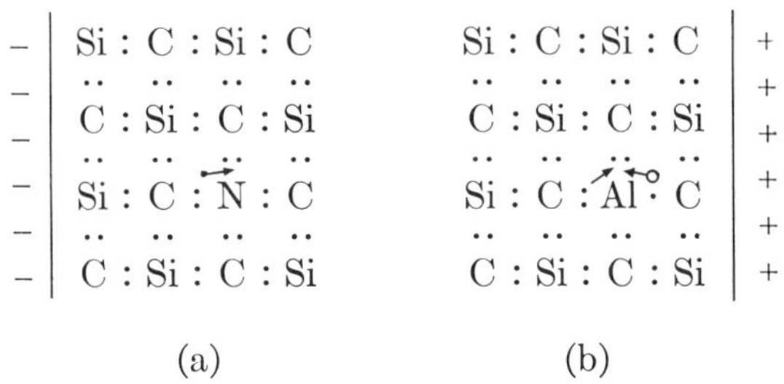

Fig. 10–5. Impurity semiconductors in SiC (schematic): (a) *n*-type conduction (excess electrons); (b) *p*-type conduction (electron holes). The proton-to-electron ratio is one to one in each case. However, the valence electron bands are not exactly filled.

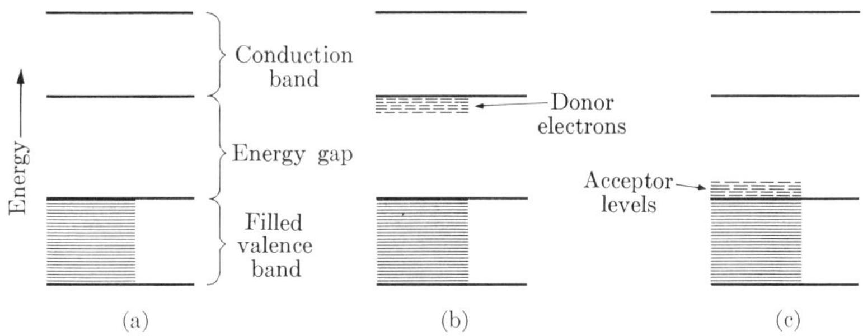

Fig. 10–6. Conduction bands: (a) intrinsic semiconductor; (b) *n*-type semiconductor (extra electrons at the donor levels); (c) *p*-type semiconductor (extra levels for accepting electrons).

semiconductors which possess extra electrons, or electron holes, in the valence-electron distribution pattern. Irregularities in the distribution of electrons introduce additional energy levels into the energy gap. The extra electron of Fig. 10–5(a) serves as a donor electron (Fig. 10–6b) which is easily energized to the conduction band. Likewise, the absence of an electron in the structure (Fig. 10–5b), opens up acceptor levels (Fig. 10–6c) or holes to which electrons may be activated and moved. The electron mobility under such conditions is commercially utilized in SiC electrical heating elements. There is enough conductivity in SiC to pass a current; however, its resistance is appreciably higher than that of metallic conductors, and considerable ohmic heating results (see Section 10–3).

Nonstoichiometric compounds are natural semiconductors. They must be classed as impurity semiconductors because either *n*-type (electron) or *p*-type (electron hole) conduction predominates. Let us consider, for example, corundum (Al_2O_3). Although at low temperatures corundum is almost perfectly stoichiometric, at very high temperatures it becomes partially reduced:

$$Al^{3+} + e^- \rightleftharpoons Al^{2+}, \tag{10–5}$$

and the nonstoichiometric compound $Al_{>2}O_3$ results. The few Al^{2+} ions which are present provide extra electrons which serve as donors (Fig. 10–6b), and limited conduction can occur. Because at low oxidation levels there is a slight reduction and the structure is more stable, the term "reduction semiconductor" has been applied when excess metallic atoms are present.

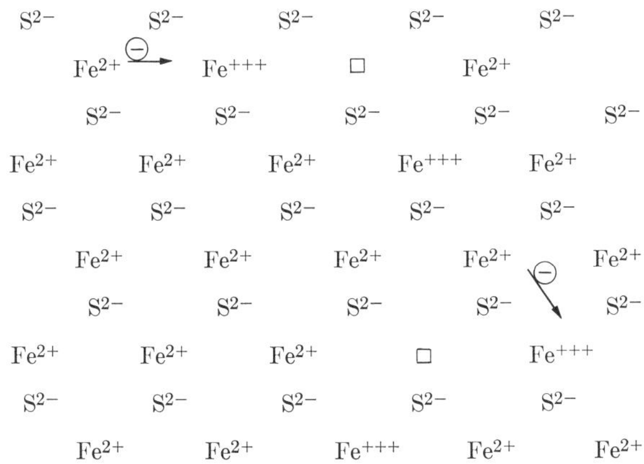

FIG. 10–7. Schematic drawing of an oxidation semiconductor: pyrrhotite ($Fe_{<1}S$). The Fe^{3+} ions provide electron holes. In an electric field, electrons can move from Fe^{2+} ions leaving electron holes. Compare with Fig. 3–21, in which there is a deficiency of cations.

Oxidation semiconductors are encountered more often than reduction semiconductors. Figure 10–7 illustrates one such example, in which both Fe^{2+} and Fe^{3+} ions are present at ambient conditions. The charge carriers are the electron holes in the 3*d*-orbital (Fig. 2–2) of the ferric ions (not the ion holes themselves). These holes provide acceptor energy levels for *p*-type semiconduction (Fig. 10–6c). This structure, and therefore electronic conduction, is also sensitive to the oxidation level; hence

$$Fe^{2+} \rightleftharpoons Fe^{3+} + e^-. \tag{10–6}$$

Effects of temperature. The electrical conductivity of ceramic materials, in contrast to that of metals, increases with increasing temperatures. This contrast is so marked that it is frequently considered to be the distinguishing difference between the two classes of materials (Fig. 10–8).

We can see the basis for this difference when we observe sources of conduction. Equation (10–1) is generally applicable, and the two variable factors are the *number* of carriers n and the *mobility* of the carriers μ. In metals the number of

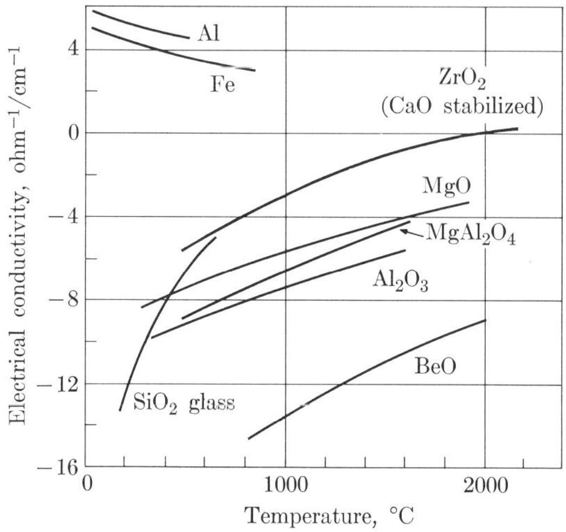

FIG. 10–8. Electrical conductivity versus temperature in ceramic materials.

carriers is large because each atom contributes one (or more) free electron. The only effect of higher temperatures is to increase the lattice vibrations, thus reducing the mobility of the electrons and shortening the mean free path of the electron wave movement through the structure. Metals, except when they are at cryogenic temperatures, have electrical conductivity inversely proportional to their temperature:

$$\sigma_{\text{metal}} \approx 1/T. \tag{10–7}$$

Conductivity in ceramic compounds increases at increased temperatures because (1) there is greater ionic diffusion, and (2) more electrons are able to enter the conduction band (Fig. 10–3). These two factors more than compensate for the shortening of mean free paths, and therefore increase the conductivity.

Effects of microstructure. On a microscopic scale, electrical and thermal conductivity are analogous, and thus Eqs. (9–9) and (9–10) may both be considered, basically, as equations for electrical conductivity. Similarly, Fig. 9–8 would also apply if the electrical conductivity of the phases is substituted for their thermal conductivity. However, a pore or a crack is relatively more important in electrical than in thermal conduction. Cracks and pores have zero electrical conductivity, although pores can still transfer heat by convection or radiation.

10–3 Ceramic conductors

Let us now consider those ceramic materials which have useful, although low, conductivities. These materials are semiconductors, and are utilized for the purpose of (1) controlled voltage drops (fixed resistors), (2) the conversion of electrical energy into thermal energy (heating elements), (3) temperature-controlled resistances (thermistors), or (4) voltage-controlled resistances (rectifiers and varistors).

Fixed resistors. Resistivity ρ is the reciprocal of conductivity and is a property of the material. It relates the voltage E and current flow I by

$$E/I = R = \rho l/A, \tag{10-8}$$

where R is resistance, and l and A are the length and cross-sectional areas, respectively. One requirement of fixed resistors is a linear relationship (Fig. 10-9) between voltage and current (where temperature is constant). Two types of fixed resistors involve ceramic insulation materials: (a) the film type and (b) the composition type. Typical of the film type of resistor is one which consists of an insulating tube, such as a glass rod, with a thin conductive coating that may be metal, graphite, or a semiconducting oxide like SnO. The resistance is controlled by the thickness, the width, and the length of the film. The width and length of the film are often established by the pitch of a helical groove around the resistor. Because it offers greater resistance control, there is lately a tendency for the film-type resistor to replace the more common composition resistor. The composition resistor consists of a mixture of carbon and insulation powders within a plastic or ceramic (porcelain) jacket, and gives definite, but low, conduction.

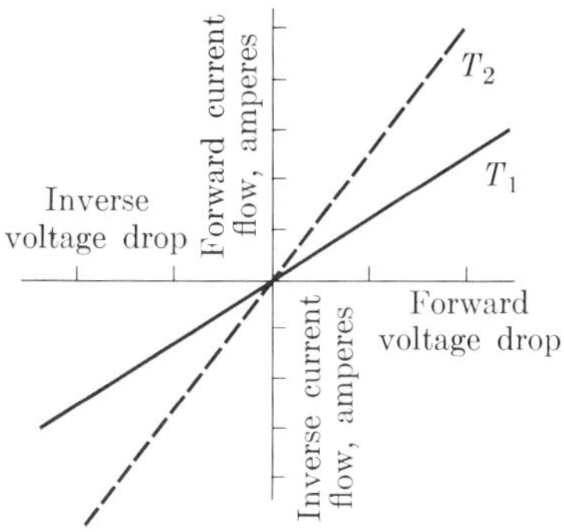

FIG. 10-9. Ohmic characteristics, showing a linear I-versus-E relationship. For ceramic materials, the proportionality constant, R^{-1}, increases with temperature.

Those resistors just described are normally used in electronic circuits where the power consumption is small (< 1 watt). However, such components are usually isolated from thermal sinks; therefore, definite attention must be given to their temperature coefficient of resistance (see Table 10-1), as well as their resistance accuracies.

Heating elements. Two widely used nonmetallic heating elements are silicon carbide and graphite. Both are semiconducting in character, with resistivities

TABLE 10-1

FIXED RESISTORS

	Film, oxide	Film, carbon	Film, metal	Composition, carbon
Manufacturing accuracy, %	1	1	1	20
Best selection accuracy, %	0.5	0.25	0.25	5.0
Percent stability after testing	2	1	1	25
Temperature coefficient, %/°C	−0.05 +0.02	+0.12	+0.02	+0.12
Maximum temperature, °C	300	115	150	115
Voltage coefficient, %/V	—	0.05	—	0.05

such that when they are used in well-designed voltage gradients they can efficiently convert energy into heat.* Both are also refractory in character (Section 9-7), which makes it possible to use them at high temperatures, thus providing rapid transfer of heat by convection and radiation mechanisms. However, they do have certain temperature limitations, which are primarily determined by oxidation (Table 10-2). Silicon carbide is more widely used than graphite in industrial furnace applications because of the protective SiO_2 glaze which develops on it.

Other semiconducting oxides and carbides are usable as high-temperature resistors. Zirconia (ZrO_2) is used because it has no limitations caused by oxidation (Example 10-3). However, as shown in Fig. 10-8, it is necessary to attain at

* Since watts, W, of the ohmic equation are

$$W = E^2/R, \tag{10-9}$$

and since industrial equipment must be capable of handling potential drops of 10-100 volts, resistances of 0.1 to 10 ohms are required. These values are easily obtained in products made of semiconductors that have bulk resistivities of 10^{-2} to 10^2 ohm·cm. Metals that are more highly conducting ($\rho < 10^{-4}$ ohm/cm) require thin, long resistors; whereas materials with high resistivities would have to be quite massive.

TABLE 10-2

TEMPERATURE LIMITATIONS AND OXIDATION PRODUCTS OF NONMETALLIC HEATING ELEMENTS

	SiC	Graphite
Maximum temperature, °F:		
25 hours in air	2800	~2000
5000 hours in air	2500	—
1000 hours in helium or vacuum	~3000	~4000
Oxidation products	SiO_2 glaze and CO gas	CO gas

least 1000°C (1830°F) by other means before low voltages can provide current flow for significant heating (see Eq. (10-9)). Furthermore, as the temperature is raised, careful voltage control must be maintained because of the continued drop in resistivity.

EXAMPLE 10-4. When ZrO_2 is used as a high-temperature resistor, it is usually preheated to 1000°C (1830°F). Above this temperature it is necessary to vary the voltage to maintain constant power consumption. Making use of the data of Fig. 10-8, show the voltage-temperature relationship for increased temperatures. (Consider the voltage at 1000°C as equal to E_0.)

Answer: Within a resistor,

$$R = k\rho = k/\sigma.$$

From Eq. (10-9),

$$W = E^2\sigma/k,$$

or, at constant W,

$$E = k'\sqrt{1/\sigma}.$$

Therefore, at 1000°C,

$$\sigma = 10^{-3}, \qquad E = E_0,$$

at 1500°C,

$$\sigma = 10^{-1}, \qquad E = 0.10E_0,$$

and at 2000°C,

$$\sigma = 10^{0}, \qquad E = 0.03E_0.$$

Thermistors. Materials with a negative temperature coefficient of resistance are highly desirable in certain electronic devices. Examples of such devices are thermistors, i.e., thermal resistors, which are used either as temperature sensors or as temperature compensators. When we are measuring temperature, we need a device made of some material which has a high temperature coefficient of resistance. These coefficients may be as high as $-4\%/°C$ at ambient temperatures, which is more than 10 times the temperature coefficient of resistance of platinum or most other metals. Thus, with precision equipment, sensitivities of millidegrees are possible. Furthermore, precision equipment is not necessary when accuracy to within 0.1°C is all that is required.

The ceramic semiconductors most widely used for thermistor purposes are the oxides. These include not only spinels such as magnetite (Fe_3O_4), which has multiple valence ions to provide intrinsic semiconduction (cf. Figs. 10–3 and 10–4), but also semiconductors containing impurities. For example, a manganosite (MnO) structure takes dissolved Li^+ ions in solid solution to give a composition of $Li_x^+Mn_x^{3+}Mn_{1-2x}^{2+}O$, so that the trivalent ions establish electron holes. The higher the temperature, the more readily electrons move into these holes or acceptor levels and the lower the resistivity.

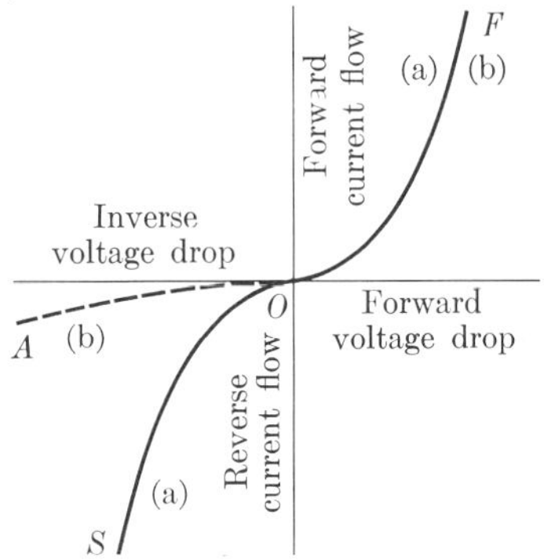

FIG. 10–10. Varistor characteristics, showing a nonlinear I-versus-E relationship. (a) FOS, symmetrical behavior; (b) FOA, asymmetrical behavior.

Rectifiers and varistors. Some materials exhibit a nonlinear relationship between voltage and amperage (Fig. 10–10). Furthermore, the behavior between forward and inverse voltage gradients may be either symmetrical or asymmetrical. A material with asymmetrical, nonlinear voltage-versus-amperage behavior (Fig. 10–10b) is usable as a rectifying material. When the voltage is in the forward direction, the current flow increases rapidly with increased voltage, and the effective resistance is low. However, an inverted voltage does not fully reverse the current, and the effective resistance is high.

Copper oxide, commonly formed by oxidizing metallic copper to Cu_2O, has had wide use as a rectifying material. Since the electron holes move away from the positive electrode, forward voltage permits the current to flow from the cuprous oxide into the metal (Fig. 10–11b). Flow in the opposite direction is blocked because the oxide becomes an insulator at the junction when the electron holes are removed from that zone (Fig. 10–11c). Copper oxide (and selenium) rectifiers are easily made, but they are often replaced by comparable silicon and germanium diodes, because these materials have a greater amperage response.

The nonlinear behavior of varistors (i.e., variable resistors) permits their use as "safety valves" or voltage surge arrestors. For example, the initial voltage differential shown in Fig. 10–10 permits negligible current flow even though high voltages will pass considerable current. Thus components which are appropriately designed may be used as insulators, except when there is sudden exposure to unusually high voltage, such as during lightning or switching surges.

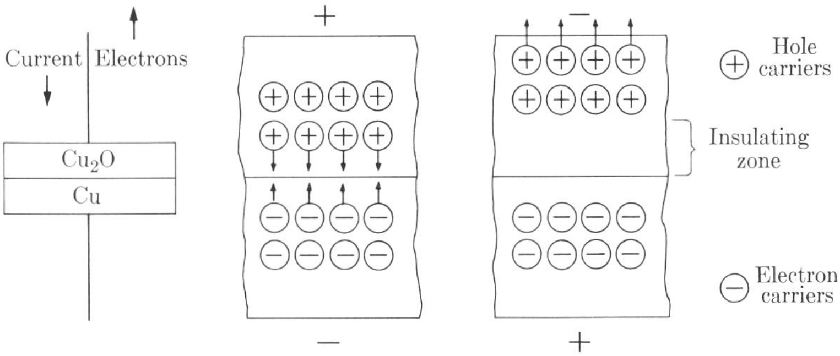

FIG. 10–11. Copper-oxide rectifier. The cuprite structure (Fig. 3–12b) is a p-type semiconductor because a few cupric ions are present. (a) The copper surface is oxidized; (b) forward voltage; (c) inverse voltage. The electron holes are removed from the junction zone so that current cannot pass.

Silicon carbide is commonly used for varistors. The dual use of SiC both as a resistor and a varistor is possible because the voltage-versus-amperage relationship for SiC is nearly linear, with a moderately high resistivity over the voltage range used for heating elements. However, with kv/in. gradients, its resistivity decreases markedly, so that very large current flow can occur.

10–4 Dielectric behavior

If an insulator is placed in an electric field, charges are not transferred; however, they are displaced. Negative charges shift toward the positive electrode, and positive charges are displaced in the opposite direction. This interaction between the electric field and the localized charges within a material has significant and useful applications for the engineer.

Dielectric constant. A major consequence of the above interaction is that a material with highly polarizable charges, when it is between the plates of a capacitor, markedly influences the charge which can be held on the capacitor's plates. The proportional increase in capacitance is a direct reflection of the dielectric constant of the material. Thus a material with a relative dielectric constant κ' of 10.0 permits the capacitor to develop 10.0 times as large a charge as it would without such material. The relationship

$$C = \kappa' A/(11.3 \times 10^6)d \tag{10–10}$$

gives the capacitance C in μf (microfarads); area A in cm^2; and thickness d in cm. Typical values of the relative dielectric constant are given in Table 10–3. Our interest centers on the origin of the relative dielectric constant κ', which

TABLE 10–3

DIELECTRIC CONSTANTS OF SELECTED CERAMIC MATERIALS†

Material	κ', 25°C, 10^6 cps	Material	κ', 25°C, 10^6 cps
NaCl	5.9	*Porcelain	5.0
MgO	9.6	Fused SiO_2	3.8
LiF	9.0	96% SiO_2	3.8
Al_2O_3($\perp c$-axis)	8.6	Borosilicate glass	4.0
Al_2O_3($\parallel c$-axis)	10.6	"E" glass	6.3
*TiO_2	96.0	Soda-lime glass	7.0
*$BaTiO_3$	1600.0		

† Adapted from Von Hippel, A. R., *Dielectric Materials and Applications,* New York: John Wiley & Sons (1954) 300ff.

* Polycrystalline.

relates the electric polarization P to the electric field strength E, as follows:

$$P = \epsilon_0(\kappa' - 1)E. \tag{10–11}$$

The permittivity of free space is ϵ_0 and constant. Polarization can arise from four sources (Fig. 10–12): (1) electronic displacement, (2) ionic displacement, (3) orientation of permanent dipoles, and (4) space charge. The total polarization, and therefore the relative dielectric constant, is the summation of these four sources:

$$P = P_e + P_i + P_o + P_s. \tag{10–12}$$

Electronic displacement, which occurs in all types of materials, involves a shift of the atom's electrons and its nucleus in opposite directions within the electric field. Because the centers of positive and negative charges are not coincident, a small dipole is formed. The total electronic polarization P_e is the summation of all of these dipole moments μ_e which are developed:

$$P_e = \sum \mu_e. \tag{10–13}$$

Although there are as many electronic dipoles as there are atoms, the moment arm (Section 2–7) is extremely short, so that the total polarization is still small. Electronic polarization responds very rapidly to changes in the electric field that occur with high frequency (up to 10^{16} cps).

Ionic polarization, which is most common in ceramic compounds, involves the relative displacement of positive and negative ions so that dipole moments are induced by an electric field. Equation (10–13) could be resubscripted for ionic polarization; and since the total displacement can be comparatively great in some structures (although still $\ll$ 1.0 A), we can obtain relatively high dielectric constants from the contribution of ionic polarization. In the presence of chang-

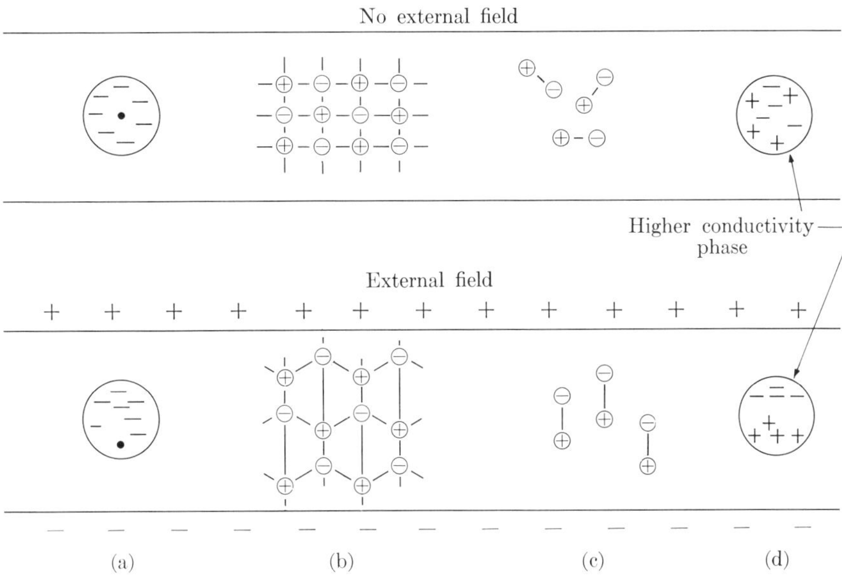

FIG. 10–12. Polarization mechanisms (schematic): (a) electronic displacements around atoms or ions; (b) ionic displacements within crystal structures; (c) dipole orientation; (d) space charge. In the latter case, the higher the conductivity of a phase, the higher the effective potential gradient within the remainder of the dielectric.

ing fields, the polarization of ions is less rapid than that of electrons. The maximum response is at about 10^{13} cps, which corresponds to the far infrared wavelengths.

Dipole orientation is relatively unimportant in ceramics (though important in polymers) because most of the permanent dipoles cannot be reoriented without destroying the crystal structure. For example, quartz (SiO_2) lacks a center of symmetry, and as a result it has a permanent dipole moment. It is impossible for an electric field to invert the existing dipoles, a fact which has useful piezoelectric consequences that will be discussed in the following subsection.

Space charges must be considered extraneous charges which collect on the interfaces. Their value and frequency response depend on geometry and on contaminants, rather than on the properties of the material itself.

Inasmuch as individual polarizations have differing frequency responses, the total polarization is a function of frequency. This is illustrated schematically in Fig. 10–13 for a wide range of frequencies. The variation in polarization is reflected by the dielectric constants in Fig. 10–14. To state the matter briefly, at high frequencies the ionic polarizations cannot keep up with the field reversals, whereas the electronic ones can. As we would expect, increased temperature

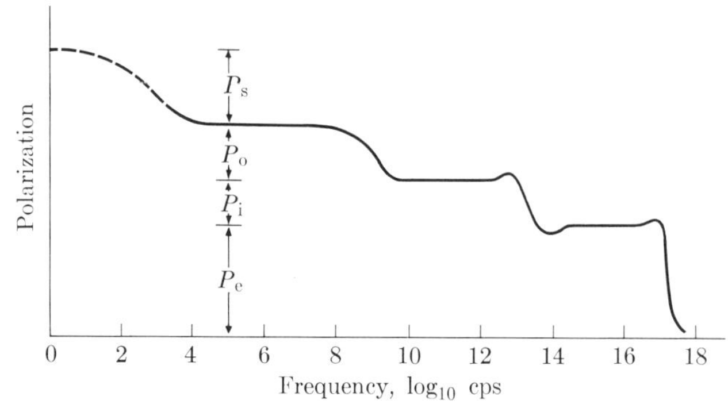

FIG. 10-13. Polarization versus frequency. The more rapidly responding polarizations continue to higher frequencies.

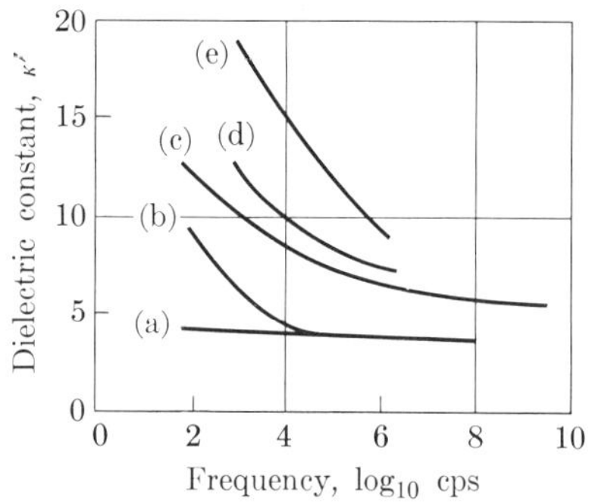

FIG. 10-14. Dielectric constant versus frequency. With higher frequencies there is a lower dielectric constant because there is less time for polarization (see Fig. 10-13). (a) Fused silica, 100°C; (b) fused silica, 400°C; (c) AlSiMag A-35, 150°C; (d) ZrO_2 porcelain; (e) Al_2O_3.

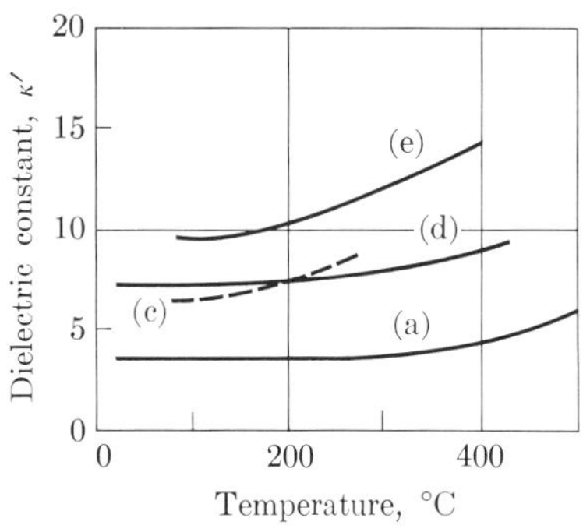

FIG. 10-15. Dielectric constant versus temperature (frequency: 10^6 cps). At higher temperatures there is more opportunity for polarization. (a) Fused silica; (c) AlSiMag A-35; (d) ZrO_2 porcelain; (e) Al_2O_3.

raises the dielectric constant because it increases ionic response to the field at any particular frequency (Fig. 10-15).

Each of the four polarizations cited is reversible and essentially proportional to the applied electric field, for small fields and low frequencies. Thus, Eq. (10-11) may be rewritten as

$$P/E = \text{constant modulus.} \tag{10-14}$$

This is analogous to the ϵ/σ = const of Eq. (8-1) (inverted). Thus the electric field corresponds to an electrical stress, and the polarization to an electrical strain. When the P/E ratio is not constant for varying fields, there are effects comparable to anelastic mechanical behavior (Section 9-6); these effects will be discussed shortly.

Piezoelectric crystals. A ceramic compound such as MgO has its overall center of positive and negative charges at the same point. When such a crystal is mechanically stressed, the charge centers are not displaced. This is not true for asymmetrical crystals where the centers of positive and negative charges are not coincident (Fig. 10-16a). In such a crystal, a mechanical stress alters the polarization of the crystal just as an electric field would do (Fig. 10-16c). Conversely, the electric field mechanically distorts the crystal.

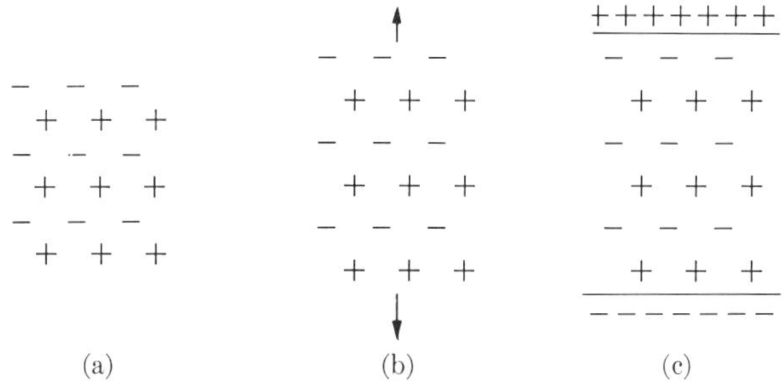

FIG. 10-16. Piezoelectric crystals (schematic): (a) unstressed, (b) mechanically stressed, (c) electrically stressed. Mechanical stressing increases the polarization; electrical polarization changes the dimensions.

The several consequences are: (1) When it is unstressed, a piezoelectric crystal has a positive and a negative "end." Actual potential differences can be measured between the ends. (2) Mechanical stresses increase or decrease the potential difference, depending on whether the stress increases or decreases the overall dipole moment of the crystal. (3) External electric fields change the dimensions of the crystal; it expands or contracts depending on the direction of the field. (4) Mechanical-electrical transducers may be made; e.g., mechanical compression may produce usable voltage changes.

Piezoelectric crystals are widely used for voltage-pressure transducers. Four types exceed the rest in present applications: (a) quartz (SiO_2), (b) rochelle salt ($KNaC_4H_4O_6 \cdot 4H_2O$), (c) barium titanate ($BaTiO_3$), and (d) lead titanates and zirconates ($PbTiO_3$ and $PbZrO_3$). Quartz is not ferroelectric, and therefore has the advantage of not having a Curie temperature, although it does lose its piezoelectricity at the $\alpha \rightarrow \beta$ displacive inversion at 575°C (1070°F), as discussed in Section 3-6. However, this is well above the Curie temperatures of common ferroelectric phases. Quartz is essentially inert to solution or environment; in fact, its main disadvantage is its low sensitivity, which means that it requires a large voltage amplification.

On the other hand, the sensitivity of rochelle salt is high. Unfortunately it is soluble in water and must be isolated from atmospheric moisture. Also, its ferroelectric (and therefore piezoelectric) behavior is confined to the temperature range of −18°C (0°F) to 24°C (75°F), where the appropriate polymorph is stable. Hence it was only after the commercial development of barium titanate that there was extensive development of piezoelectric transducers. Although the sensitivity of barium titanate is less than that of a number of semiorganic crystals such as rochelle salt, ammonium dihydrogen phosphate (ADP), and ethylenediamine tartrate (EDT), it is substantially more durable under mechanical and environmental abuse. In addition, its Curie temperature can be as high as 120°C (248°F) if no $SrTiO_3$ or $CaTiO_3$ are present in solid solution. Current applications utilize $PbTiO_3$-$PbZrO_3$ compositions which have Curie temperatures as high as 340°C (650°F).

The dimensional changes within piezoelectric crystals are anisotropic. Therefore, if there is to be maximum sensitivity, the transducer must be made of a single crystal, with the electrical field aligned with the direction of the greatest crystal asymmetry. In a tetragonal or hexagonal crystal, this would be in the direction of the *c*-axis.

When an asymmetric crystal is heated, the thermal expansion has an effect comparable to that shown in Fig. 10–16(b). The expansion so increases the dipole moment that the end-to-end voltage differential is increased. The term "pyroelectric" has been applied to this effect.

Ferroelectric crystals. In crystals which develop large dipole moments within each unit cell, there may be a spontaneous interaction between the dipoles of adjacent cells. This interaction, which is described as ferroelectric behavior,* provides a P-versus-E response which is not represented by Eq. (10–14); nonetheless the consequences are important. Let us now discuss them, with $BaTiO_3$ as the example.

Figure 10–17 shows an exaggerated sketch of {200} planes of tetragonal $BaTiO_3$. Note that there are two equivalent positions for the titanium ions with respect to the larger oxygen ions, but an ion can be in only one of the two. For example, let us assume that a $BaTiO_3$ ceramic contains randomly oriented grains, and that within these grains there are as many unit cells with their dipoles oriented in one direction as in the other. When an electric field is placed across such a material, there is an initial displacement of the ions, as indicated by Eq. (10–14). However, a relatively minor increase in the electric field serves to change the Ti^{4+} ions from one position to another if the unit cell dipole is aligned in the "wrong" direction. This produces a major change in the overall polarization. After all the unit cell dipoles are aligned, any further increase in the field simply increases the normal ionic polarization, in accordance with Eq. (10–14).

* The term "ferroelectric" arose because there is a close analogy between the electrical polarization in an electric field and the magnetization in a magnetic field (Section 10–6).

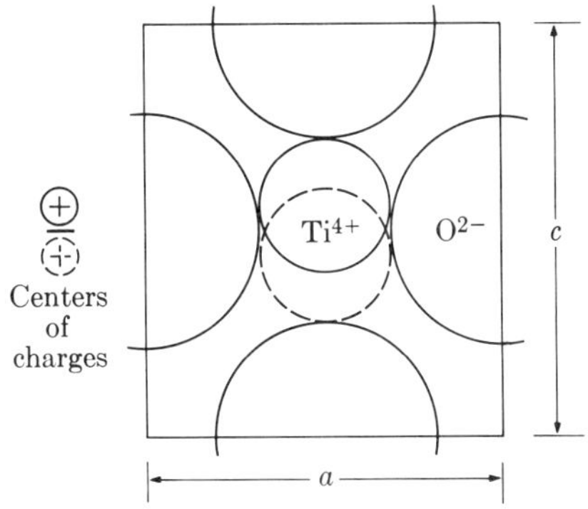

FIG. 10–17. Ferroelectric crystals: $BaTiO_3$. The {200} planes are represented with an exaggerated c/a ratio. There are two alternate positions for the Ti^{4+} ion. An activation energy is required to move it from one position to the other. This movement reverses the polarity of the unit cell (cf. Fig. 3–13).

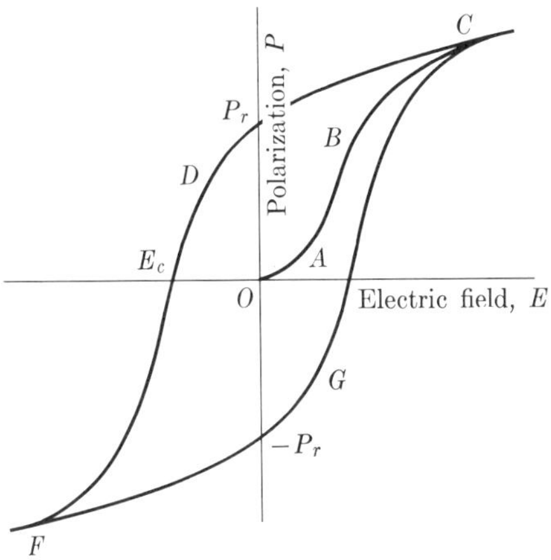

FIG. 10–18. Ferroelectric behavior: P_r = remanent polarization; E_c = the coercive field (cf. Fig. 10–23 and the text).

The previous paragraph describes the curve $OABC$ of Fig. 10–18. If the field is removed after alignment, there is a remanent polarization P_r because the dipoles of some of the unit cells do not revert to their original alignment. It is thus necessary to apply an opposing coercive field E_c to equalize the polarization. A still further reversal of the electric field will invert the polarization.

Each unit-cell dipole is influenced by its neighbors. Therefore, when a large number of unit cells mutually interact and develop the same alignment, spontaneous polarization occurs. This is indicated in Fig. 10–19, where there are several domains with many unit cells each. The major polarization of Fig. 10–18 actually involves moving the domain boundary so that there is growth of the more favorably aligned domains. In this respect, the growth of domains is somewhat comparable to the growth of grains (Section 7–3). However, the process of domain growth, unlike that of grain growth, is reversible.

The polarization possible in a ferroelectric material far exceeds that in nonferroelectric materials, and results in dielectric constants of several thousand, because there are two contributing factors: (1) the Ti ion has a charge of $4e$, and (2) the ionic displacement is very large compared to that in normal materials (Fig. 10–17).

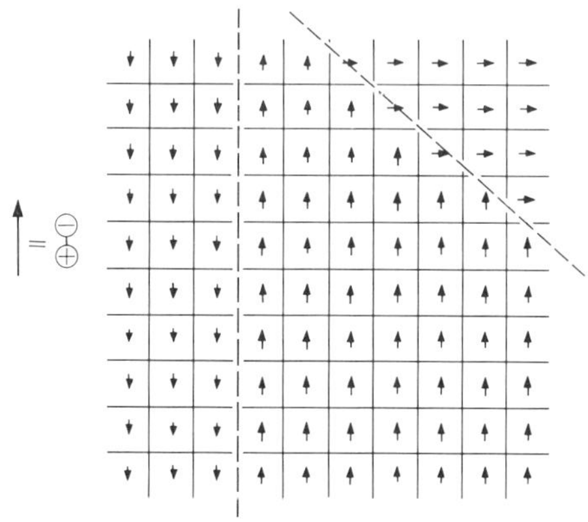

FIG. 10–19. Ferroelectric domains, whose boundaries (- - -) are interfaces of higher energy. Domains, like grains, contain many unit cells; however, adjacent domains are crystallographically related.

The polarization remanence P_r of Fig. 10–18 is a consequence of the alignment of the unit cell dipoles in domains that are more favorably oriented. In practice, this alignment is not 100 percent permanent. Occasionally Ti^{4+} ions may receive enough thermal activation to overcome the energy barrier between the two positions (Fig. 10–17) and invert the polarization. The consequent decay in the polarization has been called relaxation. Since it is a result of thermal activation, the relaxation rates have exponential characteristics:

$$dP/dt = -P/\tau, \tag{10–15a}$$

or

$$P_t = P_r e^{-t/\tau}. \tag{10–15b}$$

In this equation, P_t is the polarization after time t, and τ is a characteristic relaxation time for each material. It is a function of temperature.

Above 120°C (248°F), the ferroelectric characteristics disappear from $BaTiO_3$ because its structure changes from tetragonal to cubic (Fig. 3–13). When it is cubic, there is only one low-energy position for the Ti^{4+} ion, so that Fig. 10–17 does not apply, and there can be no polarization within the unit cell other than the normal ion displacement. Therefore the character of the P–E curves of Fig. 10–18 disappears and Eq. (10–14) applies. The temperature at which a material loses its ferroelectric characteristics is called the *ferroelectric Curie point.* It is analogous to the magnetic Curie temperature (Section 10–6) in that it is the upper limit of domain structures and is a temperature of maximum specific heat. The Curie temperature varies with composition. That is, if Sr^{2+} ions are substituted for Ba^{2+} ions, the temperature of stability for the cubic structure is lowered, and the Curie temperature drops. Other polymorphic transformations occur in $BaTiO_3$ at room temperature and at −80°C (−112°F). Between these temperatures, its tetragonal structure is displaced slightly and it takes on an orthorhombic structure. It is also ferroelectric at these temperatures, although the polarization is somewhat weaker and the dielectric constant is less than at higher temperatures.

There is an expected variation in dielectric constant with crystal orientation. The Ti^{4+} ions must be polarized in the direction of the c-axis if ferroelectric characteristics are to be realized. However, as indicated in Fig. 10–19, it is possible to interchange the c- and a-axes of $BaTiO_3$ because the c/a ratio is 4.03 A/3.99 A, or only about 1.01. A strong electric field can twin the crystal so that the c-axis is aligned with the field. Very low energies are required for this a-to-c interchange because each ion has essentially the same neighbors, the sole exception being the Ti^{4+} ion. Five of the six tetrahedral oxygens are close neighbors. In a strong electric field, one of the five moves away from its neighbors and the sixth moves closer. Therefore, the energy required for the electric field is proportionate to the polarization changes. The process appears as part of the $OABC$ change in Fig. 10–18.

The ferroelectric behavior shown in Fig. 10–18 is analogous to the anelastic behavior of a material under cycles of mechanical stress (Fig. 9–10). Since ionic displacement requires a coercive force before net polarization is removed, energy is absorbed by the material. The total energy per cycle is equal to the integrated area within the hysteresis loop.

Polycrystalline ceramics. Most of the preceding subsections concern the dielectric characteristics of single crystals of ceramic compounds, and these characteristics are important in the study of single crystals. However, most dielectric ceramics are used as part of polycrystalline and polyphase materials. The directional properties of ceramic crystals are randomized in polycrystalline materials. Furthermore, space charges, which can develop at grain boundaries and pore surfaces, alter the dielectric behavior of ceramic materials.

The mixture rules which apply for polyphase ceramics are those of a dispersion of spherical particles in the matrix of a second phase. However, rather than using the rigorous and somewhat complicated relationship of Maxwell for the bulk dielectric constant, we shall use the following empirical relationship, which is satisfactory for most purposes:

$$\log \kappa' = \textstyle\sum_i V_i \log \kappa_i', \qquad (10\text{–}16)$$

where V_i is the volume fraction of each phase. This emphasizes the significant effect of relatively small amounts of ceramic phases with low dielectric constants on the bulk constant for the ceramic product.

EXAMPLE 10–5. The mean dielectric constant for rutile (TiO_2) is 100. What is the dielectric constant for a rutile spacer with 10 percent porosity?

Answer: The pores may be considered another phase with a dielectric constant of 1.0, the same as a vacuum:

$$\log \kappa' = 0.1 \log 1 + 0.9 \log 100,$$
$$\kappa' = 63.$$

Fig. 10–20. Surface breakdown. Adsorbed moisture and contaminants provide a surface path for electrical shorting. (R. Russell, *Brick and Clay Record,* 1957, page 52.)

10–5 Ceramic insulators

Materials for electrical insulators must have: (1) high resistivity, (2) high dielectric strength, and (3) a low loss factor. The resistivity ρ is the reciprocal of conductivity, σ; the dielectric strength is the voltage gradient which is capable of producing electrical breakdown of a material; and the loss factor is the product of the dielectric constant κ' and the dissipation factor, $\tan \delta$, where δ is the loss angle ($90° -$ phase angle). Since we have already discussed resistivities at some length, we shall give no further attention to them here.

Dielectric strength. Dielectric failure can originate from several sources. Perhaps the most common cause is the arcing across external surfaces (see Fig. 10–20). Surfaces which have been subjected to contamination or adsorption particularly lend themselves to such a breakdown. Likewise, connecting pores within the bulk of the insulator provide breakdown paths. Therefore glazes and other means are used to limit permeability.

Internal failure usually originates from impurities which provide donor electrons that are rapidly accelerated along potential gradients, and in turn energize other electrons until cataclysmic electron flow results.

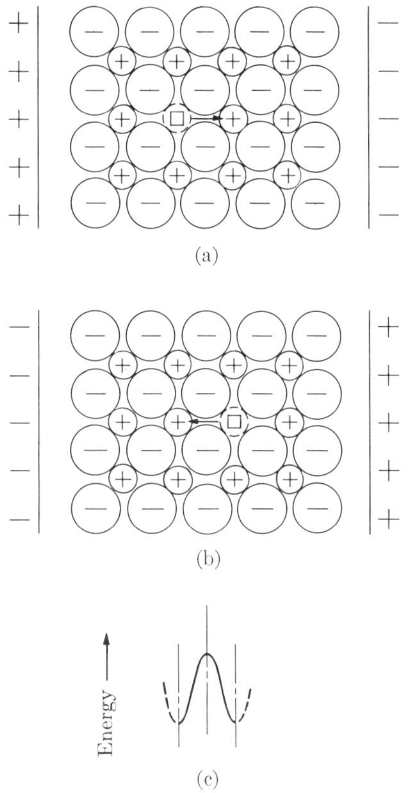

FIG. 10–21. Dielectric losses: Parts (a) and (b) show how, as the field is reversed, the movable charges are displaced across an energy barrier (c). If the frequency of the field is slow and the thermal energies of the atoms high, the charge displacement will have a relatively short lag (small loss angle). Conversely, with high frequency and/or low temperatures, polarization changes do not occur and no losses take place. With intermediate values, polarization change can occur, but with a relatively long lag, hence a large loss angle.

It is difficult to make generalizations about dielectric strength. However, some phenomenological relationships may be cited: (1) dielectric strength is related to resistivity; (2) impurities lower dielectric strength; (3) dielectric strength decreases as temperature is raised; and (4) porosity decreases dielectric strength because the number of possible paths for shorting is increased.

Loss factors. Energy losses are important in a-c circuit designs, not only because they represent a lack of efficiency, but also because electrical energy losses change the impedance of the circuit. In addition, these electrical energy losses show up as unwanted thermal energy, with consequent problems in heat dissipation and material performance.

The mechanism of energy losses is indicated in Fig. 10–21 by showing the interchange of an ion and a vacancy within a crystal as the electric field is reversed. There is a lag between the time of maximum electric field and maximum polarization, because the ion must pass across an energy barrier (Fig. 10–21c). The lag can be represented as a loss angle δ, in which case the phase angle θ is $(90° - \delta)$ rather than the 90° which occurs in a lossfree capacitive circuit. The loss angle δ, which is also equal to $(90° - \theta)$, and the relative dielectric constant κ' give a measure of the capacitive loss factor, CLF, in the form of

$$\text{CLF} = \kappa' \tan \delta. \tag{10–17}$$

This form is most commonly used because it is directly proportional to the energy loss W:

$$W = kfV^2(\kappa' \tan \delta), \tag{10–18}$$

where k is a constant, f is frequency, and V is the maximum cyclic voltage. However, in some calculations the figure of merit or Q-factor is preferred:

$$Q = 1/\tan \delta, \tag{10–19a}$$

or

$$Q = X_c/R, \tag{10–19b}$$

where R is the resistance and X_c is the capacitive reactance. Table 10–4 provides some data and Fig. 10–22 shows how temperature influences both the conductivity and the loss factor of several glasses.

The loss angle δ is directly influenced by the frequency of the a-c field. As the frequency increases, the time lag for the ion jump becomes a larger fraction of the cyclic period and the value of tan δ naturally increases. When the time

Table 10–4

Ceramic Insulation
(Typical values)

Material*	Resistivity, 25°C, ohm·cm	Dielectric strength, step-by-step, v/mil	Loss factor, κ' tan δ	
			60 cps	10^6 cps
Hard glass	10^{16}	350	0.007	0.02
Soft glass	10^{15}	250	0.4	0.08
Alumina, Al_2O_3	10^{16}	250	0.01	0.006
Silica, SiO_2	10^{19}	250	0.0025	0.0004
Mica, $KAl_3Si_3O_{10}(OH)_2$	10^{13}	~1000	0.018	0.003
Cordierite porcelain, $Mg_2Al_4Si_5O_{18}$	10^{13}	100	0.01	0.05
Steatite porcelain, $MgSiO_3$	10^{14}	300	0.007	0.005
Mullite porcelain, $Al_6Si_2O_{13}$	10^{13}	300	0.006	0.03
Zircon porcelain, $ZrSiO_4$	10^{14}	300	0.04	0.01

* The primary phase is shown. Each nonvitreous product may contain lesser amounts of other phases.

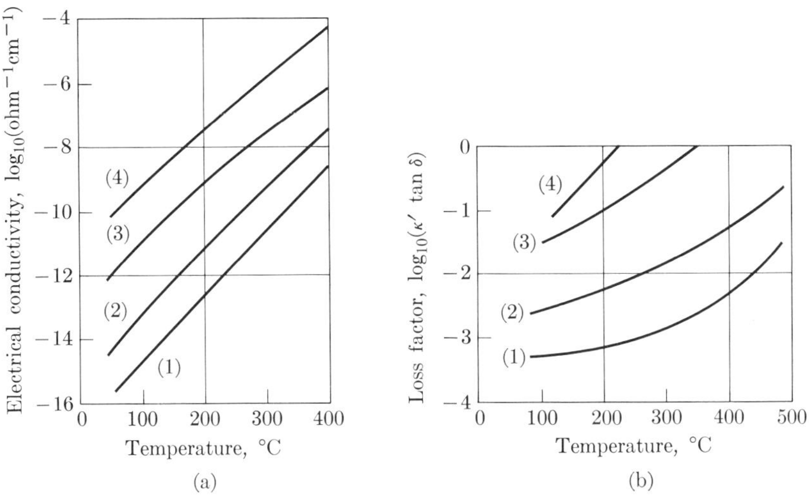

FIG. 10-22. Typical electrical properties of four glasses: (1) fused silica, (2) 96% silica, (3) borosilicate, (4) soda-lime silica, showing (a) conductivity, and (b) loss factor ($\kappa' \tan \delta$).

lag and the cyclic period correspond, energy losses are at a maximum. At still higher frequencies, the field reverses before the ion moves through the energy barrier, with the result that the phase angle returns to 90°. Hence the loss angle is greatest in the frequency limit of any particular polarization. For instance, the example of Fig. 10-13 has a large loss factor in the frequency range of 10^8 to 10^{10} cps.

The energy loss shown in Fig. 10-21 depends on the number of movable charges and the height of the energy barrier. In the crystal illustrated, the frequency range where there are high energy losses is small because all movements of ions during polarization require equal energy. In contrast, a glass has a broad frequency range because there is no uniformity in the movements of the ions as regards direction or energy requirements. Furthermore, ions vary in their displacements, according to their sizes and charges. For example, the alkalis incorporated into glasses produce a progressive increase in the energy losses as equal mole contents of smaller ions replace larger, heavier ions (Table 2-2).

10-6 Magnetic behavior

Individual atoms may or may not have a magnetic moment. They do not have a moment if their spins in opposite directions are equal in number (Section 2-2). Without a magnetic moment, the atom resists any alignment into a magnetic field; hence diamagnetic behavior results. Most atoms or ions of ceramic materials are diamagnetic, for reasons which are obvious when we examine the first 18 elements of Table 2-1. These elements include almost all of the common anions of ceramic materials, such as O^{2-}, F^-, Cl^-, S^{2-}, SO_4^{2-}, CO_3^{2-}, N^{3-}, OH^-, etc. An anion always has a full complement of 8 electrons available for the valence level. Therefore the electron spins must be balanced. In addition, the cations among these initial 18 elements, plus the heavier non-transition elements, have no partially filled orbitals (see Table 2–1) and are therefore diamagnetic.

Although minor in count, the rest of the elements are important to our subject because they have unbalanced magnetic spins (Fig. 2–1) and will respond to a magnetic field. The term "paramagnetic" applies when each atom aligns with the field individually. In some materials which have several Bohr magnetons per atom and relatively small atomic or ionic radii, there is a spontaneous alignment of the magnetic moments of adjacent atoms, which produces a ferromagnetic behavior. Although relatively few atoms have the magneton and size combination necessary for ferromagnetic behavior, we shall discuss these atoms here because the magnetic flux is greatly increased in the presence of ferromagnetic atoms.

The ratio of magnetization B to the magnetic field H is a constant for paramagnetic behavior, and is somewhat analogous to the P/E relationship of Eq. (10–14) for paraelectric behavior, and to the reciprocal of ϵ/σ in Eq. (8–1):

$$B/H = \text{const.} \tag{10–20}$$

Ferromagnetism. If moments are realigned in the process of magnetization, a nonlinear B/H relationship develops. For example, let us assume that a ferromagnetic material contains randomly oriented grains, and that within these grains there are as many unit cells with their "north poles" oriented in one direction as in the other. When a magnetic field is placed across such a material, there is an initial response, as indicated by Eq. (10–20). However, a relatively minor increase in the magnetic field serves to realign ferromagnetic atoms (or ions) in the direction of the field if they are aligned in the "wrong" direction. This produces a major change in the overall magnetization. When all of the atoms are aligned, any further increase in the field simply increases the normal paramagnetic polarization, in accordance with Eq. (10–20).

The previous paragraph describes the curve $OABC$ of Fig. 10–23. If the field is removed after alignment, there is a remanent magnetization B_r, because the north-to-south orientations of most of the atoms do not revert to the original alignment. It is thus necessary to apply an opposing coercive force H_c to remove the net magnetization. A still further reversal of the magnetic field will invert the magnetization.

Each individual atomic "magnet" is influenced by its neighbors. Therefore, when a large number of atoms mutually interact and develop parallel alignments, spontaneous magnetization occurs. This is indicated in Fig. 10–24, where there are several domains with many unit cells each. The major magnetization of Fig. 10–23 actually involves moving the domain boundary so that there is growth of the more favorably aligned domains. In this respect, the growth of

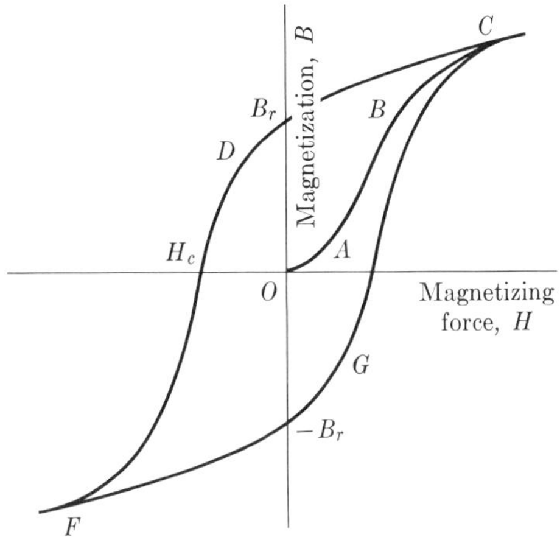

FIG. 10–23. Ferromagnetic behavior: B_r = remanent magnetization; H_c = the coercive force (cf. Fig. 10–18 and the text).

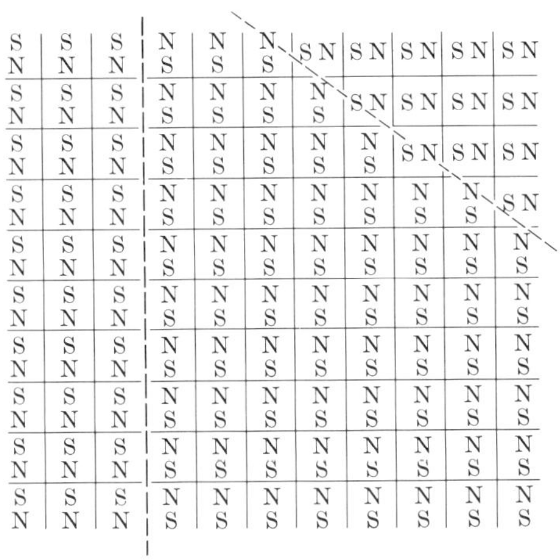

FIG. 10–24. Ferromagnetic domains, whose boundaries (---) are interfaces of higher energy. Domains, like grains, contain many unit cells; however, adjacent domains are crystallographically related.

domains is somewhat comparable to the growth of grains (Section 7–3). However, the process of domain growth, unlike that of grain growth, is reversible.

The parallelism between the above discussion of ferromagnetic behavior and that of ferroelectric behavior (Section 10–4) is intentional, since there are many similarities between the two. However, the differences should be emphasized also. Ferroelectric behavior is a result of ion displacements; ferromagnetic behavior is a result of atom orientation. The former involves asymmetric crystals; the latter involves unbalanced spins of subvalence electrons. Ferroelectric Curie temperatures occur because of an increase in entropy, which produces changes in crystal phases. Ferromagnetic Curie temperatures indicate atom vibrations which are random, and thus upset the orderly interaction between atoms which produces spontaneous magnetization.

Ceramic magnets. Since ferromagnetism originates with the subvalence electrons, it is theoretically independent of the valence state. For example, Fe^{2+} ions have the same subvalence structure as metallic atoms. However, since the behavior of domains shown in Fig. 10–23 arises from a group effect involving interactions between neighboring ferromagnetic atoms, and since the ferromagnetic ions are in effect diluted with anions so that they are less closely associated in a metal-nonmetal compound than in a metal, there is less opportunity for interaction, and the total magnetization is less. Further, all atomic positions in ferromagnetic metals are equivalent; therefore each atom makes an equivalent contribution to the magnetic domains and to the overall magnetization. This is not necessarily true in a compound. Let us now look into the consequences of these effects.

Many of the more widely encountered ceramic magnets have the spinel* structure. As we observed in Section 3–4, the spinel structure contains a basic pattern of face-centered cubic oxygens. Among the 32 oxygens per unit cell, there are two kinds of interstices: 64 small ones with four equally adjacent oxygens (i.e., tetrahedral sites), and 32 slightly larger octahedral sites, each with six surrounding oxygen ions (Fig. 10–25). The cations are located in the interstitial positions. However, there must be vacant sites because the spinel structure $[R_8^{2+}R_{16}^{3+}O_{32}]$ can have only 24 cations per unit cell of 32 oxygens. The 24 cations are not randomly located among all 96 possible sites. Two rather specific patterns result, with some intermediate gradation.

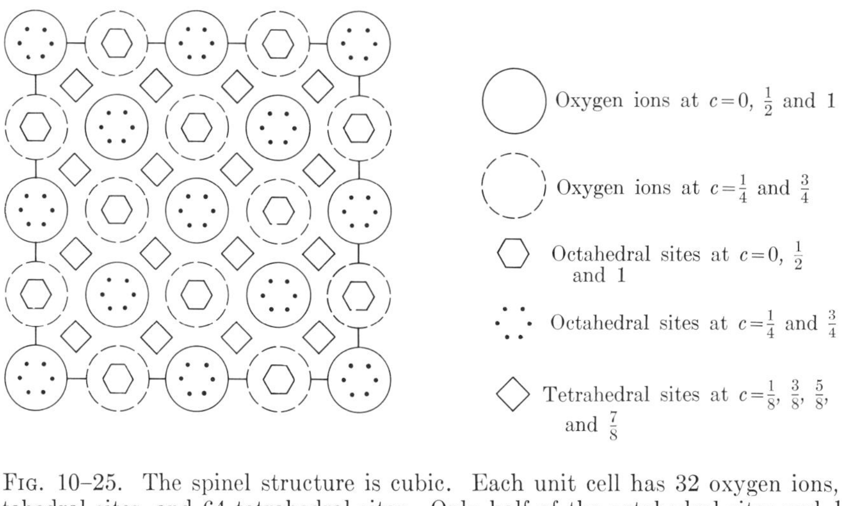

FIG. 10–25. The spinel structure is cubic. Each unit cell has 32 oxygen ions, 32 octahedral sites, and 64 tetrahedral sites. Only half of the octahedral sites and $\frac{1}{8}$ of the tetrahedral sites are occupied.

* The term "ferrite" is widely used when referring to the spinel structure. However, it is even more widely used to denote α-iron, i.e., bcc iron. Therefore either "spinel" or "ferrospinel" is to be preferred, particularly since spinel also denotes the structure (Section 3–4). The value of this distinction becomes apparent when, as a result of loose usage, garnets and other nonspinel ceramics are included under ferrites. [See Economos, G., "Ferromagnetic Ceramics," *Materials in Design Engineering*, 96–101 (September 1958), and 109–114 (October 1958).]

In a *normal* spinel, all of the R^{2+} ions are in tetrahedral sites, and all of the R^{3+} ions are in the octahedral sites. In an *inverse* spinel, all of the R^{2+} ions are in octahedral sites, and the R^{3+} ions are split equally between the two types of sites.

These two variations may be illustrated by $ZnFe_2O_4$ and $NiFe_2O_4$, respectively. In zinc ferrite, the diamagnetic oxygen and zinc ions are essentially diluents of the Fe^{3+} ions. Although the iron ions are magnetic, they are so separated that they cannot spontaneously interact to provide the ferromagnetic characteristics of Fig. 10–23. Rather, they each act independently, and only an independent paramagnetic response is observed (Eq. (10–20)). This is illustrated schematically in Fig. 10–26(a).

In the inverse spinel structure of $[NiFe_2O_4]_8$, the magnetic fields of the individual atoms interact with the tetrahedral and octahedral sites, providing *opposite* orientation (Fig. 10–26b). Although there is a partial cancellation, a net magnetic moment results. A solid solution of Zn^{2+} and Ni^{2+} for the R^{2+} component is illustrated in Fig. 10–26(c). Somewhat unexpectedly, a greater magnetism is indicated. This is substantiated by experimental data.

Ferrimagnetism, illustrated in Fig. 10–26(b) and (c), has opposing orientations for the magnetic atoms; therefore the net magnetism is a vectorial balance. Most ceramic magnets are ferrimagnetic to a greater or lesser extent, since their ions occupy nonequivalent positions (e.g., octahedral versus tetrahedral), in contrast to metals, where we usually find strictly ferromagnetic interactions.

The extreme case of ferrimagnetism is antiferromagnetism; here the two opposing magnetic vectors are equal. Although there is no external evidence of magnetism, an antiferromagnetic material in an electrical circuit does not exhibit true paramagnetic behavior. The total domain structure, rather than independent atoms, must respond to magnetizing forces. Manganese provides the most common example of antiferromagnetism among the metals, and among the compounds, MnO and $TiCl_3$ are examples of antiferromagnetism.

EXAMPLE 10–6. Estimate the number of Bohr magnetons per unit cell in the following compositions: (a) magnetite $[FeFe_2O_4]_8$, and (b) manganoferrite $[MnFe_2O_4]_8$. [*Note*: Both are inverse spinels with ferrimagnetic behavior between the two sites. See Fig. 2–1.]

Answer:

	Tetrahedral sites, 4 neighbors		Octahedral sites, 6 neighbors		Magnetons per unit cell, net
	Ions	Magnetons	Ions	Magnetons	
(a)	8 Fe^{3+}	−40	8 Fe^{3+} 8 Fe^{2+}	+40 +32	32
(b)	8 Fe^{3+}	−40	8 Fe^{3+} 8 Mn^{2+}	+40 +40	40

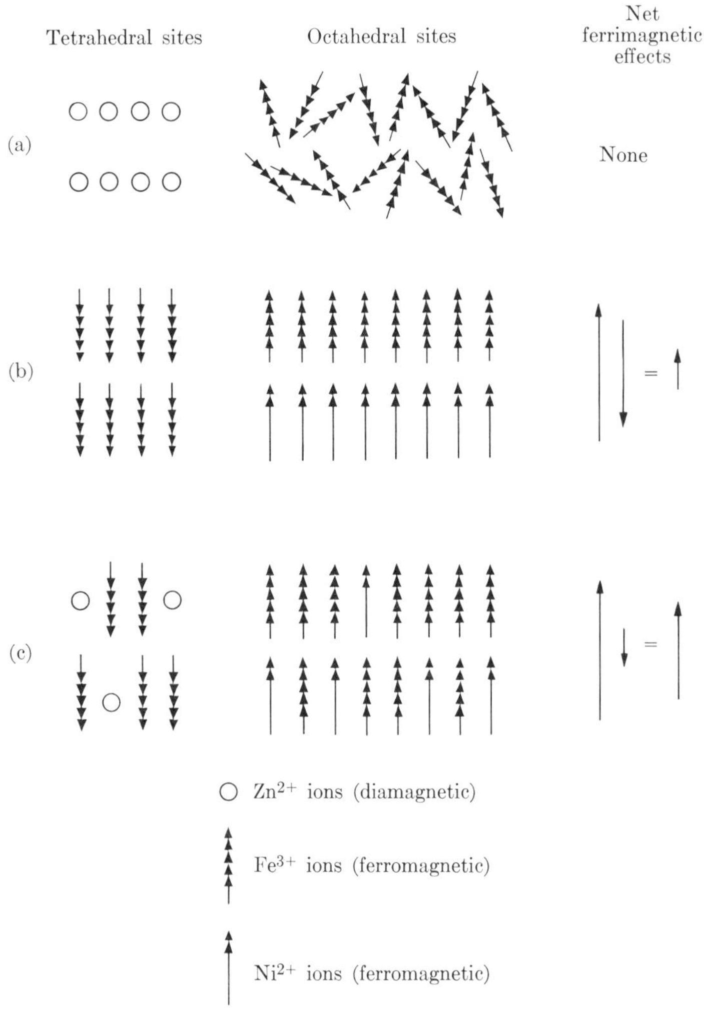

FIG. 10–26. Schematic drawing of ceramic magnets: (a) normal ferrospinel $[ZnFe_2O_4]_8$; paramagnetic behavior. (b) Inverse ferrospinel $[NiFe_2O_4]_8$; ferrimagnetic behavior. (c) Mixed ferrospinel $[(Zn, Ni)Fe_2O_4]_8$; behavior is more strongly ferromagnetic than that of $NiFe_2O_4$.

Above the Curie temperature, the alignments of Fig. 10–26(b) and (c) are so randomized that those structures lose their cooperative behavior, and there is a paramagnetic response, as shown in Fig. 10–26(a). Although the atoms do not interact cooperatively, the individual atoms still possess unbalanced electron pairs in their subvalence shells. Therefore the constant of Eq. (10–20) is higher than it would be in a spinel such as $MgAl_2O_4$, which contains only diamagnetic ions.

Magnetic remanence B_r varies markedly from material to material. As with polarization remanence (Section 10–4), there is a relaxation. Each material has a characteristic relaxation time, during which randomization of the domain alignments takes place. We can use Eq. (10–15), substituting the appropriate symbols, and apply it to demagnetization. A permanent magnet has a relaxation time which is very long, and it will maintain its magnetization until a large opposing coercive force H_c is applied. In contrast, soft magnets have extremely short relaxation times. In some cases, the ferromagnetic characteristics disappear in periods of 10^{-10} sec, and negligible remanence is observed at kilomegacycle frequencies.

The ease of magnetization varies measurably with the coordination of the atoms, which is to be expected, since the interaction between atoms differs with differing directions of the external magnetic field. For example, the $\langle 100 \rangle$ directions are most easily magnetized in metallic iron (i.e., less magnetizing force is required), and in metallic nickel and cobalt the easy alignment directions are $\langle 111 \rangle$ and $\langle 0001 \rangle$, respectively. These aspects of magnetization are used commercially in the making of transformer steels, where the manufacturers manipulate the processing in order to control the crystal alignment. The anisotropy of magnetization is not great in the spinels or other cubic ceramic structures such as the garnets.* Two other types of magnetic ceramics do possess significant magnetization preferences. For example, $R^{2+}Fe_{12}O_{19}$ is hexagonal, and is magnetized most readily in the direction of the c-axis, that is, [0001]. In addition, related but more complex compounds, typified by

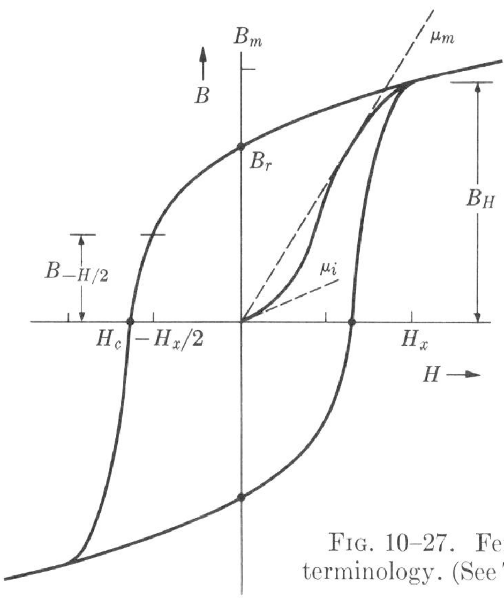

FIG. 10–27. Ferromagnetic terminology. (See Table 10–5).

* The garnets have 24 oxygens per unit cell. Because there are three types of nonequivalent interstitial sites in which the cations may be located, garnets are ferrimagnetic.

TABLE 10–5

MAGNETIC TERMINOLOGY
(See also Fig. 10–27.)

Symbol	Meaning
H	magnetizing force
B	magnetization
μ_i	initial permeability: $\mu_i = (B/H)_i$
μ_m	maximum permeability: $\mu_m = (B/H)_m$
B_m	maximum magnetization
B_r	remanent magnetization
H_c	coercive force
Q	quality factor; $Q = 1/\tan\delta$ or Q-factor: $= 2\pi fL/R$
R_s	squareness ratio: $R_s = B_{-H/2}/B_H$
f	frequency
L	inductance
R	effective resistance
B_H	magnetization for H_x of completed loop
$B_{-H/2}$	magnetization at $-H_x/2$
X_L	inductive reactance

$Ba_3Co_2Fe_{24}O_{41}$, are also hexagonal and ferrimagnetic; however, their directions of easy magnetization are in the (0001) plane.

Inductive components. Some of the magnetic terms which are of interest in circuit design are listed in Table 10–5 and described in Fig. 10–27. The items of special interest provide the basis for (1) the figure of merit, or quality factor Q, and hence the μQ product, (2) the demagnetization factor $(B)(-H)$, and (3) the squareness ratio $(B_{-H/2})/(B_H)$.

A lag in the alignment of the magnetic moments among the atoms decreases the phase angle with which the voltage leads the current. Therefore, a loss angle δ exists, which is analogous to capacitive losses (Section 10–5), and

$$Q = 1/\tan\delta, \tag{10–21a}$$

or

$$Q = X_L/R = 2\pi fL/R. \tag{10–21b}$$

The inductive loss factor ILF becomes

$$\text{ILF} = (\tan\delta)/\mu, \tag{10–22}$$

or alternatively, there is a low inductive loss factor with a large μQ product. It is this latter figure which receives the attention of the circuit designers. Several values are given in Table 10–6, where we see that the μQ product commonly

TABLE 10-6

INDUCTOR μQ PRODUCTS*

Material	60 cps	10^4 cps	10^6 cps
Iron powder	80,000	70,000	10,000
Permalloy	30,000	20,000	1,000
(Zn, Ni) spinel	20,000	20,000	15,000
(Zn, Mn) spinel	500,000	400,000	20,000

* Data from Owens, C. D., "A Survey of the Properties and Applications of Ferrites Below Microwave Frequencies," *Proc. IRE*, **44**, 1234–1248 (October 1956).

decreases (or the losses increase) at higher frequencies.

A second index of interest to the circuit designer is the demagnetization factor, which is the maximum $(B)(-H)$ product in the demagnetizing quadrant of the hysteresis loop. Such a factor indicates the amount of energy required to remove existing magnetization, and is a measure of the permanency of the magnet.

Within the past few years, with the demand for switching components for memory cores of computers, the squareness ratio has become important. As indicated in Fig. 10–28, a high squareness ratio provides a high remanent magnetization B_r and an abrupt realignment of magnetization in the opposite orientation. For this reason, a magnetic core of this type maintains the strongest "yes-no" binary signal.

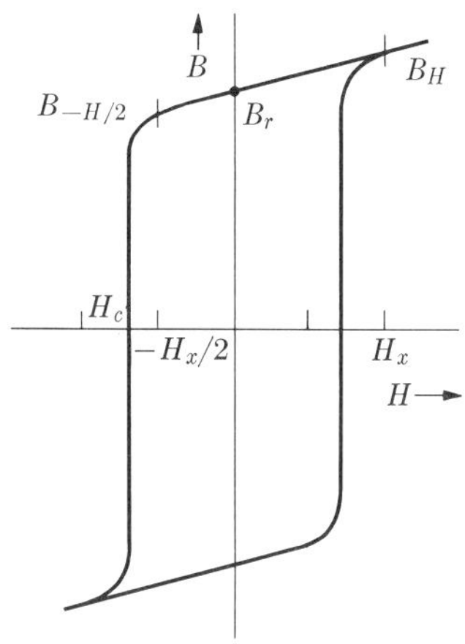

FIG. 10–28. Square loop behavior; squareness ratio = $B_{-H/2}/B_H$.

The most commonly used magnetic ceramics are the ferrospinels (or more simply the spinels). The [(Zn,Mn)Fe_2O_4] spinels are widely used in high-frequency transformer applications because they possess a high μQ product, which minimizes inductive losses, and these spinels, unlike metals, have extremely low ohmic conductivity, which all but eliminates the problem of generation of induced currents and resultant heating. The [(Mg,Mn)Fe_2O_4] spinels have found use in square-loop applications and in microwave gyrators, where the respective requirements of a high magnetic remanence and narrow resonant band width are related through the strong ferromagnetic interaction of the cations.

The $(B)(-H)$ products of spinels are not so high as they are for the anisotropic compounds, such as $BaFe_{12}O_{19}$, which is hexagonal. Therefore we find the latter type of material used for permanent magnets. In these materials the $(B)(-H)$ product is strongly affected by the processing, since, as indicated earlier, this phase is magnetically anisotropic. We shall discuss these and other processing considerations in subsequent chapters.

References for Further Reading

10–1. Azároff, L. V., *Introduction to Solids* New York: McGraw-Hill, 1960. Chapter 14 deals with electrical and magnetic properties of insulators. Chapter 12 discusses conductivity and junction properties in semiconductors.

10–2. Bozorth, R. M., "The Physics of Magnetic Materials," *The Science of Engineering Materials* (J. E. Goldman, editor). New York: John Wiley & Sons, 1957, pp. 302–335. Summary for instructors.

10–3. Caddoff, I. B., and E. Miller (editors), *Thermoelectric Materials and Devices.* New York: Reinhold Publishing Corp., 1960. For the specialist in the thermoelectric field.

10–4. Conwell, E. M., and W. J. Leivo, "Semiconductor Devices," *The Science of Engineering Materials* (J. E. Goldman, editor). New York: John Wiley & Sons, 1957, pp. 393–427. Summary for the instructor.

10–5. Dekker, A. J., *Electrical Engineering Materials.* Englewood Cliffs, N. J.: Prentice-Hall, 1959. Excellent reference for the advanced undergraduate student who has had field theory.

10–6. Economos, G., "Effect of Microstructure on the Electrical and Magnetic Properties of Ceramics," *Ceramic Fabrication Processes* (W. D. Kingery, editor). New York: John Wiley & Sons, 1958, pp. 201–213. Relates electrical and magnetic behavior to processing history.

10–7. Goldman, J. E., "The Theoretical Basis for Magnetic Phenomena," *The Science of Engineering Materials.* New York: John Wiley & Sons, 1957, pp. 267–301. Summary for the instructor; essentially nonmathematical.

10–8. Katz, H. W., *et al., Solid State Magnetic and Dielectric Devices.* New York: John Wiley & Sons, 1959. Advanced undergraduate level. Since most of the devices discussed are ceramic, this book can be said to present Ceramics for Electronic Applications. A familiarity with field theory is required.

10–9. Kingery, W. D., *Introduction to Ceramics.* New York: John Wiley & Sons, 1960. Chapter 19 discusses electrical conductivity, Chapter 20 dielectric and magnetic properties. This is the most complete reference for electrical and magnetic ceramics.

10–10. Lark-Horowitz, K., and V. A. Johnson, "The Physics of Semiconductors," *The Science of Engineering Materials* (J. E. Goldman, editor). New York: John Wiley & Sons, 1957, pp. 336–392. Summary for the instructor. This is more mathematical in approach than the three companion articles cited above.

10–11. Sinnott, M. J., *Solid State for Engineers.* New York: John Wiley & Sons, 1958. Extensive discussion of magnetic and dielectric properties in Chapters 18 and 19. Advanced undergraduate level.

10–12. Smit, J., and H. P. J. Wijn, *Ferrites.* New York: John Wiley & Sons, 1959. Specialized reference for magnetic ceramics.

Problems

10–1. A cermet is composed of 70 v/o TiC and 30 v/o Ni. The latter forms a continuous matrix. Electrical resistivities are 150 and 8μ ohm-cm, respectively. Estimate the resistivity of the cermet by an analogy with thermal conductivity.

10–2. The diffusion coefficient of a stabilized zirconia ($Zr_{0.85}Ca_{0.15}O_{1.85}$) is 10^{-8} cm^2/sec at 860°C (1580°F). Its density is 5 gm/cm^3. Determine its electrical conductivity. (Conductivity is through oxygen-ion diffusion.)

10–3. (a) On the basis of the data in Fig. 10–2 and the relationships of Section 10–2, calculate the diffusion coefficient of Na^+ ions in NaCl at 800°K and at 1000°K. (b) Calculate the diffusion activation energy for Na^+ ions. [*Note:* At 800°K, 98% of the conductivity is by Na^+ ions; at 1000°K, 80%.]

10–4. A 1 cm × 4 cm ceramic insulator is 0.5 cm thick, and produces a capacitance of $2.4\mu\mu f$ with a dissipation factor of 0.02. (a) What is the relative dielectric constant? (b) The loss factor? (c) The figure of merit? (d) The phase angle?

10–5. A forsterite (Mg_2SiO_4) porcelain contains 45% SiO_2, 5% Al_2O_3, and 50% MgO. It was fired at 1400°C (2550°F) and cooled rapidly enough to maintain the glass phase, which was equilibrated at the high temperature. The dielectric constant is 5.4. Since the dielectric constant of forsterite is 6.2, estimate the dielectric constant of the glass. (Assume that the glass has one-half the density of forsterite.)

10–6. (a) An insulator is to be made of MgO. Make suggestions for raw materials and predict processing considerations. (b) An insulator is to be 50 MgO, 40 SiO_2, and 10 Al_2O_3. Make suggestions similar to those in (a).

10–7. Magnetite $[FeFe_2O_4]_8$ and hercynite $[FeAl_2O_4]_8$ are mixed with a nine-to-one weight ratio, then ground, and sintered until a uniform solid solution is formed. How many Bohr magnetons per unit cell? (This solid solution is an inverse spinel.)

10–8. Li^+Fe^{3+} pairs replace 2 Fe^{2+} in magnetite to give $[LiFe_5O_8]_4$. Predict the number of Bohr magnetons per unit cell. (Li^+ ions locate in the sixfold sites.)

10–9. (Refer to Fig. 10–18.) Let H_1 be the field necessary to produce point C. (a) Show the hysteresis loop for cycling from $0.5H_1$ to $-0.5H_1$. (b) Show the hysteresis loop for cycling from $0.6H_1$ to $0.4H_1$. (c) Redraw the hysteresis loop for cubic $BaTiO_3$ above the Curie temperature.

10–10. By drawing an analogy with magnetic properties, describe the structure of antiferroelectric and ferrielectric materials.

Chapter 11

OPTICAL PROPERTIES AND BEHAVIOR

11–1 Introduction

Transparency, color, refraction, and reflection are among the better-known optical characteristics. When we consider these factors, we view each phase as a homogeneous mass, and disregard individual atoms. Thus we can best study these characteristics by paying attention to the wavelike behavior of light. Other optical behaviors, such as luminescence, the stimulated radiation of the newer laser materials, and photo-detectors, are best studied as interactions of photons of energy with individual atoms.

Ceramic materials are important in applications which depend on these optical properties and behaviors. In contrast to metals, ceramic materials lack the large number of free electrons which absorb light; and unlike polymers, they have considerable resistance to environmental abuse.

11–2 Refraction and reflection

The velocity of light is reduced as it moves into a material because it must go through the electrical fields which surround the atoms. The index of refraction n is the ratio of the velocity of light in vacuo, c_0, to the velocity of light in the material, c:

$$n = c_0/c. \tag{11–1}$$

Each material has an index of refraction which depends on the amount of interaction that occurs between light and internal local electric fields. As the light wave has its velocity changed it is *refracted*, and Snell's law applies at the phase boundary of condensed phases:

$$n_r/n_i = \sin i/\sin r, \tag{11–2}$$

where i and r are the incident and refracted angles respectively, and n_i and n_r are the indices in the corresponding phases (Fig. 11–1).

Although the electrons in an ionic or covalent solid are not "free" to absorb the visible radiation, they are displaced by the radiation with respect to the atom so that a small electrical dipole is formed (Fig. 10–12a). The index of refraction is related to the number and size of these dipoles per unit volume.

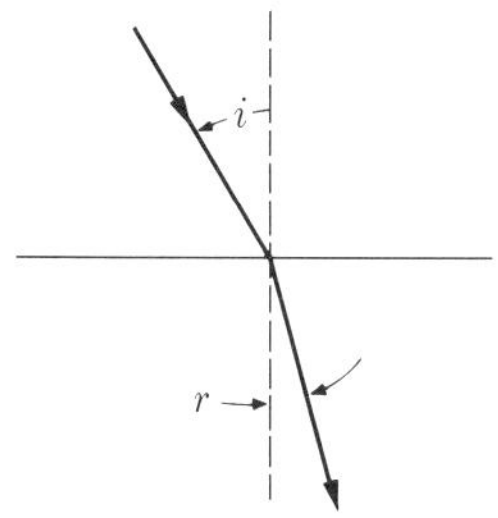

FIG. 11–1. Refraction at an interface, showing the application of Eq. (11–2). Light moves more slowly in the phase with greater electronic polarization per unit volume.

Hence two conditions contribute significantly to higher indices of refraction: (1) greater phase density (and therefore more dipoles per unit volume), and (2) the presence of atoms with higher atomic numbers (and therefore more electrons per atom). Let us illustrate these effects with ceramic examples. The four silica phases (quartz, tridymite, cristobalite and glass) have the same composition (SiO_2); however, they have different structures (Chapter 3) with different atomic packing factors. Thus the change in the index of refraction of the SiO_2 phases shown in Table 11–1 is a function of density only. Sodium chloride, shown in the same table, has a greater index than sodium fluoride because the Cl^- ion has more electrons and is larger and more polarizable than the F^- ion. Other crystals which have (1) high atomic packing factors, (2) large atomic or molecular weights, and (3) highly polarizable ions, possess appreciably higher indices of refraction than those cited in Table 11–1. Closely packed crystals such as periclase (MgO) and corundum (Al_2O_3) have indices of 1.736 and 1.76, respectively; extreme cases include crystals such as galena (PbS) at 3.9.

Birefringence. Only glasses and cubic crystals are isotropic with respect to the electronic polarization of light. Other crystalline structures (see Table 3–2) retard the velocity of light differently depending on the direction of its propagation and vibration. Thus, light vibrating perpendicular to, and light vibrating parallel with, the [001] direction of tetragonal or hexagonal crystals will have

TABLE 11–1

FACTORS AFFECTING THE INDEX OF REFRACTION

Material	Composition	Density	Index of refraction
Quartz	SiO_2	2.65	1.544–1.553
Tridymite	SiO_2	2.28	1.469–1.471
Cristobalite	SiO_2	2.32	1.484–1.487
Vitreous silica	SiO_2	2.20	1.46
		Anion radius	
Villiaumite	NaF	1.33	1.336
Halite	NaCl	1.81	1.544

two different velocities, with consequent differences in index of refraction. Because of the lack of symmetry in orthorhombic, monoclinic, and triclinic crystals, these crystals have three indices of refraction for the three orthogonal directions of vibration.

Birefringence is the difference in the index of refraction between the fastest and the slowest ray. Noncubic crystals which have almost the same atomic density in all three axial directions have very low birefringence. This is the case with cristobalite (SiO_2) of Table 11–1, with a birefringence of 0.002, that is, 1.471 − 1.469. This birefringence, which occurs because of the bond angles of low cristobalite (Fig. 3–22), disappears as the structure becomes cubic above the displacive transformation temperature of about 200°C [Figs. 3–11(a) and 9–9]. Calcite ($CaCO_3$), in contrast to cristobalite, has a large birefringence of 0.172, that is 1.658 − 1.486, because the crystal is highly asymmetrical in the [0001] and [1000] directions.

Glasses and cubic crystals can become birefringent because of stresses which change the atomic densities normal to and in the direction of the applied stress. Thus, birefringence offers a technique for evaluating residual stresses (Section 9–5).

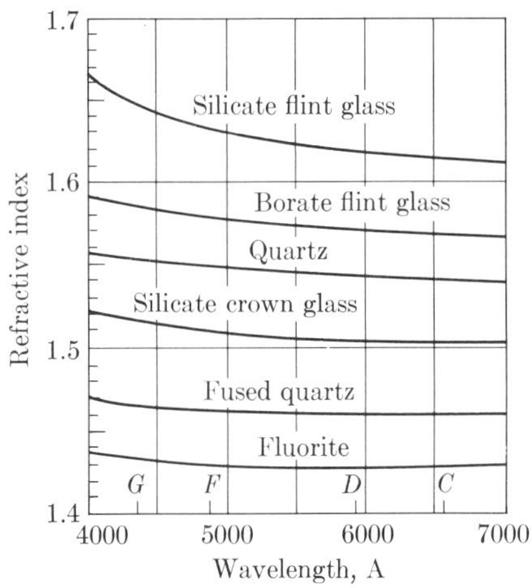

FIG. 11–2. Variation of index with wavelength. (Adapted from F. W. Sears, *Optics*, third edition. Reading, Mass.: Addison-Wesley, 1949, page 49.)

Dispersion. The index of refraction varies with the wavelength of the light. The higher the frequency (and the shorter the wavelength) the higher the refractive index (Fig. 11–2). Dispersion, $dn/d\lambda$, also varies with wavelength and is also higher for shorter wavelengths. However, in optical calculations it is convenient to use a reciprocal relative dispersion, ν, which is determined on the basis of three easily obtained indices:

$$\nu = (n_D - 1)/(n_F - n_C). \qquad (11\text{–}3)$$

The subscripted indices are identified in Table 11–2.

Phase identification. The three properties we have discussed—refractive indices, birefringence, and dispersion—can be used advantageously for phase

TABLE 11-2

SELECTED OPTICAL GLASSES*

Type of Glass	n_D†	$n_F - n_C$†	ν	Specific gravity
Borosilicate crown	1.516	0.0081	63.8	2.54
Dense barium crown	1.613	0.0109	56.3	3.58
Light flint	1.575	0.0139	41.4	3.18
Dense flint	1.647	0.0192	33.7	3.87

Type of Glass	Major components, %				
	SiO_2	B_2O	K_2O	BaO	Other
Borosilicate crown	67	7	16	7	CaO 2
Dense barium crown	37	6	1	45	ZnO 7 Al_2O_3 4
Light flint	53		10		PbO 37
Dense flint	41		8		PbO 51

* Adapted from G. W. Morey, *The Properties of Glass*, New York: Reinhold, 1947, page 408.

† Hydrogen F = 4861 A; sodium D = 5893 A; hydrogen C = 6563 A.

identification since they give a "fingerprint" which is as characteristic of the material as x-ray diffraction patterns are. The techniques of using them are well developed, and since they are described in optical mineralogy texts, no attempt will be made to present them here. Although optical procedures are somewhat more complicated than diffraction procedures, they can yield certain information not available from x-rays, e.g., information on the composition of glass.

Optical glasses. The terms "crown" and "flint" applied to optical glasses refer to the refractive index and the dispersion, which depend on the composition (Fig. 11-3). Crown glasses may have compositions similar to that of soda-lime window glasses or to borosilicate electrical glasses (Table 4-2), with low refractive index (1.50 to 1.61) and dispersion ($n_F - n_C = 0.008$ to 0.012) values. Since BaO increases the index more than it increases the dispersion, it is used to replace CaO in the higher-index crown glasses (Table 11-2). The addition of PbO to a flint glass results in index and dispersion values of (1.53 to 1.72) and (0.010 to 0.025), respectively. In practice, precise ($\pm$ 0.0001) index and dispersion values can be specified and obtained, within the limits shown in Fig. 11-3.

Because of dispersion, single-element lenses exhibit aberrations; i.e., they do not have the same focal lengths for all colors. Corrections must be made by

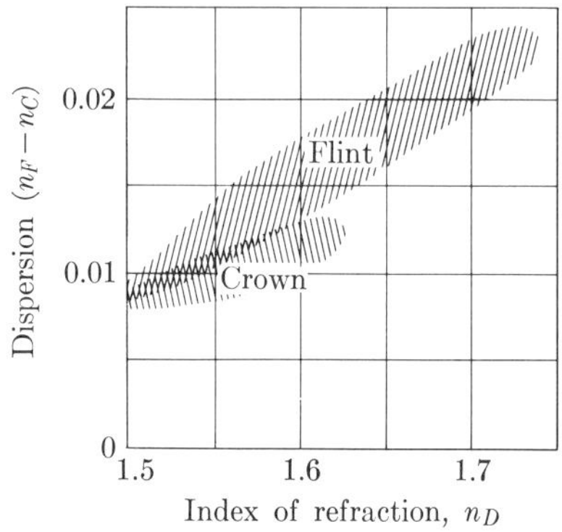

FIG. 11–3. Optical glasses: crown refers to soda-lime-silica and borosilicate glasses; flint, to lead silicate.

making appropriate combinations of crown and flint glasses. In addition to the above requirements, an optical glass must obviously be free of residual stresses, entrapped flaws, and surface imperfections. All these factors must be attended to during the processing.

Reflectivity. A smooth surface reflects a certain fraction of the incident light. The angle of spectral reflection equals the angle of incidence, and the amount is related to the index of refraction, n:

$$R = [(n - 1)/(n + 1)]^2. \tag{11-4}$$

Thus the reflectance R ranges from about 0.03 for low-index (1.4) fluoride glasses to about 0.08 for high-PbO glasses ($n = 1.8$). The extreme case, galena (PbS), has a reflectivity of 35%. In fact, the simplest way to measure the index of refraction for materials with indices in excess of 1.9 is by measuring their reflectivity.

Reflection is normally a factor to be avoided in optical systems. Therefore it is common practice to coat lenses with a thin layer ($\lambda/4$) of isotropic transparent material which provides destructive interference with the surface reflection.

Opacification. Ceramic coatings, in the form of vitreous (porcelain) enamels on metals and glazes on ceramic ware, often have a requirement of opacity or "covering power" and high reflectance. This is achieved through internal refractions and reflections. As indicated by Eq. 11–2, any internal surface changes the direction of the light path. Furthermore, if $n_1(\sin r)/n_2$ is greater than 1.0 (which requires that the light pass from the dense toward the less-dense media), all the light is reflected. Thus any surface between phases with unlike indices diverts the light. As a corollary, more scattering occurs when there is a large number of interfaces between materials with a large refractive index ratio.

White products such as enamels and glazes utilize opacifiers which have indices of refraction of 2.0 and greater in a glass matrix ($n = 1.5$–1.6). Of the common opacifiers, SnO_2 ($n = 2.0$) and $ZrSiO_4$ ($n = 2.0$) are insoluble in the glass matrix. Therefore they are considered to be inert additives to the enamel or glaze. In contrast, TiO_2 ($n > 2.5$) and ZrO_2 ($n = 2.2$) dissolve in the vitreous coating and then crystallize during the early stages of cooling. To be effective, an opacifier must be uniformly distributed in the form of very fine particles.

EXAMPLE 11-1. Corundum is hexagonal and therefore has two indices of refraction (1.760 and 1.768). A light ray passes from one grain to another. What is the maximum possible refraction at the grain boundary?

Answer: The greatest refraction $(i - r)$ occurs (1) with adjacent grain orientations which have the above limiting index values and (2) where the angle of incidence is 90° and is in the lower index grain:

$$\sin r = n_1 (\sin i)/n_2 = 1.760/1.768,$$
$$r = 85°, \qquad i - r = 5°.$$

11-3 Transparency and color

Light absorption is an exponential function of the length of the optical path. Thus,

$$dI/I = -\alpha\, dx, \tag{11-5a}$$

or

$$I/Io = e^{-\alpha x}, \tag{11-5b}$$

and

$$\ln I/I_0 = -\alpha x, \tag{11-5c}$$

where I is the intensity, x is the optical path, and α is the absorption coefficient. The absorption coefficient is a function of the frequency or wavelength. Since we see that part of the optical spectrum which is not absorbed, the color of a material is the "negative" of the colors absorbed.

Ceramic colors. The absorption, or conversely the transmission, of a ceramic material also depends on the composition. As an example, a silica glass maintains transmission through the visible range down to wavelengths of about 2000 A, thus being usable to the full spectral width of normal photographic emulsions. However, the dilution of silica glass by other oxides raises the lower limit of transmission; a borosilicate glass has a limit of approximately 3000 A, and window glass (soda-lime silica) has 3500 A as a lower limit. An Fe_2O_3-containing glass has reduced transmission in the ultraviolet, while FeO reduces the transmission of the longer red wavelengths. It thus becomes obvious that optical glasses must have compositions quite free of iron oxide.

The color of ceramic products is commonly an important characteristic; in fact, it may be the critical selection criterion for ceramic products. This is particularly true in construction materials such as (1) wall and floor tile or brick, or structural glass; (2) protective coatings of porcelain enamels; (3) tableware; and (4) certain optical devices such as light filters or colored lens covers.

Transparent colors. Agents which color transparent materials must not interfere with the light path by producing internal reflections or refractions. Thus

TABLE 11-3

CHARACTERISTIC IONIC COLORS

Ion	Atomic number	Characteristic color	
		Oxidized	Reduced
Ti	22	Yellow	None
Cr	24	Yellow-green	Dark green
Mn	25	Light purple	None
Fe	26	Light yellow	Light green
Co	27	Blue-violet	Blue-violet
Ni	28	Violet to brown	Violet to brown
Cu	29	Green-blue	Red
U	92	Yellow	Green

dissolved components are usually used in the coloring of transparent phases. An exception is a transparent glass which contains colloidal particles sufficiently small (i.e., less than the wavelength) so that their surfaces do not redirect the path of the light rays. Sometimes colloidal dispersions (e.g., gold) are used to absorb selected wavelengths. Many of the ruby glasses depend on this mechanism for their color.

It was pointed out previously that silica glasses are transparent to colors which have very short wavelengths (down to about 2000 A). Other network-forming or modifying oxides (Section 4-3) restrict this range of transparency but do not limit the transmission of visible radiation. This fact is evident not only in glasses but also in crystalline solids where (pure) compounds such as calcite ($CaCO_3$), enstatite ($MgSiO_3$), borax ($Na_2B_4O_7 \cdot 10H_2O$), corundum (Al_2O_3), and fluorite (CaF_2) serve as examples of colorless compounds. In fact, most ceramic phases do not selectively absorb visible radiation, and colored phases appear as exceptions to the general rule.

The color of many transparent ceramic phases originates from elements in the first transition row in the periodic table (Fig. 1-1). Thus we see extensive use of compounds of vanadium, chromium, manganese, iron, nickel, cobalt, and copper as colorants. Each of these elements has multiple valence states, each of which results in a variation in color (Table 11-3). The absorption characteristics of the ions vary markedly, and hence the quantities added for coloring purposes must be considered on individual bases. As a corollary, a transparent material is specifically limited as to the amounts of these various ions it can accept and still remain colorless. In addition to the compounds listed in Table 11-3, there are other transition and rare-earth elements (Section 2-2) which are also used to supply coloring ions to ceramic materials.

Pigments. Colored pigments are simply colored "opacifiers" which selectively transmit light of the appropriate color (Table 11-4). These opacifiers may be

TABLE 11–4

INORGANIC PIGMENTS

Color	For semivitrified products	
	Material	Comments
White	TiO_2	Precipitated from solution
	ZrO_2	Precipitated from solution
	SnO_2	Inert
	ZnS	Inert
Purple		
Blue	Generally obtained from cobalt-containing glasses	
Green	Cr_2O_3	
Yellow	$PbCrO_4$	
	CdS	Low temperature
Orange	$PbCrO_4 \cdot PbO$	Basic lead chromate
	$Pb(Cr,Mo)O_4$	
Red	Fe_2O_3	Brown-red
	CdSe	Low temperature
Black	MnO_2	

Color	For organic coatings	
	Material	Comments
White	TiO_2	
	$2PbCO_3 \cdot Pb(OH)_2$	White lead
Purple	$MnPO_4$	Very weak
Blue	$Fe_4[Fe(CN)_6]_3$	Prussian blue
	Ultramarine	Colored glass
Green	Prussian blue + $PbCrO_4$	
Yellow	$PbCrO_4$	
	CdS	
	$Fe(OH)_3$	Brown-yellow
Orange	$PbCrO_4 \cdot PbO$	
	$Pb(Cr,Mo)O_4$	
	Pb_3O_4	Red lead
Red	CdSe	
Black	Carbon	

added to the ceramic body, introduced into the glaze, applied as an underglaze stain, or applied as a surface glaze. Since the oxidation level affects the color of crystals as well as of glass, the color of a fired ceramic product can be partially controlled by processing steps. An example is the variation of the purple color of some manganese-containing brick according to the oxidation level of the atmosphere in which the bricks are manufactured. Either in ceramic products or in organic coatings, a pigment must be stable in the environment in which it is to serve. Paint pigments, also, may be utilized for viscosity control (fillers) or corrosion protection (metal primers).

11-4 Luminescence

Electrons may be activated to higher energy levels by photons (γ-rays, x-rays, ultraviolet rays), by electron beams (cathode rays), or by electric fields. As the electron drops back to its original energy level, it re-emits photons of energy. If these photons are in the visible range, we observe a phenomenon called luminescence (Fig. 11-4). The reradiated energy is never greater than the incident radiation (i.e., the photon never has a shorter wavelength than the initial photon).

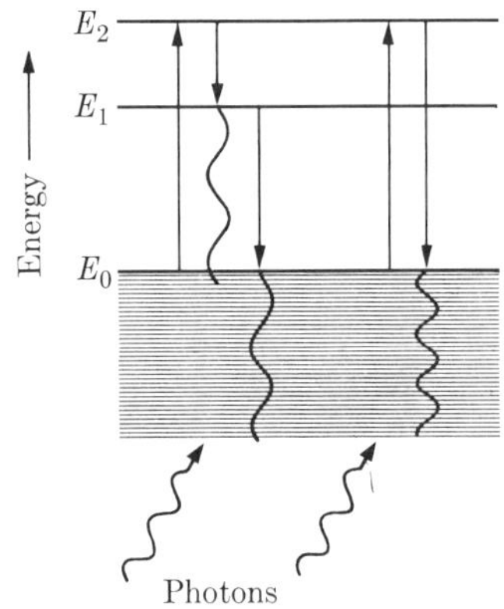

FIG. 11-4. Luminescence: photons raise the electrons from the valence band to higher energy levels. These release new photons when they return to lower energy levels. The frequency is proportional to the energy released.

Luminescence is uncommon in "pure" insulators with a large energy gap. However, if an insulator, for example calcite ($CaCO_3$), contains some impurity such as manganese which has multiple valence states, the insulator will contain additional energy levels [Fig. 10-6(b) and (c)] so that the luminescent process may exist. The term "activator" has been applied to such multivalent ions. In addition to manganese, other activators are lead, copper, chromium, titanium, silver, and tin. The color of the reradiated energy varies with the type of activator, because activation levels vary according to the nature of the impurity.

The ion that absorbs the incident energy need not be the same ion that reradiates the energy. Thus Pb^{2+} ions may absorb the energy of ultraviolet photons. In turn, these ions pass the energy to electrons of Mn^{2+} ions for re-emission. The absorbing ion is called a sensitizer. The sensitizer and activator must be adjacent in order for the transfer of energy to be efficient.

The reradiation of luminescence is not instantaneous, but is delayed on account of the residence time of the electron at the activated level. The reradiation occurs randomly and is in proportion to the number of activated electrons. Therefore, the intensity of luminescence I_t at any time t is related to the initial intensity I_0, as follows:

$$I_t = I_0 e^{-t/\tau_e}, \tag{11-6}$$

where τ_e is the median relaxation time for the reradiation. If the relaxation time for luminescence is short with respect to our visual perception, the term "fluorescence" is used. On the other hand, if the relaxation time is such that there is a noticeable delay or afterglow of luminescence, the term "phosphorescence" is used. Obviously, the distinction between these two behaviors is related to the speed of reaction of the sensing device.

Light sources. Ceramic materials are active contributors to luminescent light sources. As described above, luminescence results when excited electrons drop to lower energy levels and release a photon. Fluorescent lamps utilize phosphors which emit visible radiation after being excited by ultraviolet energy. The ultraviolet light originates when accelerated electrons hit vaporized mercury atoms. The fluorescent compounds of such fluorescent lamps can develop up to 70 lu/w, which approaches the maximum. An incandescent lamp, on the other hand, develops only about 20% as many lumens (or 14 lu/w), the rest going to heat. The phosphors which depend on ultraviolet excitation are commonly tungstates or even carbonate powders, which contain activators such as Mn^{2+} and sensitizers such as thallium or lead ions.

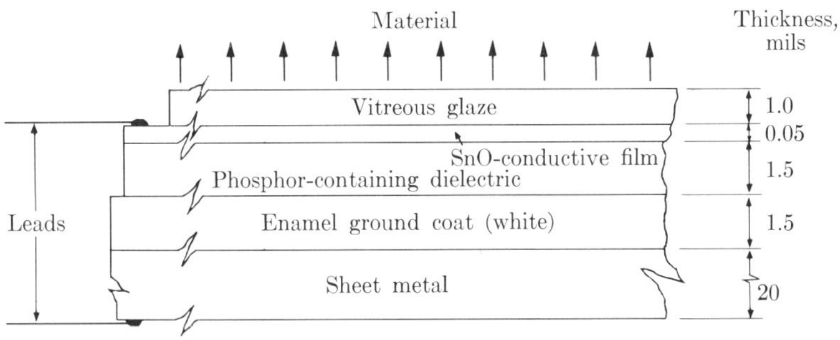

FIG. 11–5. Electroluminescent lamp: Type C (i.e., metal support). [Adapted from K. H. Butler, "Electroluminescence," *Science*, **129**, 544–550 (Feb. 27, 1959).]

Other sources of excitation are electrons (e.g., a cathode ray or a television tube), and electric fields (electroluminescence), both of which usually involve zinc sulfide as the luminescent material. Other related semiconductors are also available. Activators and other auxiliary ions are used to increase the efficiency of light emission. Electroluminescence offers challenging potentialities because of the simplicity of the lamps which use it (Fig. 11–5). By means of electro-

TABLE 11-5

PHOSPHOR CHARACTERISTICS

Type	Fluorescent color	Relaxation time
P1	Green	Medium
P2	Blue-green*	Long
P4	White*	Medium long
P7	Blue-white*	Long
P11	Blue	Short
P12	Orange	Medium

* The color shifts toward the yellow at longer times.

luminescence, wall panels and similar surfaces may be used as light sources. The efficiency of electroluminescent lamps is currently about 15% of the theoretical maximum. However, there are general expectations that further development will markedly increase this efficiency figure. The color and persistence time of phosphors have been characterized by lamp manufacturers into general categories (Table 11-5). The color is established by the activator.

11-5 Stimulated radiation (LASER)

Recently there have been additional applications of the absorption and reemission of energy by atoms. Promising among these applications is the LASER (Light Amplification by Stimulated Emission Radiation), which uses light to stimulate coherent emission. The ruby laser will serve as one of several examples. Ruby is corundum (Al_2O_3) with up to three atom percent of Cr^{3+} ions replacing the Al^{3+} ions. Photons of visible light can "pump" the Cr^{3+} ions to higher energy levels (Fig. 11-6b). It is typical of the more successful laser materials that the Cr^{3+} ions quickly lose part of their energy and fall to a metastable level (Fig. 11-6c), but normally reside in the matastable level for a significantly longer period of time. The statistics for the return to the ground level would then be the same as the statistics for luminescence. However, if the ion is stimulated with an energy equal to ΔE of Fig. 11-6(d), the metastability is removed so that the ion immediately drops to the ground level. Thus the photon emitted from the first Cr^{3+} ion serves to trigger the loss of energy from additional atoms, with a quick return of all Cr^{3+} ions to the ground state and the nearly simultaneous release of a large number of photons of identical wavelength. This optical pumping and stimulation can then be repeated.

The advantages of such stimulated emission include (1) the intense light which can be developed by the nearly spontaneous emission of photons, (2) the coherent character of the light (i.e., it is extremely monochromatic because all photons originate from the same ΔE), and (3) the directional accuracy of the

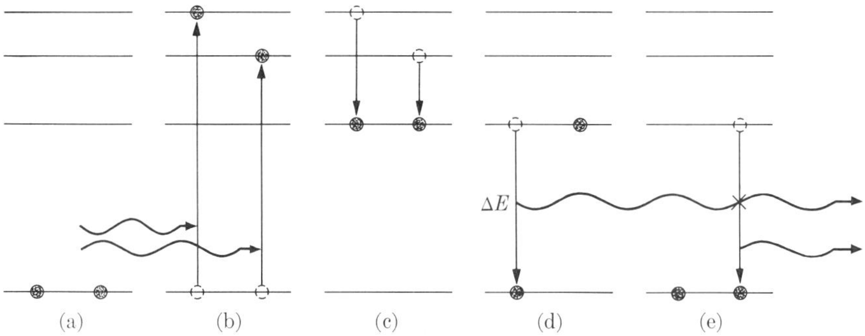

FIG. 11-6. Stimulated emission. The energy levels of chromium atoms are indicated by the black dots. (a) Ground state; (b) atoms energized by visible light; (c) return to the metastable state; (d) ΔE of energy is released as one atom returns to the ground state. (e) The photon of energy from (d) triggers the release of a coherent photon from another atom, and so forth. The pumping process of (b) can be repeated by more visible light.

light. The latter characteristic originates from the crystal structure which serves as a collimator for the light. Beam widths of one second of arc are possible, which permits accurate focusing.

The critical part of a crystal laser is the active ion. The transition and rare-earth elements are preferred; chromium, neodymium, samarium, uranium, praseodymium, holmium, thulium, and dysprosium have all been used successfully. The wavelength of the emission varies with the element. Host crystals must be inert to the "pumping" energy, and capable of taking the active ions into solid solution. The variety of such crystals is great, but they are currently limited to those which can readily be grown into large crystals. In addition to corundum (Al_2O_3), fluorite (CaF_2) and calcium tungstate ($CaWO_4$) have been utilized. Glass lasers with transition elements are also becoming important.

EXAMPLE 11-2. The wavelength of the emitted light in Fig. 11-6 is 6943 A. What is the energy of ΔE in the figure?

Answer:

$$\begin{aligned} \Delta E &= hc/\lambda \\ &= \frac{(6.62 \times 10^{-27}\ \text{erg·sec})(3 \times 10^{10}\ \text{cm/sec})(6.24 \times 10^{11}\ \text{ev/erg})}{(6943 \times 10^{-8}\ \text{cm})} \\ &= 1.79\ \text{ev.} \end{aligned}$$

References for Further Reading

11–1. AZÁROFF, L. V., *Introduction to Solids.* New York: McGraw-Hill, 1960. The optical properties of semiconductors and insulators are presented in Chapters 12 and 14. Advanced undergraduate and graduate level.

11–2. BLOSS, F. D., *Methods of Optical Crystallography.* New York: Holt, Rinehart, and Winston, 1961. Optical properties based on indices of refraction are discussed in detail.

11–3. KINGERY, W. D., *Introduction to Ceramics.* New York: John Wiley & Sons, 1960. Chapter 15 presents a summary of optical characteristics of ceramics; luminescence is not considered.

11–4. LARSEN, E. S., and H. BERMAN, *The Microscopic Determination of Nonopaque Minerals.* U. S. Government Printing Office, 1934. Standard reference for optical data of ceramic-type phases.

11–5. MOREY, G. W., *The Properties of Glass*, second edition. New York: Reinhold Publishing Co., 1954. Optical properties are correlated with glass composition.

11–6. SCHAWLOW, A. L., "Optical Masers," *Scientific American,* **204,** 52–61 (June 1961). Optical masers are now called lasers. This article is an excellent introduction to the subject. It requires the concept of quantized energy levels (Chapter 2).

11–7. SEARS, F. W., *Optics,* third edition. Reading, Mass.: Addison-Wesley, 1949. Chapter 4 presents the introductory principles of lenses.

11–8. SINNOTT, M. J., *Solid State for Engineers.* New York: John Wiley & Sons, 1958. Chapter 19 discusses optical properties on the advanced undergraduate level.

Problems

11–1. Redraw Fig. 11–3 to show the reciprocal relative dispersion as a function of the index of refraction.

11–2. Use Fig. 11–2 to check the statement: "The index of refraction varies linearly with the frequency."

11–3. With practice, the eye can distinguish between the intensity of two adjacent light sources when one is 10% brighter than the other. (a) Would you expect to distinguish between quartz and cristobalite when both are within the same reflected-light microstructure? (b) Between tridymite and cristobalite?

11–4. How thick is a quartz plate which has one additional wavelength for yellow light vibrating in one plane of polarization ($n_D = 1.553$) than in the other plane ($n_D = 1.544$)?

11–5. A leaded glass has 98% transmission through a 1.0 cm thickness. How much light would be transmitted through a 10-in. glass window in a radioactive laboratory?

11–6. A phosphor produces 0.83 as much light 30 sec after the excitation is removed as it does at the end of 15 sec. (a) What is the electronic relaxation time? (b) How much light will be produced 5 min after the excitation is removed?

Chapter 12

NUCLEAR APPLICATIONS

12–1 Introduction

Inasmuch as materials selection is a major factor in nuclear power conversion, a separate chapter on this topic is warranted. However, we shall consider only ceramic materials, except where it is necessary to make direct contrasts between ceramic and nonceramic materials.

It is assumed that the reader is generally familiar with the mechanisms of fission processes, in which energetic neutrons split heavy atoms. Released are (1) additional neutrons, (2) a variety of intermediate-size isotopes, (3) electromagnetic radiation, and (4) considerable energy. The additional neutrons sustain the process if their numbers are controlled at a steady state of production. The energy is converted to useful power within the limits of thermodynamic efficiency. The fission waste (frequently radiologically "hot") and the electromagnetic radiations present problems of protection and material behavior.

The materials specifications for nuclear reactors are more stringent than for almost any other application, for several reasons: (1) the severe safety requirements, (2) the fact that certain impurities interfere with required reactions and their control, and (3) the long life requirements.

In this chapter, we shall first discuss the behavior of ceramic materials when they are exposed to neutron radiation. These materials may be within the core proper, in structural members, or in ancillary components used in electrical control circuitry. Next we shall consider the selection of ceramic materials as related to the neutron–atom interactions. And finally we shall talk about ceramic materials as used for shielding and for waste disposal.

12–2 Alteration by radiation

All materials subject to radiation exposure are also subject to property alteration (radiation damage). This means that, in a nuclear reactor, structural members and electrical components may be altered, even though they are not directly associated with the release of fissional energy. Many treatises have been written about the effects of radiation on engineering materials. Bowman and

TABLE 12–1
EFFECTS OF RADIATION ON CERAMICS
Courtesy *Materials in Design Engineering*, **52,** 127 (July 1960)

Material	Integrated exposure*	Effects
Al_2O_3	$2 \cdot 10^{20}$ n/cm^2	No change in x-ray pattern and thermal conductivity; some color change
	10^9 ergs/gm (C), gamma	Color change
BeO	$2 \cdot 10^{20}$ n/cm^2	0.3% expansion in c_0, none in a_0 as determined by x-ray pattern; no change in thermal conductivity
MgO	$2 \cdot 10^{20}$ n/cm^2	No change in basic lattice; single crystal broken into reactive small crystals
Quartz	10^{21} n/cm^2	Decrease in density to 2.25; change in axial ratio (c/a) from 4.1 to about 1.06 at this dosage
SiO_2 (vitreous)	$2 \cdot 10^{20}$ n/cm^2	Increase in density to 2.25; heating at red heat removes increase; color changes
	$2 \cdot 10^{10}$ ergs/gm (C), gamma	No discoloration in highly purified silica; commercial silica appears dark and mottled
TiO_2	$2 \cdot 10^{20}$ n/cm^2	No change in x-ray pattern; slight decrease in thermal conductivity; color change
ZrO_2	$2 \cdot 10^{20}$ n/cm^2	Radiation-induced phase change, monoclinic to cubic; no change in cubic lattice; slight decrease in thermal conductivity
Spinel, forsterite, porcelain, steatite	$2 \cdot 10^{20}$ n/cm^2	No change in crystal structures; decrease in thermal conductivity (factor of 2); color change
	10^9 ergs/gm (C), gamma	Color change
Cordierite	$2 \cdot 10^{20}$ n/cm^2	Loss in crystallinity; decrease in thermal conductivity (factor of 4); color change
Zircon	$1 \cdot 10^{20}$ n/cm^2	1.9% expansion; large loss of crystallinity
	$2 \cdot 10^{20}$ n/cm^2	Complete loss of crystallinity; decrease in thermal conductivity (factor of 5); color change
	10^9 ergs/gm (C), gamma	Color changes
$BaTiO_3$	$2 \cdot 10^{20}$ n/cm^2	2.87% expansion in a_0; marked color change
	10^9 ergs/gm (C), gamma	Color change
Mica (muscovite)	10^9 ergs/gm (C), gamma	No effects noted

* n/cm^2 = integrated neutrons per cm^2; ergs/gm (C) = gamma-ray energy (carbon atom equivalent).

Reinsmith* provide two of the more comprehensive summaries; each includes a sizable reference list.

To state the matter most simply, when a material is exposed to radiation, neutrons energize atoms in their crystal positions. The local effect is comparable to a sudden increase in thermal energy; hence the term "thermal spike." The absorption of energy may also displace atoms and introduce crystal defects. A Frenkel defect is most common; namely, a vacancy-interstitial point (Fig. 3-18c). Extensive exposure to radiation can destroy the long-range structure of crystals to such an extent that only short-range, first-neighbor coordination remains. The existence of short-range order implies that the displaced atoms form bonds with their new neighbors; however, the energies of these bonds are not at the minimum level. Therefore, time and temperature tend to "anneal out" the changes which have developed.

Because radiation damage involves atom displacements, it is possible to predict the qualitative nature of the damage. Greater hardnesses, increased strength, and decreased ductility can be expected as a result of exposure to neutrons. Likewise, there is a reduction in density and usually a decrease in thermal conductivity. In ferroelectric materials, dielectric constants decrease, usually as a result of the decreased densities. Finally, the electrical conductivity of ceramic phases is generally increased. This latter change may be surprising when we first consider it, because it does not match the change (decrease) in thermal conductivity. However, heat is conducted by phonons (Section 9-4), and any irregularity in lattice structure decreases phonon mobility; whereas electrical conductivity of ceramic phases depends on ion movements and/or semiconducting holes and electrons, each of which is increased by exposure to radiation. In general, the consequences of neutron exposure are roughly comparable to the effects of cold work, or of disordering; however, since the temperatures required for property restoration after radiation damage are different from temperatures needed for normal annealing, it is apparent that radiation and cold work are not fully identical, and a quantitative analogy cannot be made.

Tables 12-1 and 12-2 present changes observed at selected radiation exposures. Note that many transparent ceramic materials exhibit color changes, which are generally the result of small gamma-ray dosages that change the valence level of some of the atoms. These effects can be counteracted by the addition of cerium or other materials containing multivalent ions which can provide electron-acceptor states for the electrons which have been activated by the gamma rays. In general, the other physical properties of glass are affected only slightly as compared with crystalline ceramics, in which the changes are more extensive. This might be anticipated, since glass already has a high-entropy structure.

---

* Bowman, R. E., "How Radiation Affects Engineering Materials," *Materials in Design Engineering*, **52**, 119-134 (July 1960). Reinsmith, G., "Nuclear Radiation Effects on Materials," *ASTM Bulletin*, **224**, 37-47 (September 1958).

TABLE 12–2

EFFECTS OF RADIATION ON GLASS

Courtesy *Materials in Design Engineering*, **52,** 126 (July 1960)

Glass composition or type	Type of radiation or dosage*	Property change	Remarks
K-Ba-Al phosphate	Gamma	Colors easily	High potassium favors color formation
Ca metaphosphate	Gamma	Absorption peak near 4700 A	
Silver (activated)	Gamma	Absorption near-uv	Orange fluorescence from absorption
Lead containing 60 w/o PbO	10^{14} n/cm^2	No discoloration; melts	
Pure GeO_2 glass	Gamma	Insensitive at 10^{8-9} ergs/gm (C)	
$1PbO \cdot 1.3P_2O_5$	$2 \cdot 10^{20}$ n/cm^2	Slight discoloration	
$1PbO \cdot 1.22V_2O_5$	$2 \cdot 10^{20}$ n/cm^2	No significant change	
$1PbO \cdot 1.5SiO_2$	$2 \cdot 10^{20}$ n/cm^2	Changes from yellowish-clear to amber-clear; considerable decrease in hardness	
$0.8K_2O \cdot 0.2CaO \cdot 2.75SiO_2$	$2 \cdot 10^{20}$ n/cm^2	Changes from colorless to gray	
$Li_2O \cdot 3B_2O_3$, $Na_2O \cdot 2B_2O_3$, $K_2O \cdot 2B_2O_3$, $Rb_2O \cdot 2B_2O_3$	$3 \cdot 10^{10}$ ergs/gm (C), gamma	Absorption 10 times greater than for fused B_2O_3	$Rb_2O \cdot 2B_2O_3$ has the highest absorption in visible region
Lead glass (window)	Gamma	May develop brown coloration	Color removed by annealing 2 to 3 hr at 150 to 200°C

Soft glass	$5.8 \cdot 10^{19}$ n/cm^2	Dark coloration, slight increase in tensile strength; decrease in impact strength	
Optical glass (crown and flint)	10^6 — 10^8 ergs/gm (C), gamma	Develops color at 10^6 ergs/gm (C) or less; useless at 10^8	
Optical glass (protected; containing 1–2% CeO_2)	x-ray or gamma	No coloration at 10^8 ergs/gm (C); usable at $5 \cdot 10^{10}$ ergs/gm (C)	
Soda-lime-silica glass (cobalt-containing)	Gamma	Color change sensitive to dosage; suitable to dosimeter	
Fused silica (general)	All types of high-energy radiation	Color centers form; absorption band at 2140 A	
Fused silica (pure)	Fast neutrons, 10^{11} ergs/gm (C), reactor and gamma	Density may change from 2.2 to 2.27 No change in absorption Intense absorption band at 2180 A	Gamma radiation after reactor exposure increases intensity
Fused silica (with Al)	x-ray and uv	Absorption bands at 5500, 2950, and 2150 A	
Borosilicate	10^{20} n/cm^2	Develops large cracks	Cracking presumed to be caused by (n,α) reaction
Pyrex (borosilicate)	$5.8 \cdot 10^{19}$ n/cm^2	Color darkens; slight increase in tensile and impact strength	

*n/cm^2 = integrated neutrons per cm^2; ergs/gm (C) = gamma-ray energy (carbon atom equivalent).

TABLE 12–3

NEUTRON–ATOM INTERACTION PROPERTIES

Element	σ_a Absorption cross-section, 10^{-24} cm^2 (barns)	σ_s Scattering cross-section, 10^{-24} cm^2 (barns)	δ Fractional energy loss	Moderating ratio, M
Cd	2500	5.3	0.018	0.00021
B	700	3	0.017	0.00073
Hf	120	10	0.011	0.0009
Li	64	2	0.26	0.0083
Co	33	5	0.034	0.0051
W	16	5.7	0.011	0.0039
Ti	5	6	0.041	0.0473
Mo	4	6.5	0.021	0.035
Cr	3	4	0.038	0.051
Ba	1.25	8	0.015	0.096
Zr	0.5	14	0.022	0.615
Mg	0.4	3.6	0.08	0.70
Ca	0.4	9.5	0.049	1.08
Si	0.25	1.7	0.07	0.42
Al	0.23	1.5	0.07	0.47
F	0.01	4.1	0.1	6.1
Be	0.009	6.1	0.2	133
C	0.0045	4.8	0.15	162
O	0.0016	4.1	0.12	316
H	0.3	20.0	1.0	67
D	—	—	0.72	47,500
He	0	1.5	0.425	∞

12-3 Ceramics for core components

Let us now look at the "cross-section" of various atomic and isotopic species of materials used within the core of the nuclear reactor. The cross-section, σ, is really a measure of the probability of interaction between the neutron and an atom:

$$\sigma = R/nN, \tag{12-1}$$

where R is the number of neutron reactions per second, n is neutron flux, and N is the number of atoms per unit cross-sectional area. Thus cross-section is expressed as an area per atom. The term "barn," equivalent to 10^{-24}cm^2, is commonly used.

There are several types of neutron–atom interactions, and hence several types of cross-sections. These include (1) fissioning cross-sections σ_f, (2) absorption cross-sections σ_a, and (3) scattering cross-sections σ_s. Only the latter two will be considered in further detail.

The absorption cross-section σ_a is a measure of the probability that a neutron will be captured by an atom. For this to occur, the energy of the neutron must be reasonably low (in the thermal-energy range of approximately 1 ev).

The scattering cross-section σ_s is a measure of the probability that the neutron will be deflected by an atom. When this occurs, the neutron loses a fraction of its energy. The fraction lost is greatest when the neutron collides with a light atom, because a greater portion of the energy is transferred to a light atom than to a heavy one.

Table 12-3 presents the above two cross-sections, along with δ, which is the fractional energy loss per collision.* The final column indicates the moderating ratio M:

$$M = \sigma_s\delta/\sigma_a, \tag{12-2}$$

which is a measure of the efficiency of energy reduction without neutron loss. We shall discuss this further later in this section.

Ceramic fuels. Inasmuch as the source of nuclear energy is within the atom itself and is independent of chemical combinations, nuclear fuels may be ceramic compounds, metal alloys, or enriched isotopes. The choice among these is limited only by (1) the nuclear behavior of the accompanying elements, and (2) the service stability of the material.

If we limit our discussion to uranium fuels, the most likely ceramic fuels are UO_2, UC, and UC_2 (and possibly UN, UF, and US). The oxide and carbide ceramic compounds may be considered because each nonfissionable component has a low absorption cross-section (Table 12-3) and the level of neutron efficiency may remain high. Not only that, but their physical properties render them valuable. Of the above compounds, UO_2 and UC are most used because their

---

* Wisnyi, L. G., "Ceramics in Nuclear Reactors, Part 2," *Ceramic Industry*, **74** [3], 79 (1960).

Table 12–4 Ceramic Fuels

Composition	Density, gm/cm³ Total	Density, gm/cm³ Uranium	Approximate melting point, °C	Thermal conductivity, $\frac{\text{cal}\cdot\text{cm}}{\text{cm}^2\cdot\text{sec}\cdot°\text{C}}$
UO_2	10.0	8.8	2800	0.02
UC	13.7	13.0	2400	0.08
UC_2	11.7	10.6	2500	0.08
U	18.1	18.1	1130	0.07

stability is relatively good (Table 12–4). The stability of both UO_2 and UC is better than that of metallic uranium for two reasons: (1) Orthorhombic uranium undergoes thermal growth as a result of its anisotropic structure (Section 9–2), and (2) the oxide has internal vacancies within its CaF_2-type structure (Fig. 3–10 and Section 3–4). This open structure can accommodate fission fragments with little distortion of the crystal structure; UO_2 does have the disadvantage of low thermal conductivity, which decreases further as the temperature increases.

In some reactor designs, it is advantageous to use either binary solid solutions or compounds of UO_2 with other oxides, so that there are greater fuel burnups with less structural deterioration due to radiation. The compound ZrO_2 forms a suitable solid solvent when it is stabilized in the cubic form. Possible binary compounds include the alkaline earth uranates.

Control components. Cadmium, with its high absorption cross-section, is currently used as a damping or control element. However, as temperatures are increased to achieve more efficient energy conversion, other elements will have to be considered more seriously. Boron carbide and hafnia are possibilities. The lanthanum-series elements (not listed in Table 12–3) also have high absorption cross-sections (>1100 barns), so that their oxides can possibly be used for control components.

Moderators. Reactors which depend on slow (thermal) neutrons for operation require the presence of moderators to reduce the neutrons' energy. The moderating ratio (Table 12–3) identifies the most suitable materials. The advantage of D_2O (heavy water) for low-temperature reactors is obvious, whereas high-temperature reactors make use of purified graphite or BeO. In general, graphite is favored because it has better thermal conductivity at higher temperatures. However, care must be taken in fabrication to produce a random crystal orientation, since individual crystals have a marked anisotropy in many properties. Thermal expansion is a particularly significant factor, because unequal dimensional changes can disrupt close structural tolerances. By virtue of its use, a moderator is subjected to severe radiation damage; this factor must be considered in design.

12-4 Shielding and waste disposal

Neutron shielding is best achieved by materials which have high absorption cross-sections, particularly if there is a space limitation. Gamma-radiation, on the other hand, is best absorbed by dense materials of heavy atoms. Where there are no limitations as to the shielding mass, material bulk is simply increased. Where there are limitations, lead oxide finds wide use in glasses, and barium sulfate (barite) is used as concrete aggregate. Of course, metallic lead, iron, tungsten, and other metals are also used.

A perplexing problem in nuclear technology is disposal of atomic wastes, because many of the by-products of nuclear fission have exceedingly long half-lives, and must be made as inert as possible. There is a potential solution among the principles mentioned earlier in this book (Chapter 5). Since some of the more damaging radioactive fission products are alkalis and alkaline earths, such as Cs^{127} and Sr^{90}, it is possible to adsorb them onto the surfaces of base-exchange materials (Section 5-3). Montmorillonitic clays (Table 3-5 and Fig. 5-2b) are utilized for this purpose. After ion exchange has taken place, these clays are dried and fired to produce a chemically inert ceramic of glass and nonexchangeable crystals.

References for Further Reading

12–1. Bowman, R. E., "How Radiation Affects Engineering Materials," *Materials in Design Engineering*, **52,** 119–134 (July 1960). Introductory presentation; includes metallic, ceramic, and organic materials.

12–2. Fan, H. Y., and K. Lark-Horovitz, "Effects of Radiation on Semiconductors," *Effects of Radiation on Materials* (J. J. Harwood, *et al.*, editors). New York: Reinhold, 1958, pp. 159-190. Summary article on the subject.

12–3. Kling, H. P., and B. Kopelman, "Ceramic Fuels," pp. 90–109, and H. Shapiro, "Ceramic Cores," pp. 199–206; from *Materials for Nuclear Reactors* (B. Kopelman, editor). New York: McGraw-Hill, 1959. Brief summaries of the indicated topics.

12–4. Kriedl, N. J., and J. R. Hensler, "Special Glasses for Nuclear Engineering Applications," *Modern Materials*, Vol. 1 (H. H. Hausner, editor). New York: Academic Press, 1958, pp. 217–241. Excellent summary for the specialist.

12–5. Norton, F. H., *Refractories*, third edition. New York: McGraw-Hill, 1949. Chapter 24 presents an introduction to refractories for nuclear power production.

12–6. Primak, W., and M. Bohman, "Radiation Damage," *Progress in Ceramic Science*, Vol. 2 (J. E. Burke, editor). New York: Pergamon Press, 1962, pp. 103–180. Nonmathematical presentation, with emphasis on ceramic materials. Extensive bibliography.

12–7. *Radiation Effects on Materials*, Vol. 1–3. Philadelphia: American Society for Testing and Materials, 1957–58. Papers from three symposia on the subject. The last two articles of Vol. 3, on radiation damage of piezoelectric crystals and of optical glasses, are of direct interest to the ceramist.

12–8. Reinsmith, G., "Nuclear Radiation Effects on Materials," *A.S.T.M. Bull.*, **224**, 37–47 (September 1958). Excellent introductory summary.

12–9. Smoluchowski, R., "Effects of Radiation on Electronic and Optical Properties of Inorganic Dielectric Materials," *Effects of Radiation on Materials* (J. J. Harwood, *et al.*, editors). New York: Reinhold, 1958, pp. 144–158. Production and annealing of color centers. Advanced undergraduate and graduate presentation.

Problems

12–1. Provide explanations for the neutron-induced alterations of ZrO_2 in Table 12–1.

12–2. Suppose that $BaTiO_3$ is being considered for use as a piezoelectric crystal in a neutron-radiation environment. Discuss possible complications, if any.

12–3. Account for the variations in the "fractional energy loss" values of Table 12–3.

12–4. (a) Which will undergo a greater change in thermal conductivity when exposed to neutron radiation: quartz or fused silica? Why? (b) Explain the opposing effects of exposure to neutron radiation on the densities of these two phases.

Part 3

THE PROCESSING OF CERAMIC PRODUCTS

Because this book is primarily concerned with physical ceramics, we shall not go into any great detail on the subject of the processing of ceramics. However, the processing of ceramics is important to the physical ceramist and the design engineer, since processing procedures frequently influence properties, and there are limitations encountered in the processing of a material. A design engineer who ignores either of these considerations cannot make the most advantageous selection of material. Thus Part 3 will deal with the Processing of Ceramic Products because the choice of processing procedures is dependent on the Nature of Ceramic Materials (covered in Part 1) and on the Properties and Behavior of Ceramic Materials (Part 2). We shall discuss both the forming (or shaping) of ceramic products and the thermal treatments required to achieve the final specifications.

The reader must devote further study to processing principles and requirements if he wishes to enter the field of ceramic engineering, in which material production is the center of technical interest. Also, the nonceramist will undoubtedly want to acquaint himself with the production details and variables for the specific ceramic materials which he incorporates into his designs.

Chapter 13

CERAMIC PRODUCTS: FORMING

13–1 Introduction

Ceramic products available to the engineer all require some type of shaping or forming procedure, and here we find specific dissimilarities between metals and ceramics. Most metal-forming operations depend on the casting of a molten liquid and/or mechanical working. These procedures are rarely used in ceramic-forming operations where hydroplastic forming of solids is of major importance. Powder metallurgy, which employs sintering for purposes of agglomeration, is the procedure which is most comparable to the forming procedures in ceramics, but it is used relatively seldom with metals, in comparison with its use in ceramics. The plastic-forming and vitreous-shaping operations of ceramics manufacture are quite similar to some of the polymer production methods. However, let us forgo further analogies with plastics and metals, and devote ourselves to the processing requirements of ceramic products.

The forming of ceramic products starts with the preparation of the raw material. The final shape of the product depends mainly on (1) particle agglomeration and (2) viscous deformation, plus a few other methods. For convenience, we shall subdivide agglomeration into (a) dry pressing, (b) hydroplastic forming, and (c) slip casting. We shall discuss each subject in turn. Properties of the ceramic materials (Chapters 8–12) govern the choice of forming methods. The nonductile, brittle character of ceramic materials is, of course, of prime importance. Also, the nature of the surface becomes important in colloidal-sized raw materials.

13–2 Raw materials

The raw materials needed for ceramic processing all have their prime origin in nature. Some raw materials are incorporated into ceramic products in their natural form, while others are beneficiated and processed before being used.

Two generalities apply here: (1) Raw materials for large-tonnage products (such as brick, concrete, refractories, etc.) receive little or no preliminary processing, while those for low-tonnage ceramic products (such as ceramics used for electrical circuits, ceramic cutting tools, and optical glass) receive extensive beneficiation. This is a consequence of both economics and property requirements. (2) There is a general trend toward additional processing of all raw materials, even for large-tonnage applications, because there are more severe requirements nowadays as to the properties of ceramic products, and because the best natural deposits are being depleted.

Natural raw materials. Many clays, limestones, quartz, and related minerals are used directly in the forming process. The first three of these are widely distributed in nature, so that selection is made on two bases: (1) economy of mining and transporting, and (2) the absence (or presence) of impurities. Raw materials other than these generally are restricted in occurrence.

The geological processes which lead to the concentration of raw materials include (1) mechanical separation, (2) chemical precipitation, (3) chemical leaching, and (4) thermal changes. Quartz sand is an example of a raw material which has been extensively washed and *mechanically separated* by the geological processes of wave and wind action. Since quartz (SiO_2) is much harder and less soluble than most other common minerals, it may become highly concentrated. The sand grains have been rounded during the geological processing; the more thoroughly worked sands are quite uniform in size.

The limestones are typical of geological processing which involved *chemical precipitation.* The reaction

$$Ca^{2+} + 2(HCO_3)^- \rightleftarrows CaCO_3\downarrow + H_2O + CO_2 \qquad (13\text{–}1)$$

is responsible for limestone formation, and is just as responsive to chemical conditions in nature as the same reaction would be in the chemistry laboratory. The solubility of the Ca^{2+} and $(HCO_3)^-$ is increased by higher CO_2 contents. Therefore, carbonic acid-containing ground waters dissolve $CaCO_3$ and carry the Ca^{2+} and $(HCO_3)^-$ ions to areas where CO_2 and the acidity are lost, inducing precipitation of $CaCO_3$ (calcite). The purest limestone was formed in waters free from fine sediment (e.g., clay). Also, as in the chemical laboratory, the conditions for $CaCO_3$ solution and precipitation are sufficiently different from those for other ions that a "quantitatively clean" separation is possible. Most precipitated limestones are relatively free from alkalis, iron, manganese, and most other cations. They may contain a considerable amount of Mg^{2+} and in fact may have a Mg^{2+} content sufficiently high to form a dolomite [$CaMg(CO_3)_2$] or even a magnesite ($MgCO_3$). All of these raw materials have refractory applications (Section 9-7). Limestones were also formed through the accumulation of shells and similar biological residues. Here, the animal aided the chemical precipitation process by accelerating the reaction (13-1) to obtain calcite ($CaCO_3$) for its own use.

Over many centures, geological processing which involved hydrolysis and *chemical leaching* was utilized in the formation of primary clays from the feldspars of granites:

$$2KAlSi_3O_8 + 11H_2O \rightarrow (OH)_4Al_2Si_2O_5 + 2KOH + 4Si(OH)_4. \qquad (13\text{–}2)$$

$$\begin{bmatrix}\text{orthoclase} + \text{water} \rightarrow \text{kaolinite} + \text{soluble} \\ \text{feldspar} \qquad\qquad\qquad\qquad\qquad\qquad \text{compounds}\end{bmatrix}$$

Some of these clays remained in place as residual kaolins, and some eroded and mechanically separated into sediments composed of finely divided particles. The purest clay sediments were formed in quiet bodies of water which lacked sea life. Geologists associated these waters with times between coal-formation periods.

Pegmatite deposits illustrate the final mechanism of geologic processing; namely, *thermal alteration*. Pegmatites are geologic structures which were enriched by minerals that commonly develop large-sized grains. Pegmatites vary considerably in composition, but in general contain minerals which have relatively large quantities of easily ionizable elements. Alkali-containing minerals are common, and fluorine ions are often substitutes for $(OH)^-$ ions. Small ions, such as Be^{2+}, are also found. In brief, pegmatites are the products of geologic diffusion and extensive crystal growth. Undoubtedly temperatures were elevated; however, there is considerable evidence to indicate that the crystallization and grain growth occurred primarily as solid-state reactions or hydrothermal processes (Section 14–5). Although less common than sedimentary, igneous, or metamorphic rocks, pegmatites are important sources of raw material for ceramic purposes because they provide commercially usable concentrations of many of the less readily available elements (such as Li, Be, K, etc.).

We shall not attempt here to enumerate the geologic occurrences of all the other raw materials. In general, these other materials have been concentrated by one of the geological processes outlined above.

Processed raw materials. Geological processes, with their nearly infinite times and their widespread action, have supplied us with many usable raw materials. However, technological requirements as to quantity and composition have induced the engineer to beneficiate many of those available raw materials. Beneficiation and additional processing will, in fact, undoubtedly increase in importance in the future.

It is interesting to note that many of the processes are similar, whether they are slow geologic processes occurring in nature or the faster artificial processes. This is not surprising, since the same physical and chemical principles apply. The only significant difference is the limitation which man encounters on account of the shortness of time available for processing.

Mechanical separation is accomplished in a variety of ways. Once the raw material has been extracted from its natural source, it must be crushed and sized so that impurities can be removed. Since most ceramic raw materials are brittle, crushing does not present too formidable a problem. Even abrasive materials such as emery (Al_2O_3 with Fe_2O_3 in solid solution) may be crushed if care is given to provide wear-resistant equipment. For effective separation, the material must be crushed until the particles are smaller than the grains to be liberated. For this reason, it is impossible to use mechanical equipment to achieve separations such as the removal of adsorbed manganese ions from clay particles. However, the desirable, or undesirable, phase can be liberated in many coarser-grained raw materials.

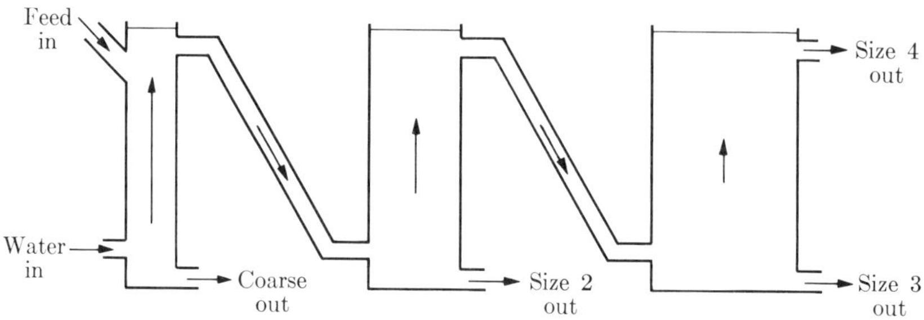

Fig. 13-1. Schematic drawing of the elutriation process. Crushed particles of various sizes settle at different velocities against the rising stream of water [Eq. (13-3)].

Once we have liberated the desired phase, we can utilize a variety of mechanical or physical operations. These include separation procedures such as elutriation, where particles of various gradations in size and weight settle against a rising water stream (Fig. 13-1). Magnetic separation is widely used when there are iron impurities; this process may require preliminary heating to transform the iron oxide to magnetite (Fe_3O_4). Flotation is increasing in use, particularly when minerals must be separated from coarse pegmatitic raw materials. Flotation works by the principle of surface wetting: a phase which is not wetted by a liquid is carried upward with a rising air bubble, while the particles that *are* wetted sink. Selective wetting and nonwetting can be achieved by minor additions of certain highly polarized molecules to the liquid (Section 5-4). These polar molecules are surface-active, and therefore can change selected surface energies drastically. To date, it is difficult to predict the flotation agent which will be most suitable in a given case, without resorting to experimentation. However, there are some general principles to guide us, so that we do not have to proceed by trial-and-error methods.

Example 13-1. A particle of anorthite ($CaAl_2Si_2O_8$, $\rho = 2.75$ gm/cm^3) settles in water of 20°C at the rate of 1μ/sec. What is the equivalent spherical diameter D of the particle?

Answer: For viscous flow,

$$V_m = (\rho_s - \rho_f)\, gD^2/18\eta, \tag{13-3}$$

where V_m is the maximum settling velocity, ρ_s and ρ_f are the densities of the solid and fluid respectively, and η is viscosity. From handbook data:

$$\begin{aligned} D &= \left[\frac{(10^{-4}\text{ cm/sec})(18)(0.01\text{ gm/cm}\cdot\text{sec})}{(981\text{ cm/sec}^2)(2.75 - 1.0\text{ gm/cm}^3)}\right]^{0.5} \\ &= 1.02 \times 10^{-4}\text{ cm} \\ &= 1.02\mu. \end{aligned}$$

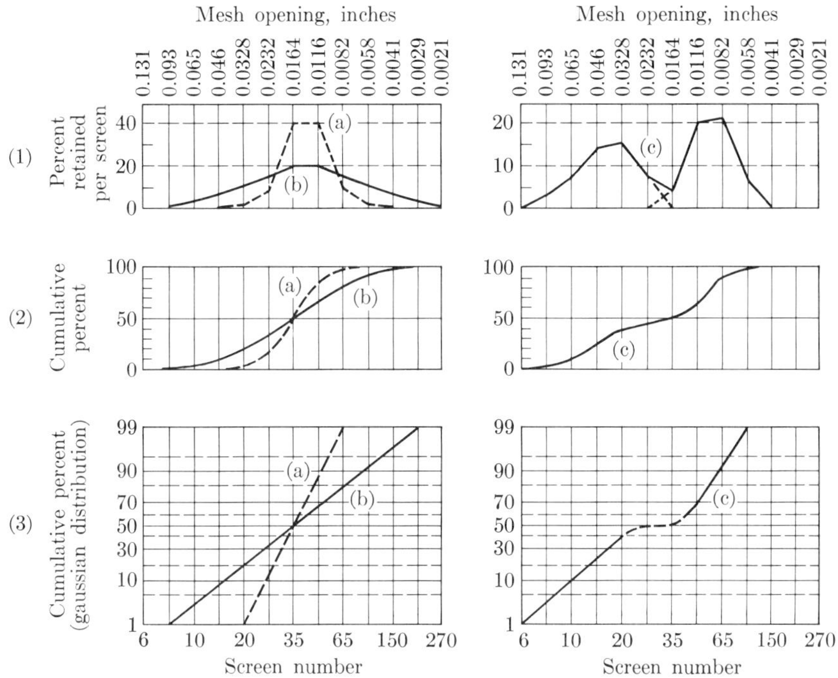

FIG. 13-2. Sizing distributions, in w/o: (a) closely sized, (b) broadly sized, (c) mixed sizing, equal weights. The three plots are (1) class frequency, (2) cumulative percentage (non-gaussian), and (3) cumulative percentage (gaussian).

We shall see in the next section that the sizes and size distributions of raw materials are quite important. The Tyler mesh sizes, used almost universally to classify raw materials, is a standard which presents an index of cross-sectional areas. These sizes progress downward on the scale on a basis of $\sqrt{2}$ (upper abscissa of Fig. 13-2). As sizes decrease below the lower limit of mechanical screening, they are indexed according to settling velocities, and these small sizes are described on the basis of equivalent spherical diameters.

Figure 13-2(a) and (b) show distribution plots for closely and for broadly sized materials, respectively. It is significant that if the crushing operation yields a statistical distribution of sizes it can be plotted as a straight line on a cumulative plot with a gaussian ordinate (see Fig. 13-7). Similarly, a bimodal mixture such as is frequently required in pressure fabrication appears as shown in Fig. 13-2(c).

We shall discuss *chemical leaching and precipitation* together. The purification of bauxite [$Al(OH)_3$] to form an Al_2O_3 raw material is an example which typifies the procedure. The seven general steps which enter into this process

TABLE 13-1

Step	Purpose
1. Dry and calcine alumina at the mine	Reduces weight for shipping Removes any organic material
2. Grind to −70 mesh	Accelerates the solution in Step 3
3. Dissolve (leach) in an NaOH solution [2–8 hours at 160°C (320°F) and 60 psi]	Forms soluble Na^+ and AlO_2^- ions, $Al(OH)_3 + NaOH \rightleftarrows AlO_2^- + Na^+ + 2H_2O$
4. Settle any precipitate and insoluble residue, and filter	Removes $Fe(OH)_3$ precipitate and SiO_2
5. Cool to reverse the reaction of Step 3. [Fresh $Al(OH)_3$ particles are added to nucleate further $Al(OH)_3$ precipitation.]	Precipitates $Al(OH)_3$
6. Filter and wash. (The NaOH solution is recycled to Step 3.)	Removes $Al(OH)_3$ and salvages NaOH
7. Calcine the $Al(OH)_3$ at 1150°C (2100°F)	Forms Al_2O_3

(called the Bayer process) involve solution and leaching as well as mechanical separation (see Table 13–1).

Thermal preparation may be an end in itself, or it may simply be a step in a more complicated procedure (e.g., Step 7 in Table 13–1). The calcining process was originated to attain the reaction:

$$CaCO_3 \rightarrow CaO + CO_2\uparrow . \qquad (13\text{–}4)$$

The raw material is dissociated into a crystallographically simpler solid and a gas. The term "calcination" is now used to describe any dissociation operation which releases gas as a product. As an example,

$$(OH)_4Al_2Si_2O_5 \rightarrow Al_2O_3 + 2SiO_2 + 2H_2O\uparrow \qquad (13\text{–}5)^*$$

is the calcination reaction for kaolinite [$(OH)_4Al_2Si_2O_5$]. Also, magnesite ($MgCO_3$) is calcined to produce periclase (MgO) refractories (Chapter 9).

* Again, a more complicated structure dissociates (incongruently) into simpler structures; x-ray evidence indicates that corundum (Al_2O_3) and amorphous silica (SiO_2) are the initial products. However, since these two phases are incompatible, they will react slowly at higher temperatures to form mullite ($Al_6Si_2O_{13}$).

Since cement is so important in our technology, special mention should be made of clinkering (Fig. 13-3), a thermal treatment which initially involves calcination, but which is carried far enough so there is chemical reaction. For example, the calcined CaO, Al_2O_3, and SiO_2 from limestone and clay react to form dicalcium silicate (Ca_2SiO_4), tricalcium silicate (Ca_3SiO_5), and tricalcium aluminate ($Ca_3Al_2O_6$), plus a number of other minor compounds. These ingredients hydrate in the forming operation at the construction site (Section 13-6). Phase diagrams such as Fig. 6-12 give an excellent guide to the high-temperature reactions encountered in clinkering.

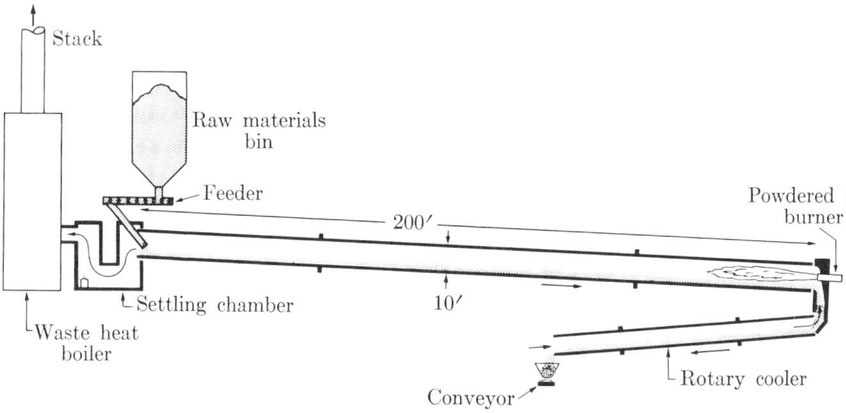

FIG. 13-3. Cross section of a rotary kiln for manufacturing portland cement clinker. (From Norton, F. H., *Elements of Ceramics*. Reading, Mass.: Addison-Wesley, 1952, page 211.)

Almost identical in purpose to calcination and clinkering, but different in name, is the presinter step used in preparing electrical and magnetic ceramics. Such materials as highly purified carbonates, hydroxides, etc., are mixed and heated until they react and homogenize the oxides into the desired phases. The "presinter," like the cement clinker, is then reground before forming it into the final shape desired for the product.

Another thermal homogenization process is the fritting operation, used to make the raw materials that go into glazes and vitreous enamels. The raw materials are heated above their melting temperatures, to form a liquid, then cooled. The resulting glass is crushed to a powder, so that it can later be applied as a surface coating on a ceramic or metal base. The final operation is firing (Section 14-3) to produce an impermeable surface.

Some raw materials are made coarsely crystalline by arc fusion, which results in crystallization accompanied by extensive grain growth. The abrasive industry (Chapter 8) utilizes this method to a considerable extent.

EXAMPLE 13-2. A porcelain body is to have the following weight composition: SiO_2, 59; Al_2O_3, 32; K_2O, 3; Na_2O, 1; and CaO, 5. The following are available as raw materials:

1. Quartz	SiO_2		
2. Kaolinite	$(OH)_4Al_2Si_2O_5$	or	$Al_2O_3 \cdot 2SiO_2 \cdot 2H_2O$
3. Potash feldspar	$KAlSi_3O_8$	or	$K_2O \cdot Al_2O_3 \cdot 6SiO_2$
4. Soda feldspar	$NaAlSi_3O_8$	or	$Na_2O \cdot Al_2O_3 \cdot 6SiO_2$
5. Calcite	$CaCO_3$	or	$CaO \cdot CO_2$

Assuming pure raw materials, indicate the number of pounds of each material needed to make 100 lb of body.

Answer:

$$\begin{aligned}
CaO &= 5 = X_5(56.1/100.1)\\
Na_2O &= 1 = X_4(62.1/524.6)\\
K_2O &= 3 = X_3(94.2/556.7)\\
Al_2O_3 &= 32 = X_4(101.9/524.6) + X_3(101.9/556.7) + X_2(101.9/258.1)\\
SiO_2 &= 59 = X_4(360.6/524.6) + X_3(360.6/556.7) + X_2(120.2/258.1) + X_1
\end{aligned}$$

Solving, we have

$$\begin{aligned}
X_5 &= 8.9 \text{ lb calcite}\\
X_4 &= 8.4 \text{ lb soda feldspar}\\
X_3 &= 17.7 \text{ lb potash feldspar}\\
X_2 &= [32 - (8.4)(101.9/524.6) - (17.7)(101.9/556.7)](258.1/101.9)\\
&= 69.0 \text{ lb kaolinite}\\
X_1 &= 59 - (8.4)(360.6/524.6) - (17.7)(360.6/556.7) - (69)(120.2/258.1)\\
&= 9.5 \text{ lb quartz}
\end{aligned}$$

13-3 Pressure fabrication

"Dry" pressing has become an important means of forming ceramic products. To describe the method briefly, we can say that particles of raw materials are compressed into as small a volume as possible by the application of pressure, with two results: particle packing and particle bonding. Packing is a juxtaposition of particles (aggregate) without coherence. Subsequent bonding of the particles is required to bring the particles into a coherent mass (agglomerate).

Particle packing. It is impossible to completely fill space with irregularly shaped particles. Therefore any aggregation of odd-shaped particles into a bulk implies porosity, or a packing factor of less than unity. The two are related as follows:

$$\text{Porosity} = 1 - \text{PF}. \tag{13-6}$$

The packing factor varies with (1) particle shape, (2) particle arrangement, and (3) size distribution. It does not vary with the size, *per se*, unless small containers with large surface-to-volume ratios are used.

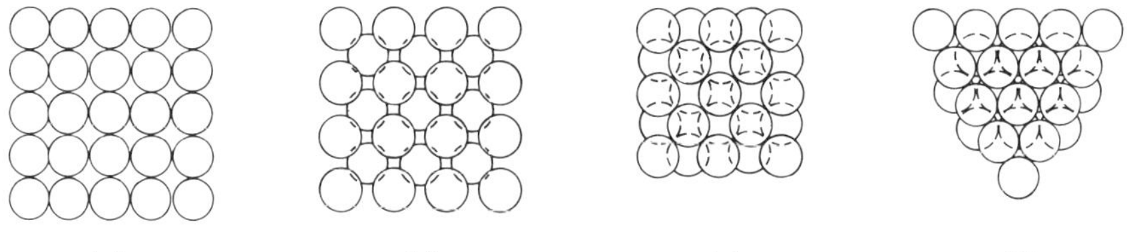

FIG. 13-4. Packing arrays: (a) Simple cubic, 0.52 PF. (b) Body-centered, 0.68 PF. (c) Face-centered: 0.74 PF. (d) Hexagonal close-packed, 0.74 PF.

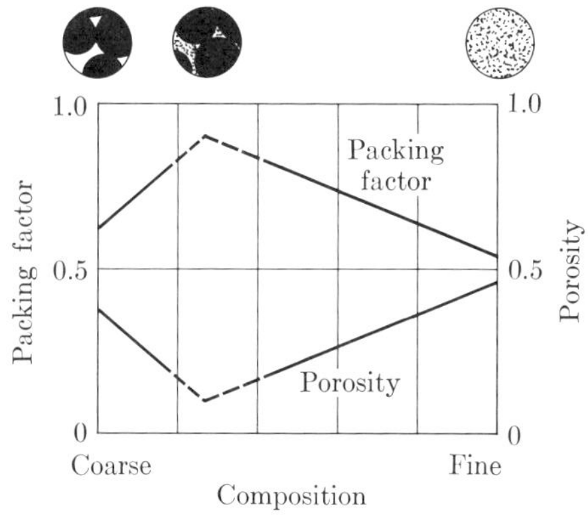

FIG. 13-5. Mixed sizing. The fine particles fill the pores in the coarse aggregate.

The simplest packing involves spheres in a simple cubic array (Fig. 13-4a). The packing factor of such a pattern is easily calculated to be $\pi/6$ or 0.52. Higher packing factors and therefore greater bulk densities may be achieved with body-centered or face-centered arrays (Fig. 13-4b and c). Here the packing factors are 0.68 and 0.74, respectively. The latter is the highest possible packing factor for uniformly sized spheres, and also applies to the hexagonal close-packed array (Fig. 13-4d). A reader who tries to verify these figures for packing factors will realize that packing factors are independent of the sphere diameter.

Other shapes have other packing factors, which may be higher if the units are correctly aligned. For example, cylindrical rods can have a packing factor as high as 0.91, and brick may have a PF of nearly 100%.

The packing factor may be increased markedly by mixed sizing, or grading. This is illustrated in Fig. 13-5, showing that small particles can locate themselves in the interstices among large particles. The concrete producer is highly aware of this fact, and of course utilizes sand/gravel mixtures to fill space more effectively. For the highest packing factor and the lowest porosity, the mixture should contain a coarse/fine bulk volume ratio of 1/(porosity of the coarse fraction). In other words, there must be just enough fines to fill the pore spaces among the coarse aggregate particles. This assumes comparable shape factors for the two fractions. It also assumes that the size ratio between the two is sufficiently high so that there are negligible edge effects at the boundaries between the two sizes.

Edge effects reduce the packing factor because the particles cannot fill space efficiently. However, these effects are usually less significant at the external surface of the product than at the boundary between coarse and fine particles, as illustrated in Fig. 13-6. The reduction in the packing factor, ΔPF, is a function of the diameter d of the small particles and the ratio of the surface area to the volume of the large particles. Thus the following relationship is appropriate:

$$\Delta PF = kd/D, \tag{13-7}$$

so that bulk density is harder to achieve when the large-particle diameter D and the small-particle diameter d are comparable. In typical aggregates k, the proportionality constant, is approximately -2 when the size ratio of the two aggregates is greater than 5/1.

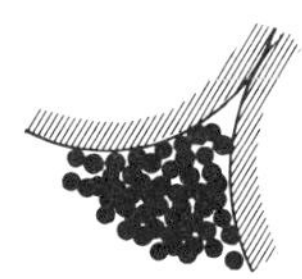

FIG. 13-6. Schematic drawing showing edge effects. Small particles cannot be packed efficiently adjacent to large particles. The width of the lower-density boundary zone is dependent on the size of the fine aggregate. The area of the boundary is related to the size of the coarse aggregate.

EXAMPLE 13-3. Coarse and fine fireclay grog are mixed, with the following characteristics.

	Fine grog	Coarse grog
Mean diameter	0.01 in.	0.25 in.
Bulk density	100 lb/ft^3	105 lb/ft^3
Specific gravity	2.65	2.7

Estimate the maximum possible packing factor.

Answer: Use as a basis 1 ft^3, with the fine grog filling the pores of the coarse grog, and figure the packing factor of the coarse grog as

$$PF_c = \frac{105 \text{ lb/ft}^3}{(2.7)(62.4 \text{ lb/ft}^3)} = 0.623.$$

The packing factor of the fine grog is

$$PF_f = \frac{100 \text{ lb/ft}^3}{(2.65)(62.4)} = 0.605.$$

$$\text{Maximum true volume} = \text{PF mix} = 0.623 + (0.605)(1.0 - 0.623) = 0.85.$$

The edge correction is

$$\Delta PF \approx (-2)(0.01)/(0.25) = -0.08.$$

$$\text{Maximum PF} \approx 0.77.$$

As already indicated in Eq. (13–6), porosity is the complement of the packing factor. There are two types of pores, open and closed. Unless it contains internal voids, an aggregate has only open pores. During firing or sintering processes (Section 14–3), there is often a closure of pores because firing or sintering reduces the total surface area. Also, when concrete is poured specific attempts are made to entrain air, so that internal (or closed) pores remain after setting.

The apparent porosity (that is, the open porosity) can be determined by the absorption method. Usually this involves measuring the weight of a certain volume of a material both dry and saturated. If *water* is used to fill the voids,

$$\frac{\%\ \text{porosity}}{100} = \frac{\text{saturated weight} - \text{dry weight}}{\text{bulk volume}}. \tag{13–8a}$$

When other liquids are used in order to avoid solution or slaking (i.e., disintegration), appropriate specific-gravity corrections must be made. When bonding has taken place, the volume of irregularly shaped pieces may be obtained by buoyancy measurements, and the following equation applies:

$$\frac{\%\ \text{porosity}}{100} = \frac{\text{saturated weight} - \text{dry weight}}{\text{saturated weight} - \text{suspended weight}}. \tag{13–8b}$$

Three types of density figures (and also specific gravities) are encountered. *True density*, as the name implies, is the mass per true volume; whereas *bulk density* is mass per total volume, which includes all pores as well as occupied space. *Apparent density* is mass per apparent volume, which includes actual material and closed pores. The sum of the apparent volume and the apparent pore space equals the total or bulk volume.

EXAMPLE 13–4. An insulating brick (true specific gravity = 2.58) weighs 3.90 lb dry, 4.77 lb saturated in kerosene, and 2.59 lb suspended in kerosene (specific gravity = 0.82). (a) What is the apparent porosity? (b) Apparent density? (c) Bulk specific gravity? (d) Closed porosity?

Answer:

$$\begin{aligned}
\text{True volume} &= (3.90\ \text{lb})/[(2.58)(62.4\ \text{lb/ft}^3)] &&= 0.0242\ \text{ft}^3.\\
\text{Apparent pore volume} &= (4.77 - 3.9)/[(0.82)(62.4)] &&= 0.0170\ \text{ft}^3.\\
\text{Bulk volume} &= (4.77 - 2.59)/[(0.82)(62.4)] &&= 0.0426\ \text{ft}^3.\\
\text{Apparent volume} &= (3.9 - 2.59)/[(0.82)(62.4)] &&= 0.0256\ \text{ft}^3.
\end{aligned}$$

(a) Apparent porosity $= 0.0170/0.0426 = 39.9\%$.

(b) Apparent density $= 3.9/0.0256 = 152.3\ \text{lbs/ft}^3$.

(c) Bulk specific gravity $= 3.9/(0.0426)(62.4) = 1.47$.

(d) Closed porosity $= (0.0256 - 0.0242)/0.0426 = 3.3\%$.

If the range of size distribution and other factors are equal, a change in the median size of the particles will not affect the porosity of an aggregation, but *will* affect the permeability. The net velocity v of gas or liquid flow through a porous aggregation is

$$v = kD_p^2(-\Delta P)F_p/\eta L \tag{13-9a}$$

where k is a proportionality constant, D_p is a characteristic particle dimension such as diameter or mesh size, η is the fluid viscosity, $-\Delta P/L$ is the pressure gradient, and F_p is a packing coefficient which varies with the apparent porosity and particle sphericity. If either porosity or sphericity is decreased, F_p decreases. Thus we note that the particle size, as a result of its second-power relationship, is highly significant.

In Eq. (13-9a) k, D_p^2, and F_p are sometimes combined into a permeability coefficient, K. Thus,

$$v = K(-\Delta P)/\eta L. \tag{13-9b}$$

The size of the particles is a result of the treatment given to the raw material (Section 13-2). For coarsely aggregated products, crushing and screening are used. For a mixture containing a certain percentage of particles in a number of different sizes, the best procedure is separation of the various sizes, followed by a reassemblage, so that the mixture contains the desired proportions of each component size. As many as three or more size fractions may be used. When finely divided particles are required, so that the end product will have low permeability or high strength, broadly sized aggregations are used. In such a case, there are interstices of widely varied sizes, which are filled with a gradation of particles in many sizes.

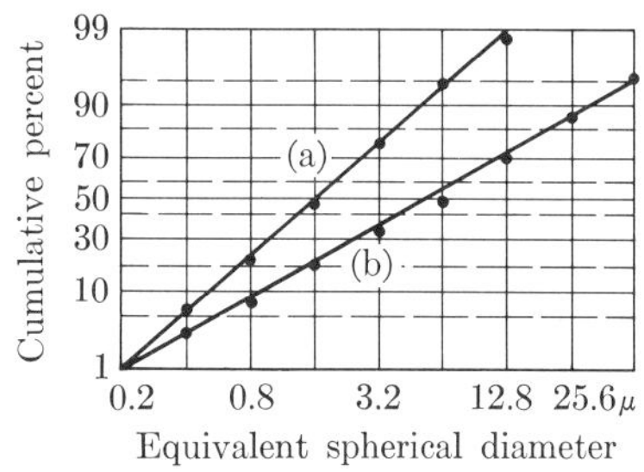

FIG. 13-7. Diagram of sizing control of ball-milled alumina by grinding: (a) wet grinding; (b) dry grinding. (Adapted from non-gaussian plots by Kingery, W. D., "Pressure Forming of Ceramics," *Ceramic Fabrication Processes*. New York: John Wiley & Sons, 1958, page 56.)

Ball-milling is a procedure commonly used for producing fine sizes. This rotating mill contains a wear-resistant lining and a charge of wear-resistant balls which provide grinding surfaces. Figure 13-7 shows variations in the size distribution of ground particles of alumina as a result of the milling procedure. The coarser, larger-sized particles from dry grinding probably result from dust suspensions within the mill. During wet grinding, the particles are brought between the grinding surfaces more frequently. In each case a statistical distribution is obtained, as indicated by the straight line on the gaussian plot.

Pressed density. The object of pressure fabrication (or vibratory compaction) is to force the aggregate particles into closer contact and therefore increase the

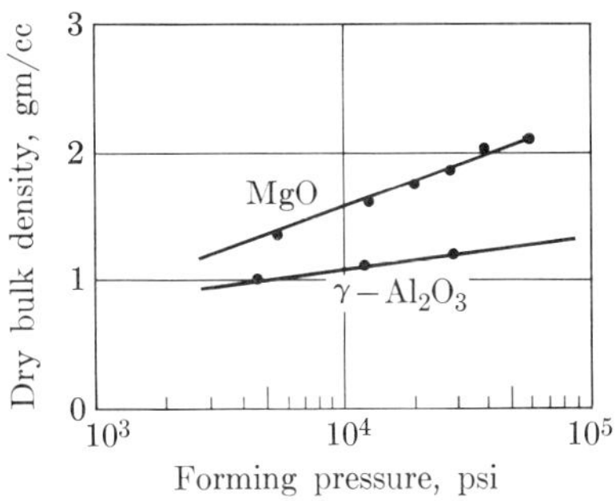

FIG. 13-8. Bulk density versus forming pressure. (Adapted from Kingery, W. D., *Ceramic Fabrication Processes*. New York: John Wiley & Sons, 1958, page 60.)

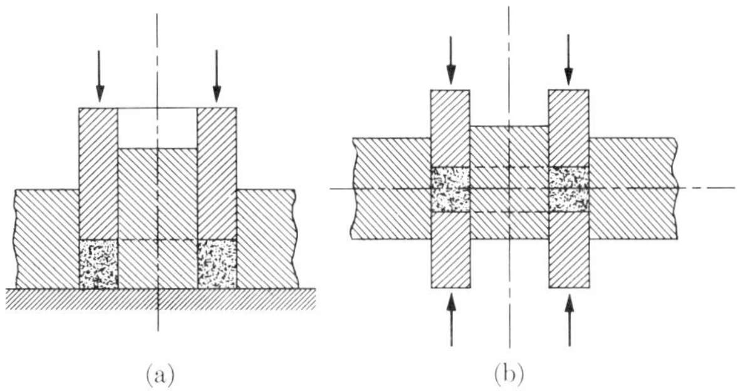

FIG. 13-9. Pressure fabrication of a magnetic toroid: (a) single-end pressing; (b) double-end pressing.

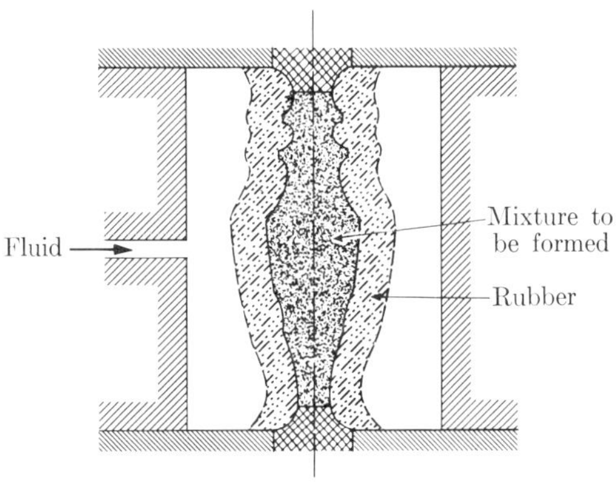

FIG. 13-10. Schematic drawing of hydrostatic molding of a spark plug. In the actual process, the design of the mold is somewhat more complicated so that the pressure distribution can be accurately controlled. (See Jeffery, B. A., U. S. Patent 1,863,854, Champion Spark Plug Co., June 21, 1932.)

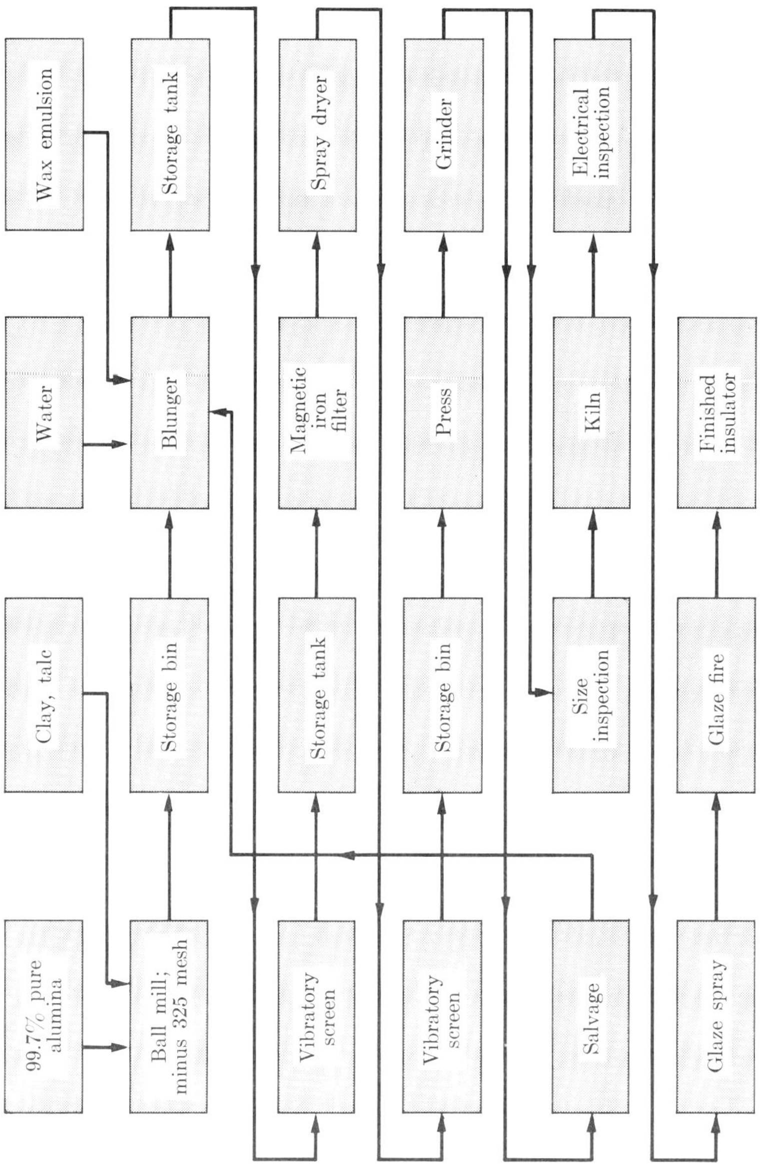

FIG. 13–11. Flow sheet for spark-plug manufacture. (Courtesy A. C. Spark Plug Company.)

density for better bonding and less shrinkage in subsequent firing processes. The bulk density ρ_B increases from the initial bulk density ρ_i with the log of the pressure P,

$$\rho_B = \rho_i + A \log P. \tag{13-10}$$

The constant A is empirical, but is probably related to friability in the cold-pressing of most ceramic materials because much of the increase in density results from crushing of the brittle particles. From Fig. 13-8, which shows data for MgO and Al_2O_3, it is apparent that excessive pressures would be required to even begin to approach the true density; therefore, sintering or some similar treatment is required to reduce the porosity to desirable levels (Chapter 14).

The two main procedures for applying pressure are sketched in Figs. 13-9 and 13-10. The directional pressing of Fig. 13-9 is simpler in operation than the hydrostatic molding, but is limited to simple shapes which have equal compaction ratios in all longitudinal sections. The hydrostatic (or isostatic) molding of Fig. 13-10 is more complicated, but is more popular because there is negligible wall friction to produce pressure gradients. This procedure has been used extensively in spark-plug manufacture (Fig. 13-11).

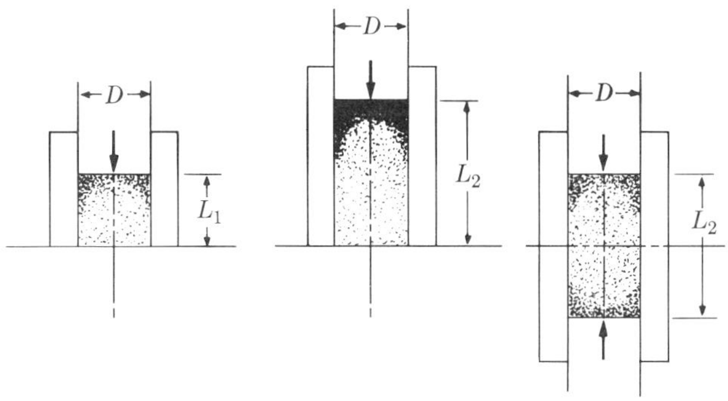

FIG. 13-12. Bulk density distribution: (a) $L/D = 0.9$, (b) $L/D = 1.8$, (c) $L/D = 1.8$, double-end pressing.

The importance of wall friction is shown in Fig. 13-12, where bulk densities are shaded differently for various parts of the compact. Because part of the compressive load is transferred to the mold wall, larger length/diameter ratios (L/D) provide greater variations in applied pressures (and therefore differing densities resulting after pressure). This was demonstrated in two ways by Duwez and Zwell.* (1) If the manner of loading is that shown in Fig. 13-12 (a) and (b), the pressure ratio between the upper (moving) piston and the lower

* Duwez, P., and L. Zwell, "Pressure Distribution in Compacting Metal Powders," *Trans. AIME*, **185**, 137-144 (1949).

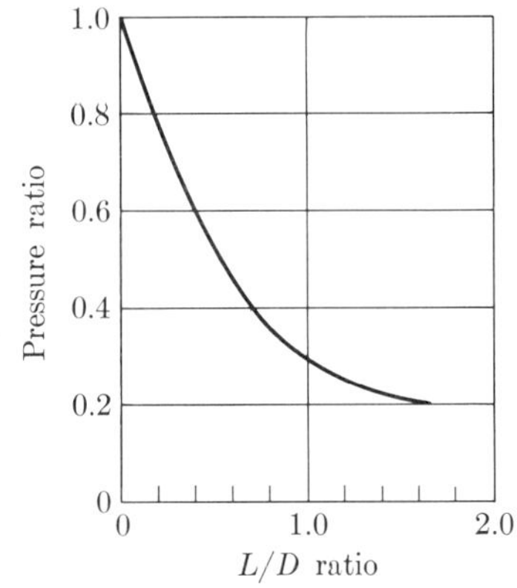

FIG. 13-13. Pressure ratio versus length/diameter ratio. The pressure ratio is the ratio between the pressure on the stationary piston and the pressure applied through the compacting piston. Applied pressure = 100,000 psi. (After Duwez, P., and L. Zwell, "Pressure Distribution in Compacting Metal Powders," *Trans. AIME*, **185,** 137–144, 1949.)

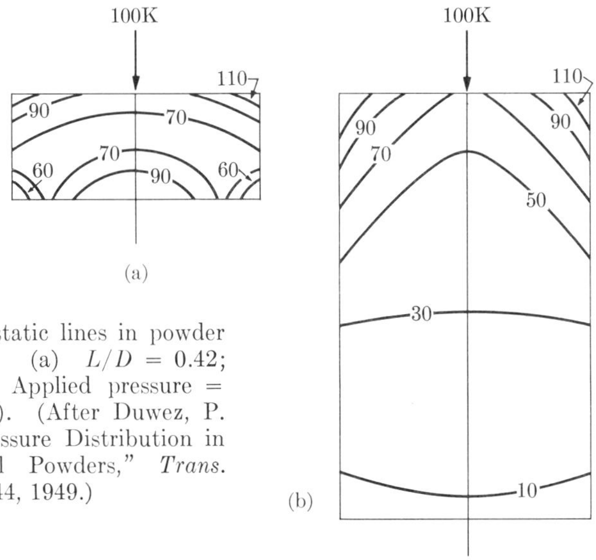

FIG. 13-14. Isostatic lines in powder metal compacts: (a) $L/D = 0.42$; (b) $L/D = 1.66$. Applied pressure = 100,000 psi (100 K). (After Duwez, P. and L. Zwell, "Pressure Distribution in Compacting Metal Powders," *Trans. AIME*, **185,** 137–144, 1949.)

(fixed or supporting) piston increases with an increase in the L/D ratio (Fig. 13–13). (2) The pressure distribution within the compacts ranges as indicated in Fig. 13–14. Each demonstration of pressure loss to the wall emphasizes that bulk density differences in the aggregate do result in subsequent differences in shrinkage and strength of the product.

Common procedures used to avoid the above heterogeneities are (1) double-end pressing (Fig. 13–12c) and (2) die lubricants. Concerning lubricants, Kamm† and his coworkers have shown that if the die wall is lubricated, pressure and packing homogeneity is obtained with single-end pressing, even though the

† Kamm, R., M.A. Steinberg, and J. Wulff, "Lead Grid Study of Metal Powder Compaction," *Trans. AIME*, **185**, 694–706 (1949).

particle surfaces are not lubricated. However, if there are lateral movements among the particles (as there would be with an irregular shape), the particle surfaces must also be lubricated.

A third procedure, vibratory compaction, may be used to avoid the bridging and frictional difficulties cited above. This principle is well known, and is widely used in the forming of large concrete structures, where vibration is applied directly. However, the whole mold is vibrated when smaller shapes are formed. Vibratory compaction utilizes kinetic energy rather than pressure as the mechanism for overcoming the resistance to compaction. In general, engineers choose vibrational amplitudes comparable to the particle dimensions. Larger amplitudes disrupt compaction which has already been accomplished. When fine particles are used, it is usually desirable to generate the vibrations by electrical or magnetic transducers instead of by mechanical means. The number of vibration cycles, and not the vibrational time *per se*, is important. Therefore, it is desirable to use relatively high frequencies, particularly when compacting fine powders which can quickly receive inertia from the vibration. It is interesting that resilient rubber molds resist the wear of abrasive vibrating particles better than the harder metal molds. As yet, the use of vibratory compaction has attained only limited acceptance in ceramic production.

Hot pressing. Greater densification occurs in the pressing of oxides and carbides when pressing is carried out at high temperatures, probably because of the increased slip and ductility of materials at high temperatures. Densification by hot pressing is in contrast to densification by firing or sintering (Section 14–3), where diffusion and grain growth reduce porosity. In general, hot pressing requires temperatures of greater than $\frac{2}{3}$ of the absolute melting temperature. Since many of the oxides and carbides have a high melting temperature, this procedure requires special molds and heating techniques. Graphite is the material most widely used for molds, but graphite molds are costly because of their rapid wear.

Particle alignment. Certain ceramic products can be made to have better properties if their particles are specifically aligned prior to pressing. For example, magnetic ceramics provide greater magnetic permeability if the easy direction of magnetization is aligned with the direction of the magnetic field (Section 10–6). Therefore, pressing is sometimes performed within a magnetic field. Likewise, higher dielectric constants may be developed in ferroelectric components which are pressed in an electric field, so that the individual particles are most favorably oriented.

13–4 Hydroplastic forming

Ceramic raw materials with layered structures are amenable to plastic forming because they provide a paste with a low initial shear strength. There are several industrial modifications of plastic forming. These modifications include repressing, which is equivalent to the procedures described in the previous section, but which utilizes larger amounts of liquid as a plasticizer. Extrusion is also widely used, for products which have regular cross-sections; this method finds use for large-tonnage items such as porcelain and various construction materials. Jiggering, which involves a rotating mold to form one surface of the product, is used for products such as tablewares which have axes of circular symmetry. The opposite surface is formed when the plastic material rotates under a contoured tool. Other, relatively less important, methods of plastic forming include hand pressing, throwing, and certain techniques used by artists working with ceramics.

In general, plastic forming requires a fine-particle paste with sufficient yield strength so that the formed product will maintain its shape during handling and during the period of time prior to drying (Section 14–2). Excessive yield strengths mean that there are greater opportunities for rupture. Yield strength is most readily controlled by regulating the amount of water or other plasticizer which is used (Table 8–7). It is further influenced by the structure of the colloidal paste (Section 8–8). During plastic forming, the layered colloidal particles are oriented by the shear stresses. In some respects, this may be compared to the preferred orientation developed in metals during metalworking operations. In each case, a critical shear stress is exceeded (Section 8–6). A direct comparison of ceramic materials with cubic metals cannot be made, because these metals have several equivalent planes on which shear can occur. In contrast, the plastic forming of two-dimensional ceramic particles must be compared more directly with the plastic deformation of zinc, magnesium, or other metals with hexagonal close-packed structures that possess only one major slip plane. Because a preferred orientation is developed, it is to be expected that anisotropic behavior can occur with respect to drying shrinkage and to various properties.

The reader can gather, from the emphasis on two-dimensional particles in this section, that plastic forming is most widely used with products containing clays (Fig. 3–17), and related structures (Table 3–5). However, other colloidal materials can be plastically formed, although more difficulty is encountered when the "ready-made" planes, present in clays, are missing. In no case does slip occur between the primary bonds within the crystal structure, as it does in metals. Ceramic bodies usually contain mixtures of clay and nonclaylike particles. As we shall see in Chapter 14, these mixtures frequently have better dimensional stability and mechanical strength than the pure-clay ceramic bodies.

A practical problem may arise during plastic forming if there are any air pockets in the original paste. These pockets are compressed by the forming pressures, but after the pressures are removed, they reopen to form internal fissures or laminations, which follow the shear stress patterns that were developed. These air pockets must be avoided if strength is required in the finished product. Current technology avoids this problem by deairing. In extrusion, this is accomplished by reducing the pressure to about 30mm Hg in the pugmill operation. Similarly, in dry pressing operations, vacuum techniques are applied.

13-5 Slip-casting

Slip-casting is a long-established process which utilizes a colloidal suspension. The suspension, also called a slip, is dewatered by means of a porous mold. After the formation of a shell deposit of a desirable thickness, the remaining slip is drained, the shell dewatered still further, and the mold removed. At this point the green ware (unfired product) has properties somewhat like a low-water colloidal paste.

The primary advantage of slip-casting is that rather intricate shapes may be formed by using this method. Further, it has economic advantages for limited production requirements, which makes it a favorite forming operation for the ceramic artist. It also has industrial applications when thin-walled products are to be made from nonplastic raw materials, as in the case of high-alumina combustion and thermocouple tubes.

We can discuss the principles governing slip-casting most cogently if we compare the process to filtration. In both cases there are pressure differentials which cause diffusion of water through the accumulating solids. Filtration utilizes either a positive pressure on the suspension side, or a partial vacuum on the exit side (or both). The pressure differential in slip-casting results from the capillary action which draws the water into the mold.

The depositing shell increases in thickness at a rate proportional to the velocity of water that flows through it. This may be shown* to be approximately

$$dL/dt = -\Delta P/2CL, \tag{13-11}$$

and

$$t = CL^2/-\Delta P. \tag{13-12}$$

In these equations, t is time, L is the thickness of the deposit, and ΔP is the pressure drop. The constant C includes liquid viscosity, liquid content of the slurry, the packing factor within the deposit, and the reciprocal of the permeability coefficient [Eq. (13-9b)]. Therefore attention must be given to (1) the nature of the colloidal suspension, (2) the structure of the deposit or shell, and (3) the absorption behavior of the mold.

A colloidal suspension with a high liquid content forms a deposit slowly because considerable water must be filtered through the deposit, and the absorption gradient in the mold must be reduced. Conversely, a low-viscosity slip is required to ensure filling of the mold contours and to avoid the entrapment of air bubbles. It is therefore apparent that deflocculation is important in order for there to be maximum fluidity with minimum liquid content. One means of

---

* See Brown, G. G., *et al.*, *Unit Operations*. New York: John Wiley & Sons, 1950, pp. 242-243. This calculation assumes a noncompressible deposit. Since this is not strictly true, the thickness of the deposit will accumulate at a slower rate than is implied by Eq. (13-11).

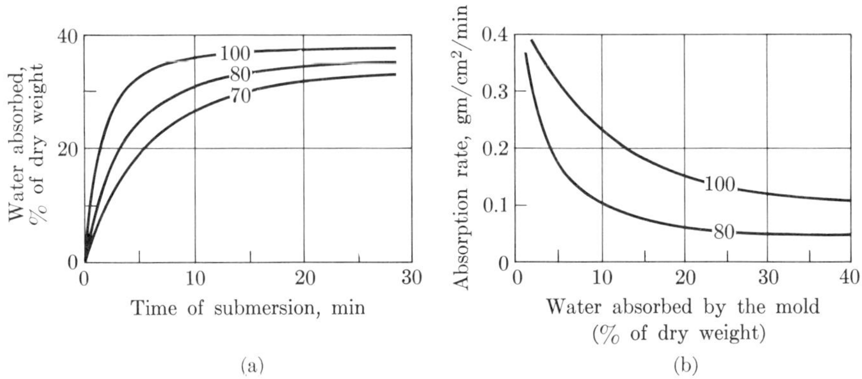

FIG. 13-15. Absorption rate versus water absorbed. The numbers 70, 80, and 100 indicate the pounds of water added to 100 lb of gypsum plaster in the original mix. More water produces a more porous mold and therefore faster dewatering of the suspension in slip casting. (After Lambe, C. M., "Preparation and Use of Plaster Molds," *Ceramic Fabrication Processes*. New York: John Wiley & Sons, 1958, pages 31-40.)

control is through additions of sodium silicate or sodium carbonate to adjust the pH (Fig. 8-12).

Equiaxed particles (as contrasted to layered or two-dimensional particles) form a relatively permeable deposit. It is for this reason, among others, that it is desirable to include crushed silica or other raw materials in slips, if they are not already present as impurities in the clay. Without the addition of such materials, a highly oriented nonpermeable deposit would form (Section 7-6).

Molds. Many compositions are potential materials for molds, but gypsum plaster (plaster of Paris) is used almost universally. Its advantages are: (1) relatively low cost, (2) simple procedures for producing molds from master patterns, (3) the fact that its porosity may be controlled, and (4) the fact that it may be dried and reused. Only the third item will be given further attention here. The porosity of a plaster mold, and hence its ability to dewater slip, is in proportion to the original water content of the plaster mix (Fig. 13-15a). The strength of a mold is of course decreased by high absorptivity; however, the rate of absorption is rapidly increased when there is greater pore space within the mold (Fig. 13-15b).

13-6 Cementation

The forming processes described above require enough agglomeration for the product to maintain its shape and dimensions until it is sintered and gains its useful strength. This early coherency depends on the electrostatic attraction between colloidal particles, or on intentional additions of minor amounts of waxes, oil, water, or other "fugitive" binders.

In contrast to these temporary binders, cements are materials which directly agglomerate aggregates into their final usable product. A chemical reaction is always involved, and chemical bonds are formed. Although cements may be either organic or inorganic, we shall discuss only the latter.

Physical and structural requirements. The cement must first of all have cohesive strength within itself. Probably equally important, the cement must adhere to the aggregate. Wygant* summarizes the structural requirements of cements by pointing out that the best ceramic cements are those whose products are *not* highly crystalline. Although this seems surprising at first, there are theoretical justifications for the experimental evidence. (1) An amorphous or finely colloidal cement is more likely to match the structure of the aggregate and develop good adhesion than a highly crystalline cement is. (2) Interatomic variations do not produce long-range boundary mismatches. (3) A variation of stoichiometry is permitted in an amorphous material. (4) An amorphous structure is somewhat flexible, so that it does not develop severe stress concentrations when loads or dimensional changes are encountered.

Wygant also notes that the best cements contain glass-forming oxides, and Zachariasen's rules for glass formation (Section 4-3) are usable for predicting the best cement compositions. Of course, this follows from the above discussion, because glass has a noncrystalline structure. We therefore find SiO_2 and P_2O_5 among the more common cement oxides; Al_2O_3 is also encountered in both glasses and cements. Further, the network modifiers of glass, such as Na_2O, CaO, and BaO, are also utilized in cements. It is true that a cement makes less use of primary ionic and covalent bonds than does a glass, partly because of the fact that cements are not melted, but depend on water or other ambient liquids to attain a workable consistency. Therefore, primary bonds in a cement can occur only when atoms from the cement are chemically adsorbed onto the aggregate surface. The van der Waals bonds of the stronger cements utilize hydrogen (and hydroxyl) bridging and do not depend on the weaker dipole moments as a source of surface adherence.

The chemical reactions which produce ceramic cements fall into two main categories: (1) water-releasing and (2) water-consuming. Although CO_2 and other reactants can be used in cements, water is by far the most common additive, and a cement made with water illustrates the chemical behavior of cementitious materials. The hydraulic cements, which are among those with water-*consuming* reactions, include not only the portland cements of highway and building construction, but also gypsum plaster. The water-*releasing* reactions may involve polymerization by condensation (silicic acid, for example). Also, water is released by reaction-type cements, as in the case of two compounds such as Al_2O_3 and phosphoric acid, which react to give aluminum phosphate and water.

---

* Wygant, J. F., "Cementitious Bonding in Ceramic Fabrication," *Ceramic Fabrication Processes*. New York: John Wiley & Sons, 1958, pp. 171–188.

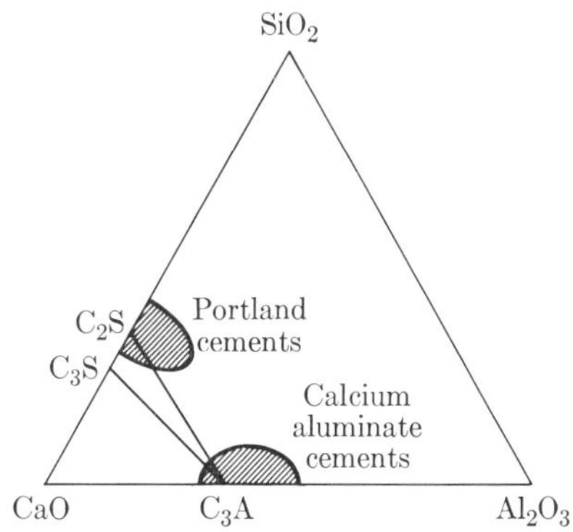

FIG. 13-16. Cement compositions. Compare with Fig. 6-12 for phase details.

Hydraulic cements. The portland cements are the most common of the hydraulic cements (Fig. 13-16). Their chief ingredients are di- and tricalcium silicates and tricalcium aluminate. Their hydration reactions, although somewhat complicated, are indicative of the principles governing the reactions of hydraulic cements in general:

$$Ca_3Al_2O_6 + 6H_2O \rightleftarrows Ca_3Al_2(OH)_{12}, \tag{13-13}$$

$$2Ca_2SiO_4 + (5 - y + x)H_2O \rightleftarrows Ca_2[SiO_2(OH)_2]_2 \cdot (CaO)_{y-1} \cdot xH_2O + (3 - y)Ca(OH)_2, \tag{13-14}$$

$$2Ca_3SiO_5 + (5 + x)H_2O \rightleftarrows Ca_2[SiO_2(OH)_2]_2 \cdot (CaO) \cdot xH_2O + 3Ca(OH)_2. \tag{13-15}$$

In Eq. (13-14), y is approximately 2.3. The value of x in the hydrated calcium silicate (called tobermorite) varies from zero to more than one, depending on the vapor pressure of the available water. In each case the hydrated form is slightly less soluble than the anhydrous phases. Therefore solution and precipitation occur through the liquid medium.

Tobermorite (hydrated calcium silicate) is typical of materials which do not maintain highly ordered crystal structures. That tobermorite has a variable composition is verified by x-ray diffraction and by electron microscope analyses which show that the tobermorite is in a colloidal gel form.*

Hydrated tricalcium aluminate, on the other hand, has a structure which is more regular and crystalline. Other contrasts between these two structures are their rates of setting, their ultimate strengths, and their dehydration when exposed to heat (Table 13-2). Hydration, as reflected by the length of time taken to reach 30% of strength, occurs comparatively rapidly in the more crystalline calcium aluminate. Calcium aluminate also loses its water of hydration at a

* Brunauer, S., "Tobermorite Gel—The Heart of Concrete," *Am. Sci.*, **50**, 210-229 (1962).

TABLE 13-2

CEMENT PROPERTIES IN RELATION TO CRYSTALLINITY

(Collected from several sources, but primarily from Bogue, R. H., and W. Lerch, *P.C.A.F. Reports*, 1936–1945.)

Characteristics	Hydrated tricalcium aluminate	Hydrated dicalcium silicate
Structure	Crystalline	Colloidal and nearly amorphous
Approximate strength of paste (after 1 year)	1000 psi	10,000 psi
Time to achieve 30% strength	12 hours	60 days
Temperature range of dehydration	550–580 °F	200–1000 °F

rather specific temperature 290–305°C (550–580°F). The dicalcium silicate hydrates to greater strength (but does so more slowly) by producing a structure with a wide range of interatomic bond strengths.

Commercial cements which are to be of the "high early strength" type may have relatively large quantities of tricalcium aluminate added in order to capitalize on the latter material's initially rapid hydration kinetics. In this case, maximum strength is sacrificed. An alternative means of accelerating the rate of strength attainment is finer grinding of the cement. Since the setting operation is one of solution and reprecipitation, setting is a factor related to surface area. If the average particle size can be reduced by a factor of two, the surface area and hydration reaction increase by a factor of four.

The dehydration of a hydraulic cement is comparable to the removal of water from other colloids (Section 6–2). The free energy of the water to be removed varies with the actual bond strength, and influences the dehydration temperature. Since a cement with large amounts of dicalcium silicate involves a dehydration range starting at relatively low temperatures, it is not usable to the same temperature level as a hydrated tricalcium aluminate cement.

The moisture content of a colloidal cement varies with the humidity of the environment. Furthermore, any change in moisture content produces a volume change. Theoretically, the volume change is reversible, but in practice, reversibility is not complete because volume expansions occur somewhat unevenly, and this produces local displacements that are not fully recoverable. As Fig. 13–17 shows, there is a net dimensional growth in colloidal cements that may be compared to thermal growth in polycrystalline materials which have anisotropic thermal expansions (Section 9-2).

Polymeric inorganic cements. The prototype reaction for this type of cement is silicic acid:

$$x\mathrm{Si(OH)_4} \rightarrow \left(\begin{array}{c} \mathrm{OH} \\ | \\ -\mathrm{Si} \\ | \\ \mathrm{OH} \end{array} - \mathrm{O} - \right)_x + x\mathrm{H_2O}. \tag{13-16}$$

Other polymeric cements are the methyl and ethyl orthosilicates, as well as comparable titanium and zirconium compounds. The resulting structures are those described in Section 1-7.

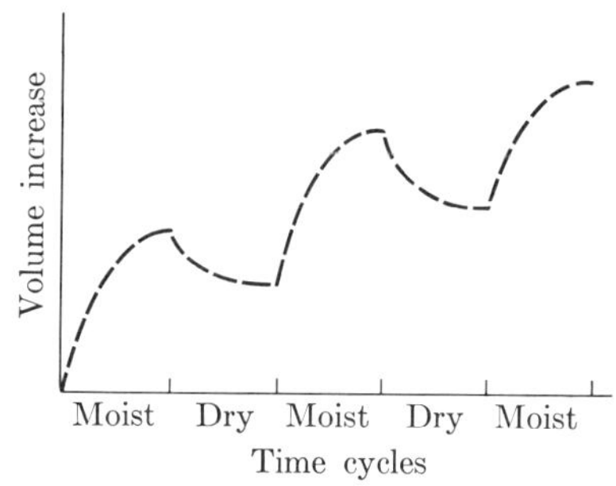

FIG. 13-17. Growth in colloidal cements. Nonuniform adsorption produces volume changes which are not completely reversible in the return cycle.

For the most part, these cements require the existence of a small atom coordinated tetrahedrally with larger oxygen atoms. Both ionic and covalent bondings are common. In addition, polymeric reactions are encountered in sodium and potassium silicates, where the sodium ion serves as an ionic bridge between one polysilicate anion and another. These react with many aggregate materials in the same way that reaction cements behave (see below).

Polymeric cements are often used as cold-setting cements for high-temperature applications. This is possible because the water from the reaction is released during setting. Once this water is removed by drying, the reaction is not reversible. As sintering temperatures are reached, the cement forms a low-melting liquid and a vitreous bond.

Reaction cements. Phosphoric acid reacts with many oxides to produce metal and acid phosphates. One such reaction will serve as an illustration:

$$\mathrm{Al_2O_3 + 2H_3PO_4 \rightarrow 2AlPO_4 + 3H_2O}. \tag{13-17}$$

Other oxides which produce similar reactions include Fe_2O_3, ZnO, PbO, CaO, CdO, etc. We can compare $AlPO_4$ and SiO_2 in properties and behavior, especially as regards their crystal structures, which are analogous (Fig. 13-18). [$AlPO_4$ may be considered an isomorph of $SiSiO_4$ (that is, SiO_2) in which an SiSi pair is replaced with AlP.]

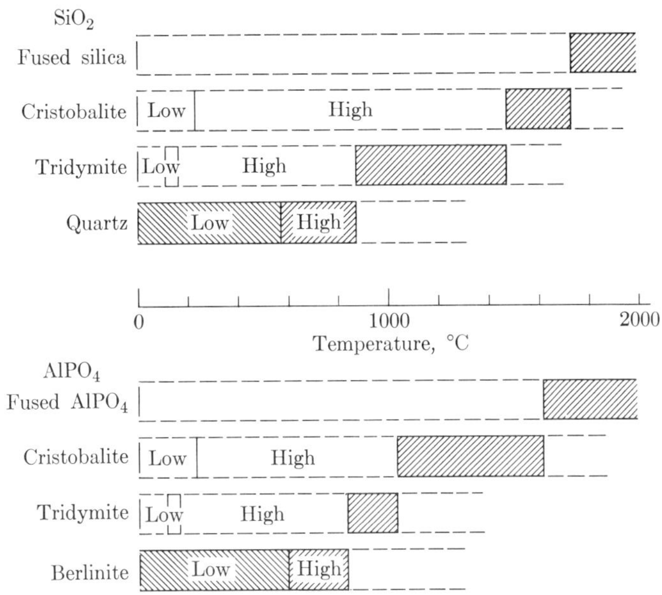

FIG. 13-18. $AlPO_4$ versus SiO_2. The $AlPO_4$ may be compared structurally to SiO_2 (or $SiSiO_4$). The stable temperature ranges are indicated by solid outlines. (Data from Beck, W. R., "Crystallographic Inversions of the Aluminum Orthophosphate Polymorphs and Their Relation to Those of Silica," *Journ. Amer. Ceram. Soc.*, **32,** 147-151, 1949.)

The above reaction (13-17) is particularly interesting because it yields an air-setting cement which is very refractory in behavior. When Al_2O_3 is used as the aggregate as well as a reactant, the resulting product can withstand temperatures in excess of 3000°F (1650°C).

Other reaction-type cements employ sulfuric and nitric acids. The reactions are similar to those indicated in (13-17); however, the bonds are somewhat less permanent in the face of environmental conditions.

13-7 Vitreous shaping

The forming operations for glass products are performed at high temperatures where the desired shapes can be obtained by viscous deformation. The sintering procedures which are required after pressure fabrication, plastic forming, and slip casting are not necessary for glass; however, subsequent stress relief may be required. Vitreous shaping operations include (1) pressing, (2) drawing, (3) blowing, and (4) sagging. These operations are controlled by selecting the temperature, pressure, and time variables appropriate to the composition of the glass. The temperature controls the viscosity of the glass (Section 8-7), and the pressure influences the deformation rate.

TABLE 13-3

TYPICAL CALCULATION FACTORS FOR GLASS VISCOSITY*

Glass Type	A†	B†
Fused silica	6.9	28,200
96% silica	2.9	18,800
Borosilicate	5.0	13,900
Soda-lime	4.6	11,200
Lead	3.9	10,100

* Applicable only in the melting and working temperature ranges. H. A. Robinson and C. A. Peterson, in "Viscosity of Recent Container Glass," *Journ. Amer. Ceram. Soc.*, **27**, 129–38 (1944) use the equation $\log \eta = B/(T - T_0) - A$, where T_0 is a characteristic temperature for each glass. If $T \gg T_0$, T_0 may be ignored. This is possible in the working range but not in the annealing range (Fig. 9–11).

† $\text{Log}\ \eta = B/T - A$, where T is in °K.

Viscosity control. The relationship between temperature and viscosity was shown in Fig. 9–11 for several common glass types. These relationships are also given in Table 13–3 for use above the fictive temperature, where

$$\log \eta = B/T - A. \tag{13-18}$$

The constants A and B are empirical; however, B may be related to the activation energy of diffusion. Temperature T must be in °K. Obviously, variations in glass composition within any one type lead to variations in the constants of Table 13–3; therefore more specific data about composition need to be obtained before one can utilize these constants. The sensitivity of viscosity to temperature may be demonstrated by a calculation. A change of 20°C (36°F) will change viscosities (and therefore deformation rates) of some glasses by a factor of two in the working range where η is approximately 10^5 poises. This requires a temperature control of ± 10°C for satisfactory process control.

Forming procedures. Pressing operations require viscosities of 10^6 to 10^7 at the low-temperature end of the working range, because external pressure is applied to the glass. A carbon-coated, cast-iron mold of the desired shape is used. Examples of pressed glass include lens blanks and tablewares (see Fig. 13–19).

Drawing is used where thin but long continuous shapes are to be formed, such as sheet glass, glass tubing and continuous glass fibers. The latter deserve additional mention because of the specialized procedures. In brief, glass is heated in a "bushing" made of platinum or some similar material. This bushing contains small orifices, through which flows the relatively fluid glass. The glass is drawn away from the bushing rapidly as fibers onto a rotating drum or mandrel. Four factors control the fiber diameter: (1) glass composition, (2) glass temperature (these two factors influence the viscosity), (3) orifice diameter, and

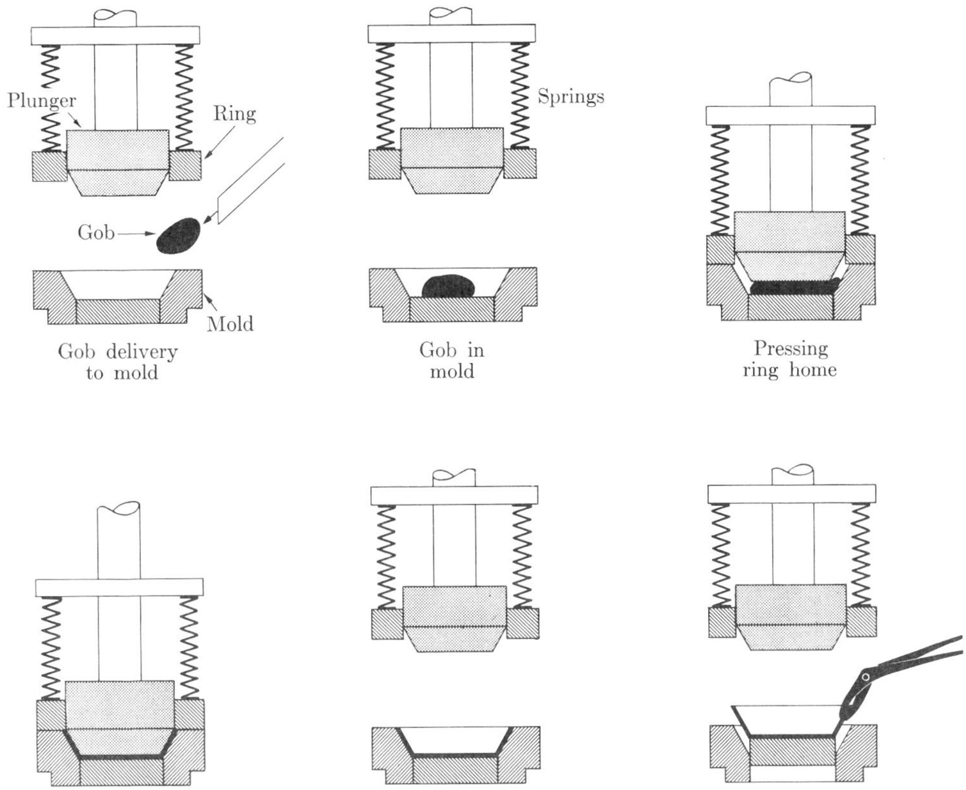

FIG. 13-19. Steps in pressing a glass dish. (From Norton, F. H., *Elements of Ceramics*, Reading, Mass., 1952, page 163.)

(4) mandrel velocity. The latter is usually 8,000 to 10,000 surface feet per minute. Therefore, in practice, there is a balance between the rate of flow through the orifice and the rate of drawing of the fibers from the bushing.

The velocity at which sheet and tube drawing take place is much less than that of fiber drawing, but the weight is greater. These operations are facilitated by pressures provided by exit rolls which aid in dimensional control as well as regulating the drawing rate (Fig. 13-20).

Blowing of glass, long regarded as a craftsman's art, is now almost completely mechanized to shape products such as containers and bulbs. It is possible to make 600-800 light bulbs per minute by pneumatic machines which are fed a ribbon of drawn glass. Mechanical glass-blowing requires molds (usually cast iron) into which a controlled quantity of semiviscous glass is fed, after which air pressure is applied to an internally formed cavity. The air pressure must supply

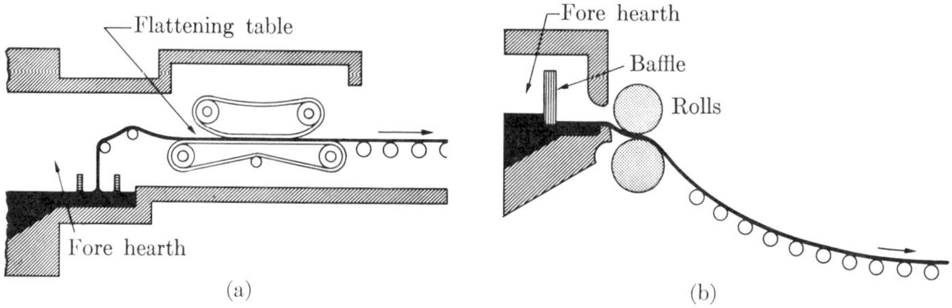

FIG. 13-20. (a) Drawing sheet glass. (b) Rolling sheet glass. There is also a new "float process" which utilizes molten tin after drawing or rolling. As the hot glass passes over the molten tin, it acquires a surface so smooth that the product competes directly with ground and polished plate glass. (From Norton, F. H., *Elements of Ceramics*, Reading, Mass., 1952, page 166.)

the energy to make the viscous glass flow, as well as overcome the surface energies of the glass.

Sagging is a special subject which has considerable commercial importance because the principle is used in the production of curved sheet glass products. Car windshields, for example, are made from two pieces of precut plate glass which are heated while supported on appropriately shaped metal frames. The two sheets sag together and therefore match perfectly when they are later laminated, with a plastic sheet between them, to form safety glass. This shaping operation requires the highest viscosities (and hence lowest temperatures) of any of the vitreous forming procedures. (It also requires a separation agent such as $MgSO_4{\cdot}7H_2O$ to prevent the two glass sheets from sticking while they are being heated.)

Product defects. For glass to be flawless, there must be no entrapments of air or other substances, no surface blemishes or residual stresses. The latter may be removed by annealing (Section 14-4). Surface blemishes are usually consequences of poor molding practices; for instance, defects result when cold, viscous glass does not flow into the exact contours of the mold, or when hot glass sticks to the mold surface. Entrapments may arise from several sources, such as refractory occlusions (stones) and gas bubbles (seeds).

13-8 Miscellaneous forming procedures

There are several other forming methods which are important but which are not as generally used as the foregoing procedures. The other methods include grinding, spraying, and dipping. Grinding, widely used in the glass industry after initial vitreous shaping, is used on such products as plate glass and optical lenses. There are also electrical and structural components which attain the required close dimensional tolerances by means of grinding and polishing.

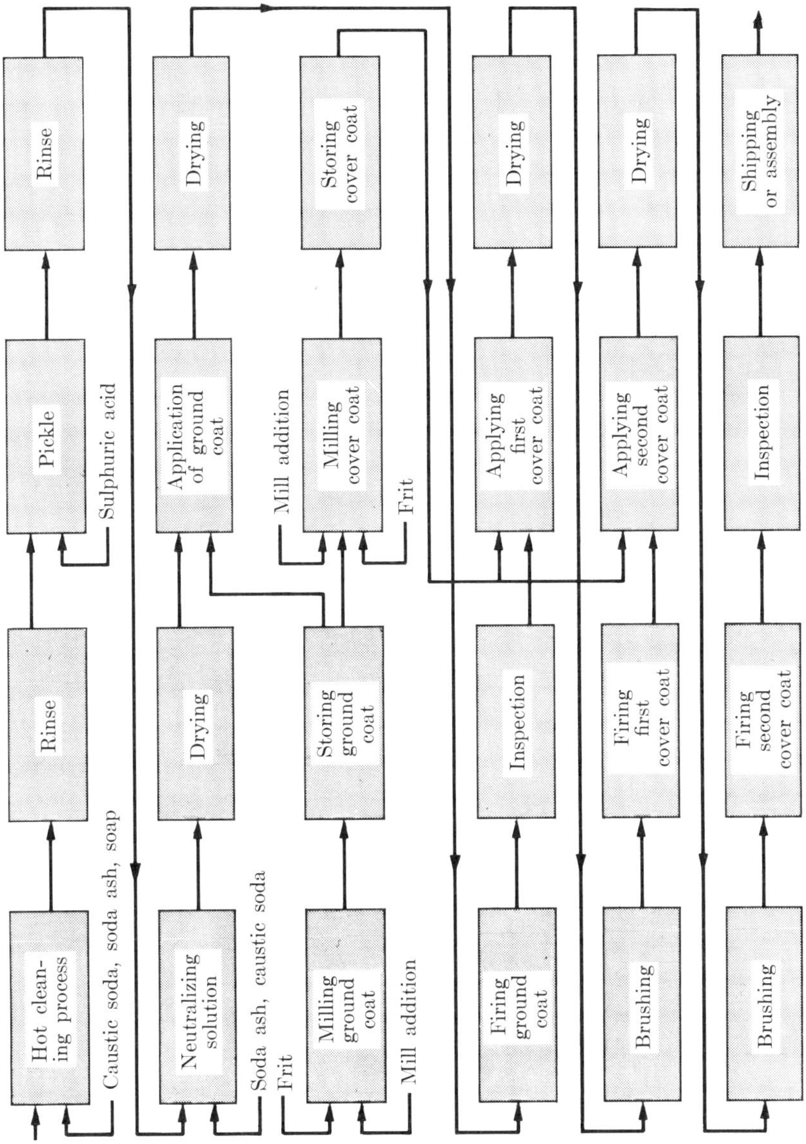

FIG. 13–21. Flow sheet for the enameling of sheet steel. (From Norton, F. H., *Elements of Ceramics*. Reading, Mass., Addison-Wesley, 1952, page 187.)

There is a variety of applications of the spraying technique. Enamel and glaze frits (low-melting glass) are sprayed, as finely divided suspensions, onto ceramic and metal surfaces. (Dipping is also used to achieve the same end.) There is partial evaporation of the liquid from the atomized spray so that the built-up coating does not flow after it lands on the surface. This spraying is followed by a heating or firing which sinters and remelts the coating into a homogeneous and impermeable coating. The total process is summarized, for the enameling of sheet steel, in Fig. 13–21.

A recent variation of the spraying method involves melting of refractory materials (such as Al_2O_3) in a high-temperature flame or arc; these materials turn into spray as they melt. The liquid spray solidifies when it comes into contact with the surface of a previously shaped product. Although the resultant surface is somewhat porous, it does act as a highly refractory and inert protective veneer.

Atomization is used as a process for making noncontinuous glass fibers, which are used for heat and sound insulation as well as for filtering media. Since the glass has a viscous network structure, the falling stream of molten glass may be elongated into strands or fibers by an atomizing air blast. Because cooling is rapid, the relatively low surface tension (approximately 300 dynes/cm) cannot overcome the viscosity to spheroidize the resulting fibers. Continuous glass fibers, described previously, are more commonly used for plastic reinforcement and for electrical insulation fabrics.

References for Further Reading

13–1. A.I.M.E., *Industrial Minerals and Rocks.* New York: Amer. Inst. Min. Metal. & Petro. Engrs., 1949. Includes the sources and processing for all common ceramic raw materials.

13–2. Bogue, R. H., *The Chemistry of Portland Cement,* second edition. New York: Reinhold Publishing Corp., 1955. A thorough treatment of the chemistry of portland cement. However, some of the studies on hydration have been superseded. (See Brunauer, 13–3, and Taylor, 13–9.)

13–3. Brunauer, S., "Tobermorite Gel—The Heart of Concrete," *Amer. Scientist,* **50,** 210–229, March 1962. Emphasis on the hydration of portland cement.

13–4. Huber Corp., *Kaolin Clays and Their Industrial Uses,* second edition. New York: J. M. Huber Corp., 1955. This book, published by a clay supply company, contains considerable useful data for the whiteware producer.

13–5. Kingery, W. D., (editor), *Ceramic Fabrication Processes.* New York: John Wiley & Sons, and Cambridge, Mass.: The Technology Press, 1958. This book contains the most complete current source of information on the principles of ceramic processing. Chapter 1, "Ion Exchange in Clay Minerals," by G. W. Brindley. Chapter 2, "Rheological Properties of Aqueous Clay Systems," by A. S. Michaels. Chapter 3, "Preparation and Use of Plaster Molds," by C. M. Lambe. Chapter 4, "Controls Required and Problems Encountered in Production Slip Casting," by H. S. Magid. Chapter 5, "Slip Casting Nonclay Ceramics," by P. D. S. St. Pierre. Chapter 6, "Pressure Forming of Ceramics," by W. D. Kingery. Chapter 7, "Controls Required and Problems Encountered in Production Dry Pressing," by H. Thurnauer. Chapter 8, "Hydrostatic Molding," by W. D. Kingery. Chapter 9, "An Introduction to the Vibratory Compaction of Powders," by W. C. Bell. Chapter 10, "Claywater Pastes," by F. H. Norton. Chapter 11, "Particle Orientation in Clays and Whitewares and Its Relation to Forming Processes," by W. O. Williamson. Chapter 12, "Some Experiences with the Control of Plastic Bodies for Automatic Jiggering," by R. E. Gould and J. Lux. Chapter 13, "Vertical Extrusion of Nonclay Compositions," by C. Hyde. Chapter 18, "Cementitious Bonding in Ceramic Fabrication," by J. F. Wygant. Chapter 20, "Some Possible Applications of Metalworking Processes to Ceramics and Cermets," by P. Loewenstein.

13–6. Norton, F. H., *Elements of Ceramics.* Reading, Mass.: Addison-Wesley, 1952. May be read by a student who has had college chemistry. The first 11 chapters present raw materials and forming processes.

13–7. Norton, F. H., *Ceramics for the Artist Potter.* Reading, Mass.: Addison-Wesley, 1956. Provides details of "How We Do It" with some explanation of "Why We Do It." Nontechnical.

13–8. Shand, E. B., *Glass Engineering Handbook.* New York: McGraw-Hill, 1958.

13–9. Taylor, H. F. W., "The Chemistry of Cement Hydration," *Progress in Ceramic Science,* Vol. 1 (J. E. Burke, editor). New York: Pergamon Press, 1961, pp. 89–145. A thorough presentation of the subject. For the specialist.

13–10. Tooley, F. V., *Handbook of Glass Manufacture.* New York: Odgen Publishing Co., 1953. Presents detailed descriptions of glass-forming operations.

Problems

13–1. A ceramic body contains the following ingredients: kaolinite 65.0 lb, quartz 12.6 lb, feldspar 14.5 lb, calcite 7.8 lb. What will be the chemical composition of the final ceramic product? [Assume pure raw materials except for the feldspar, which is a 35/65 combination of albite ($NaAlSi_3O_8$) and orthoclase ($KAlSi_3O_8$).]

13–2. Experiments indicate that a satisfactory ceramic body can be made from the following ingredients, with batch and mineral analyses as indicated.

		Kaolinite	Quartz	Kspar	NaSpar	Calcite
Clay	100 lb	92%	7%			1%
Flint	11 lb	4%	96%			
Feldspar	21 lb		10%	60%	30%	
Chalk	10 lb		5%			95%

(a) What is the chemical composition of the raw material batch? (b) What will be the chemical composition of the fired ceramic product?

13–3. Frit for a sheet-steel cover-coat enamel is to have the following composition: Al_2O_3, 2–4; B_2O_3, 11; K_2O, 2; Na_2O, 9; SiO_2, 48–50; ZnO, 6; AlF_3, 8; NaF, 12. (a) Calculate the batch composition if the following raw materials are available: feldspar, flint, borax, soda ash, cryolite (Na_3AlF_6), zinc oxide. The feldspar and the flint have the composition indicated in problem 13–2. (Assume no loss of SiF_4.)

13–4 A laboratory sample of a dried but unfired whiteware body has the following weights: dry, 873 gm; saturated (with kerosene), 898 gm; suspended (in kerosene), 587 gm. What are (a) the apparent porosity, (b) the apparent density, (c) the bulk density, (d) the apparent volume, and (e) the bulk volume? (170 cm^3 of kerosene weighs 141 gm.)

13–5. (a) Why was kerosene used in obtaining the data of problem 13–4? (b) If water could have been used, what would have been the saturated weight? (c) The suspended weight?

13–6. Water elutriation is used to separate magnetite (Fe_3O_4) from quartz (SiO_2) sand. If their densities are 5.18 and 2.65 gm/cm^3 respectively, what size magnetite particles will remain with (a) 1.0 mm quartz sand? (b) 0.1 mm quartz sand?

13–7. Sand composed of uniform spherical grains (specific gravity = 2.65 and PF = 0.7) has water added to it. Plot the bulk density versus (a) weight percentage of water, (b) bulk volume percentage of water.

13–8. Sketch a flow sheet for Al_2O_3 purification.

13–9. Referring to Table 4–2, specify raw materials of low cost, but with sufficient purity for each of these glasses.

Chapter 14

CERAMIC PRODUCTS: THERMAL TREATMENTS

14–1 Introduction

Elevated temperatures play an important part in ceramic production. High temperatures are used (1) to dry previously formed products, (2) to bring on agglomeration and reduce porosity, (3) to modify residual stresses, and (4) to produce crystallization. Thermal treatments result in weight and volume changes, promote atom movements and accentuate diffusion rates, reduce porosity and surface areas, and change properties. In the following four sections, we shall consider each of the above effects.

14–2 Drying

Drying is a major step in those processes which utilize hydroplastic forming (Section 13–4) and slip-casting (Section 13–5). When a product is fabricated by means of pressure, it usually contains small amounts of liquid or fugitive lubricants and binders, and although these may not be water, they must be removed. Products formed by vitreous shaping and grinding, of course, do not have drying steps.

Drying, as an engineering process, must be rapid to be economical. At the same time, it must not be so rapid that the product is damaged by cracks or warping as a result of volume changes. To design an efficient drying process requires a familiarity with the distribution of the liquid within the product, and also with the kinetics of liquid movements. With this knowledge it is possible to predict various dimensional and property changes with better accuracy.

Liquid distribution. Water is the liquid most commonly used in ceramic production; therefore we shall give it most of our attention. The general principles which apply to water also apply to other liquids such as oil, carbon tetrachloride, etc. Water may be considered a fugitive vehicle because it can be used for purposes of suspension or plasticity, and then can be removed by evaporation because of its relatively high vapor pressure.

Water distribution can be categorized into several types. (Not all types are present in all ceramic products prior to drying.) The various categories include:

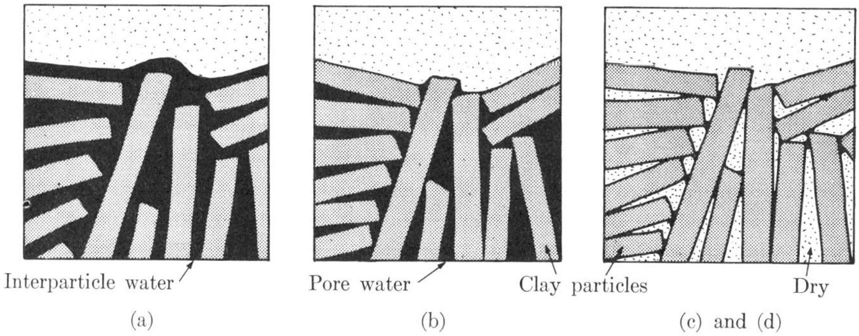

FIG. 14–1. Schematic of location of water in plastically formed ceramic products. Shrinkage occurs when the interparticle water is removed, but not when the pore water is removed. See Fig. 14–2. (After Norton, F. H., *Elements of Ceramics*. Reading, Mass.: Addison-Wesley, 1952, page 109.)

(1) water of suspension, (2) interparticle or interlayer water, (3) pore water, (4) adsorbed water, and (5) lattice water. A sixth type, hydration water, can arise in some materials as a result of thermal decomposition during presintering, and we shall discuss this hydration water in the next section, under firing operations. We shall not talk further about the water of suspension, because it must be removed either during or prior to the formation of the shape of the ceramic product.

Figure 14–1(a) shows interparticle (or interlayer) water in products made of claylike materials. After the water of suspension has been removed by filtration, decantation, or mold absorption, a considerable quantity of water in a relatively free state can remain between the particles. This interlayer film may be as much as 500 A thick, which is comparable to the colloidal dimensions of some ceramic raw materials. When the water is removed, the layered particles move closer together (probably by capillary action) and the shrinkage of the previously formed product is readily apparent. During this stage of drying, the volume shrinkage directly equals the volume of interparticle water removed [Fig. 14–2 (a) to (b)].

Pore water is shown in Fig. 14–1(b) as the water in the interstices among the particles after the particles have come into contact. Removal of this water produces very little if any additional shrinkage (Fig. 14–2, B to C). In fact, in special cases there is even some evidence that a slight expansion occurs because the capillary forces of the liquid disappear.

Plastically formed products generally contain both interparticle and pore water because they usually contain clays or other two-dimensional raw materials. This is always true for slip-cast ceramics. Pressure-fabricated ceramics contain very little if any interparticle water and therefore do not exhibit the large shrinkages depicted in Fig. 14–2. This factor constitutes an advantage for pressure molding which partially offsets the other limitations of pressure forming.

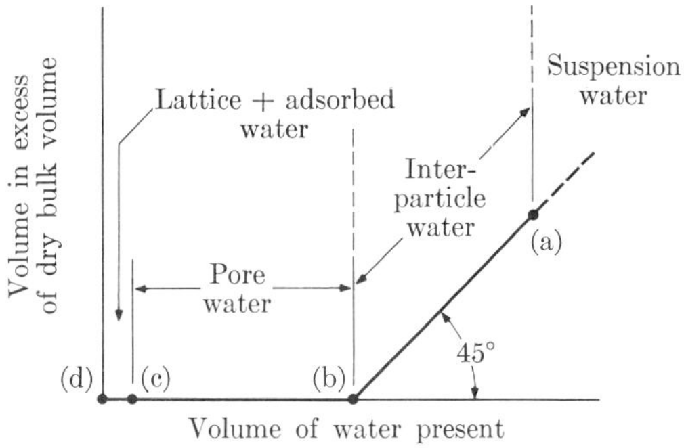

FIG. 14-2. Schematic of shrinkage versus water content. The letters refer to Fig. 14-1. (After Norton, F. H., *Elements of Ceramics*. Reading, Mass.: Addison-Welsey, 1952, page 109.)

Lattice water and adsorbed water are similar in that each has molecular dimensions only (in contrast to the 300-500 A films of interlayer water), and each has lower free energy than interlayer or pore water as a result of the attraction of water to the particle itself. Because lattice water enters the crystal structure at specific points, these molecules are not randomly positioned, and their presence is evidenced in x-ray diffraction measurements. Lattice water, which is not to be confused with the so-called water of hydration that forms through crystal dissociation, is not present in all phases. However, it is found between the layers of montmorillonite (Table 3-5), which is a significant phase in bentonite and some ball clays, within attapulgite, which is a clay with a fiberlike structure, within the tobermorite phase in concrete [Eq. (13-14)], and within some other less commonly encountered phases.

Adsorbed water is that water which is held on the surface by the physical forces of van der Waals attraction (Section 5-2). Although the adsorbed water is only a few molecules thick at the most, its quantity can be significant in ceramic products formed from colloid-sized particles. Both adsorbed and lattice water can be removed from a material by increases in temperature or by reduction in the partial pressure of the surrounding atmosphere. This was demonstrated in Chapter 13 for the tobermorite of hydrated portland cement, where the composition was determined as $Ca_2[SiO_2(OH)_2]_2 \cdot (CaO) \cdot xH_2O$, and x varied from zero at a dew point of $-78°C$, to 0.5 for a dew point of $-60°C$, to more than 1.0 in normal atmospheric humidity. The volume changes which accompany the removal of lattice or adsorbed water vary significantly. Montmorillonite shrinks a good bit, because its lattice-water content is quite high. Other phases exhibit much less shrinkage as a result of this final drying than as a result of the removal of the interparticle water (Fig. 14-2); however, sometimes the volume changes resulting from the removal of lattice or adsorbed water are important because they account for long-time growth or shrinkage; for example, the case of cement (Fig. 13-17).

TABLE 14-1

WATER VISCOSITY VERSUS TEMPERATURE

Temperature*		Viscosity, poises
°C	°F	
20.2	68	1.00×10^{-2}
25	77	0.89×10^{-2}
50	122	0.55×10^{-2}
75	167	0.38×10^{-2}
90	194	0.31×10^{-2}
100	212	0.28×10^{-2}

* Equation (14-2) requires absolute temperature for calculation purposes.

Liquid movement. The drying process requires a movement of liquid from the interior of the shaped product to the surface, where volatilization can occur. While the drying is limited by liquid movement within the product, the rate of water volume decrease dV/dt may be expressed as

$$\frac{dV}{dt} = \frac{AK(C_d - C_w)}{\eta L}, \tag{14-1}$$

where K is the permeability coefficient of Eq. (13-9b), A is a constant, η is viscosity, and C_d and C_w are the concentration-related activities of the liquid in the drier and wetter zones, respectively. The "half-thickness" of the product is L, if we assume that drying is permitted to occur equally rapidly from both surfaces of a slablike shape. In general, product requirements limit the processing variables to (1) the control of the water viscosity (through temperature) and (2) the control of the humidity of the drying air. We shall shortly see that the former provides a more satisfactory method of control in the early stages of drying, when shrinkage takes place. Between 20 and 100°C, the viscosity of water may be approximated as

$$\log \eta = -4.56 + \frac{750}{T}. \tag{14-2}$$

Therefore, viscosity drops by a factor of 4 as temperature is increased from 20 to 100°C (Table 14-1). Rather than handling these calculations mathematically, let us look at Fig. 14-3, which indicates schematically the effects of temperature and atmospheric humidity.

As we mentioned before, the maximum amount of shrinkage during drying takes place while the interparticle or interlayer water is being removed. Thus it is desirable that the water content within a product should be as uniform as

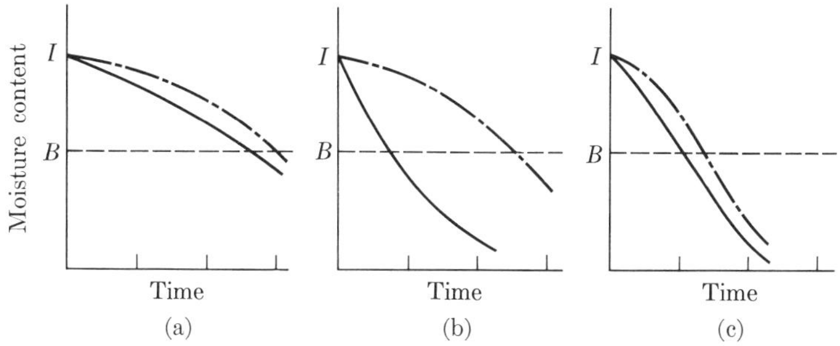

FIG. 14-3. Drying of ceramic products. Solid line indicates surface moisture; dot-dashed line, center moisture; I = initial moisture content; B = moisture content at the end of shrinkage [cf. point (B) of Fig. 14-2]. (a) Drying with low temperature and high humidity. (b) Drying with low temperature and low humidity. (c) Drying with high temperature and high humidity.

possible during the early stages of drying. Uniformity can best be maintained and cracking avoided by initially heating the product in a saturated atmosphere before water is removed. Then, as the water is removed from the surface, a steep moisture gradient is not established, because the remaining water of low viscosity readily moves outward. Once the shrinkage has ceased, we can use drier air to remove the pore water. Finally we must employ temperatures in excess of 100°C to remove the lattice water and the adsorbed water.

Dimensional changes. Volume and linear changes during drying are important in ceramic production. If the shrinkage is uniform, the chief concern is to establish a shrinkage allowance which will enable the product to assume the required final dimensions. However, shrinkage is not always uniform; the surface may shrink faster than the subsurface, which presents the danger of potential cracking. Tension stresses are established in the more rapidly dried zones and compression stresses in the higher-moisture zones. If these stresses arise while the product is still plastic, warpage can occur; also, dry strengths are never exceptionally high in "greenware," which is another cause of rupture. Since drying cracks form in the direction of the steepest moisture gradient, they are usually perpendicular to the external surface; however, their direction may vary somewhat with variable cross sections.

Shrinkage is anisotropic when there is anisotropy in the microstructure of the product. Thus it is common to find greater shrinkage *across* a thin slip-cast wall than parallel to it, because of the orientation which develops during deposition. As a consequence, there are differential shrinkages at corners and reentrant angles, where there are changes in orientation.

Shrinkage can be reduced by the addition of noncolloidal raw materials. The best example is the addition of finely crushed grog (previously fired clay) to an unfired clay. The grog particles reduce the amount of interparticle or interlayer water which is retained in the undried product because the structure of grog is not two-dimensional. Furthermore, even though the grog may be −200 mesh, it is coarse in comparison with the colloidal clay, and this coarse fraction increases the packing factor of the mix (Section 13-3).

The limit in additions of nonplastic materials such as quartz, feldspar, etc., is usually the workability of the raw material for the forming operation. Thus it is essentially impossible to reduce drying shrinkage of an extruded or similar product to near-zero levels, as is done in pressure fabrication.

The addition of coarse, nonplastic components to an otherwise plastic ceramic product introduces a possibility of microcracks. The nonplastic material serves as a source of stress when the plastic materials shrink around it. The effects which such microcracks can have on the strength of the final fired product are of course pronounced, unless the firing permits a healing of the cracks.

EXAMPLE 14-1. The following data were taken in a control laboratory: initial weight, 33 gm; dry weight, 22 gm; initial volume, 24 cm^3; final volume, 16 cm^3. From this data supply all the information you can which would be pertinent to the drying process.

Answer:

Plastic bulk specific gravity $= 33/24 = 1.38.$

Plastic apparent porosity $= (33 - 22)/24 = 0.46\ cm^3/cm^3.$

Linear drying shrinkage $= (\sqrt[3]{24} - \sqrt[3]{16})/\sqrt[3]{24} = 0.125\ cm/cm.$

Dry apparent porosity $= [11 - (24 - 16)]/16 = 0.187\ cm^3/cm^3.$

Dry apparent specific gravity $= 22/13 = 1.69.$

Properties after drying. Few ceramic materials are used in a dried but unfired condition. Therefore the chief property we must consider for a dried product is the mechanical strength, i.e., the green strength, required for handling prior to firing.

Dry strengths vary markedly with the nature of the ceramic product. Pressure-fabricated ferroelectric and magnetic ceramics usually are very weak, because their composition intentionally omits plasticizing clays or other colloidal materials. Fortunately, these ceramic products are usually small, so that careful handling can prevent rupturing stresses. These products may not even be completely dried before they are introduced into the furnace because their volume changes during drying are insignificant. Thus they may be handled while some moisture (or volatile wax) is present to give handling strength. In contrast, dried ceramic products containing a high percentage of clays with large base-exchange capacities (Section 5-3) have dry strengths amounting to several hundreds or even thousands of psi.

The mechanisms of dry strength involve both van der Waals and ionic forces of attraction. Whenever the surface structure possesses a large number of ions (i.e., a large base-exchange capacity) it is possible to have a considerable number of primary bonds. Further, the surface ions may serve as bonds between colloidal plastic particles and nonplastic particles such as fine quartz (SiO_2) grains.

14-3 Firing (sintering)

An important step in the production of most ceramic products is firing, where the purpose is to agglomerate particles into a coherent mass by sintering. The sintering operation brings about several significant changes in the ceramic product: (1) the total surface area is reduced, (2) the bulk volume is reduced, and (3) the strength is increased. Although some sintering operations produce no phase changes, it is common to encounter changes such as liquid formation.

The dimensional changes which occur during firing are just as important as those which occur during drying; in fact, probably more so, because shrinkage takes place in almost all aggregated materials during firing. The firing operation generally bears a much closer relationship to property control than other thermal treating procedures do. We shall consider this relationship at the end of this section.

Presintering reactions. Formed and dried ceramic products which contain relatively complex phases may undergo presintering decomposition as they are fired. The most familiar examples are the calcination reactions of Eqs. (13-4) and (13-5), where calcite ($CaCO_3$) and kaolinite [$Al_2Si_2O_5(OH)_4$] are dissociated. Just as calcite does not contain molecules of CO_2 but dissociates to form CO_2, so also do kaolinite and related structures dissociate to form H_2O. The water thus formed is called the water of hydration. Its formation is endothermic, and therefore energy must be supplied to decompose this clay mineral. Similar reactions are observed when heat is applied to pyrophyllite [$Al_2Si_4O_{10}(OH)_2$], talc [$Mg_3Si_4O_{10}(OH)_2$], muscovite mica [$K_2Al_4(Si_3AlO_{10})_2(OH)_4$], hydrous tricalcium aluminate [$Ca_3Al_2(OH)_{12}$], gypsum [$CaSO_2(OH)_4$ or $CaSO_4 \cdot 2H_2O$], and amphibole asbestos [$Ca_2Mg_5(Si_4O_{11})_2(OH)_2$], to name but a few.

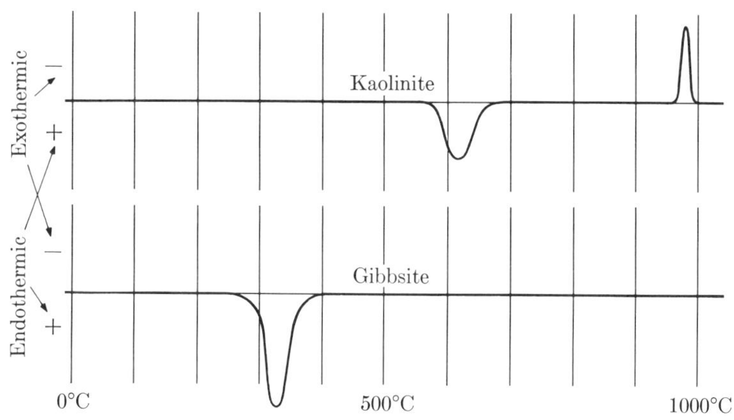

FIG. 14-4. Differential thermal curves of kaolinite [$Al_2Si_2O_5(OH)_4$] and gibbsite [$Al(OH)_3$]. The endothermic arrests occur at the dissociation temperatures. The exothermic peak results from the initial step in the recombination of the SiO_2 and Al_2O_3 that were dissociated from the kaolinite.

Each of the previous phases has its own characteristic temperature at which the endothermic dissociation reaction occurs. These and other reactions, endothermic and exothermic, may be used to identify ceramic phases through differential thermal analyses. Figure 14-4 shows thermal curves for kaolinite [$Al_2Si_2O_5(OH)_4$] and gibbsite [$Al(OH)_3$]. The endothermic "peak" is that point at which there is an arrest in the heating of the sample as compared to the heating of an inert (usually Al_2O_3) reference. The area under the peak is an index of the energy required. The endothermic reactions shown in Fig. 14-4 are a consequence of the previously mentioned dissociation and release of the water of hydration. An exothermic peak is also encountered in the heating of kaolinite (although it is absent in the heating of gibbsite). Dissociation releases Al_2O_3 and SiO_2, which do not, however, immediately recombine into phases stable in the Al_2O_3-SiO_2 system (Fig. 9-14). Partial reaction does occur at about 1000°C (depending on the rate of heating), and this releases heat. Metastable products of dissociation are common during presintering reactions.

Oxidation and reduction also occur as presintering reactions. One example of each reaction will be cited. (1) Oxidation: A clay used in the manufacture of brick commonly contains minor amounts of pyrite (FeS_2), which oxidizes to Fe_2O_3 and SO_3 during firing, giving a red color to the brick. (2) Reduction: Cupric oxide is reduced to cuprous oxide at high temperatures under the partial pressure of oxygen in the air (0.21 atm). This reaction is also evidenced by a change in color.

EXAMPLE 14-2. Equal parts by weight of kaolinite [$Al_2Si_2O_5(OH)_4$], calcite ($CaCO_3$), quartz (SiO_2), and corundum (Al_2O_3) are mixed. Predict (qualitatively) the characteristics of the thermal analysis between 20° and 1000°C (1830°F).

Answer: Corundum, with no phase changes, will be inert to thermal analysis.

Kaolinite will follow Fig. 14-4, with an endothermic peak at 600°C (1100°F) and an exothermic peak near 1000°C (1830°F). The area under the peaks will be about $\frac{1}{4}$ of the area for 100% kaolinite.

Quartz will have an endothermic peak at 573°C (1060°F) when it has the displacive transformation between the low and the high form (Fig. 9-4). The reconstructive transformation to tridymite at 875°C (1610°F) will probably be undetected because of its sluggishness.

Calcite will have an endothermic peak at its dissociation temperature for 1 atm CO_2 at 900°C (1650°F).* However, dissociation will start slightly below this temperature if the CO_2 can escape readily.

Sintering mechanisms. During sintering, particles are brought into close contact and porosity is reduced. The driving force is the decrease of total surface energies (Section 5-4) which comes about because of grain contact and grain growth (Sections 7-3 and 7-4). The atoms in the small grains are transferred to

* These data are from chemical handbooks.

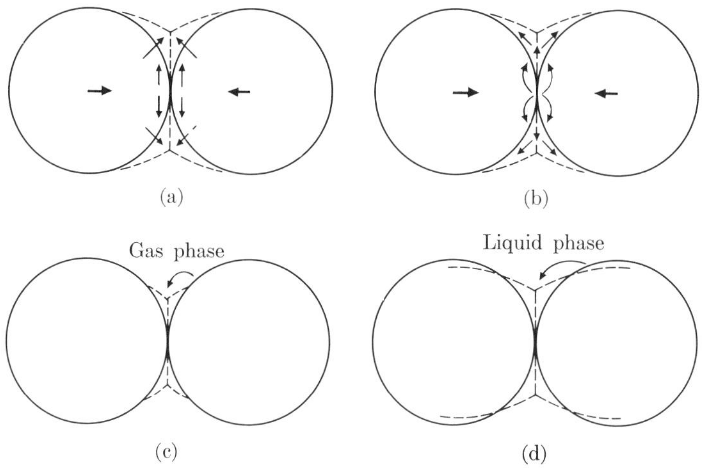

FIG. 14-5. Schematic of sintering mechanisms: (a) viscous or plastic flow of particles; (b) diffusion; (c) vaporization and condensation; (d) solution and precipitation.

the larger grains, and the pores are replaced by solid material. This mass transfer can be accomplished by four distinct but sometimes concurrent mechanisms, which are illustrated schematically in Fig. 14-5.

(1) Sintering can proceed by viscous or plastic flow.

(2) Diffusion can occur, not only along the grain boundaries between the particles in contact, but also within the grains.

(3) Highly volatile materials evaporate from convex surfaces and condense on concave surfaces because of vapor-pressure differentials.

(4) Similarly, the surface energy of a solid in a liquid is greater on convex surfaces than on concave surfaces. Therefore, liquid-phase sintering results in a reduction in surface energy.

Each of the above mechanisms may predominate within certain materials. The remaining mechanisms may be present, but to a lesser and oftentimes even insignificant extent.

Viscous flow predominates in the sintering of glassy particles. Two examples may be cited. The first involves the rather common operation of spraying glass frits (Section 13-8) as coatings onto previously shaped ceramic or metal products. The firing operation sinters the glass particles into a nonporous glaze or vitreous enamel. The only driving force available to achieve the required material flow is the surface energy (because gravity works against flow, as well as for it). The second example involves the sintering of slip-cast glass products. Small glass particles are agglomerated into a less porous product through viscous flow as a result of surface stresses. Higher temperatures increase the sintering rate quite markedly, because the flow rate, like the viscosity, is logarithmically related to the reciprocal of the temperature [see Eqs. (8-11) and (13-18)]. Hot-pressing (Section 13-3) can be considered comparable to viscous flow as a means of thermally agglomerating particles; here the surface forces are supplemented by externally applied forces.

Diffusion (Fig. 14-5b) is apparently the major mechanism in the sintering of single-phase ceramic products, e.g., ferroelectric and magnetic ceramics. Such diffusion involves the movement of atoms and the countermovement of lattice vacancies. There is an accompanying shrinkage of bulk volume which maintains contact of the particles and brings the grain centers closer together. This shrinkage, $\Delta L/L_0$, may be expressed as

$$\Delta L/L_0 = kt^n. \tag{14-3}$$

In this equation, t is time, n is a constant which is very close to 0.4, and k is a temperature-sensitive proportionality constant that behaves as

$$k/k_0 = e^{-Q/RT}, \tag{14-4}$$

where Q is the activation energy for diffusion, R is the gas constant and T is temperature. Combining the above two relationships, Burke* obtains

$$\ln \frac{t_1}{t_2} = \frac{Q}{R}\left(\frac{1}{T_1} - \frac{1}{T_2}\right). \tag{14-5}$$

With these equations, we can compare time requirements for specified firing shrinkages at various temperatures. Although these equations are theoretically sound and have been corroborated experimentally, we must be aware of their limitations so that we will not extrapolate them too far. Specifically, the above relationships are not applicable when theoretical density (0% porosity) is approached; nor are they applicable during the very slow process of eliminating internal pores which become isolated from grain boundaries. The above relationships *are* useful, however, in the early stages of sintering where the initial fired strength is developed.

The role of volatilization and condensation in ceramic materials has not been widely studied. It apparently is significant in ZnO and SiC sintering. In SiC, very high temperatures are developed on account of resistance heating. In other cases, reducing atmospheres are used to develop species of suboxides which are volatile; for example, SiO from SiO_2, and possibly Al_2O from Al_2O_3. The principles of sintering by vapor transport are identical with those of sintering by liquid transport. Therefore we shall consider them together in the following paragraph.

---

* Equations (14-3), (14-4), and (14-5) are from J. E. Burke, "Recrystallization and Sintering in Ceramics," *Ceramic Fabrication Processes*. New York: John Wiley & Sons (1958), p. 123.

Liquid-phase sintering (Fig. 14-5d) involves the removal of ions or atoms from surfaces of higher energy and depositing them on surfaces of lower energy. The highest-energy surfaces are those with small radii of *convex* curvature; the lowest-energy surfaces are those with small radii of *concave* curvature. Large-radii surfaces and plane surfaces are intermediate. Two situations which may arise are illustrated in Fig. 14-6. (1) The solid grains are always separated by a liquid film because the dihedral angle of liquid penetration along the solid grain boundary is 0° [that is, $2\gamma_{L/S} < \gamma_{S/S}$, Eq. (7-3) and Fig. 5-5d]. (2) The solid grains form solid-to-solid grain boundaries so that finite dihedral angles have the lowest-energy configurations. In Example (1) above, the strength to resist external stresses at high temperatures is solely a consequence of capillary forces and liquid viscosity. In contrast, Example (2) maintains strength by solid-to-solid contact even when a liquid is present. At low temperatures, solidification or increased viscosity of the liquid contributes to the strength in either case.

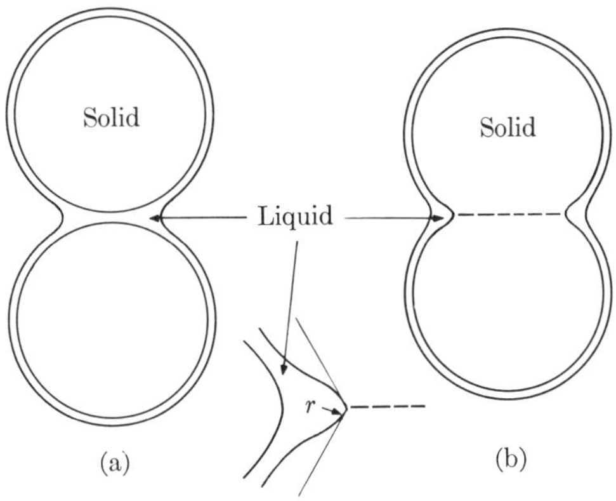

FIG. 14-6. Schematic of liquid-phase sintering: (a) dihedral angle = 0°; (b) dihedral angle ≠ 0°. [Center enlargement shows sketch at grain junction of (b).]

The rate of densification by liquid-phase sintering has been formulated by MacKenzie and Shuttleworth.* The somewhat complex relationship may be summarized by noting that densification increases as the liquid viscosity decreases at higher temperatures. Densification is also affected by the liquid's surface tension and by the initial shear stress required to effect deformation. Further, it is ultimately limited by the pore:solid ratio. Figure 14-7 illustrates these limitations for periclase-liquid compositions.

Dimensional changes. The densification described above induces shrinkage. It is apparent from Fig. 14-7 that the amount of shrinkage depends on both

---

* MacKenzie, J. K., and R. Shuttleworth, "A Phenomenological Theory of Sintering," *Proc. Phys. Soc. of London,* **B62**, 833 (1949).

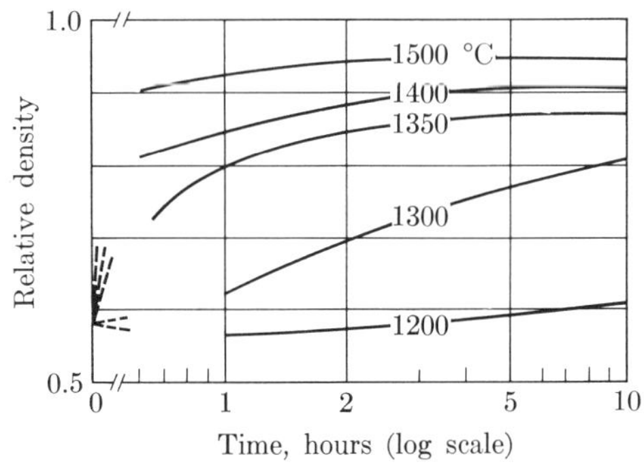

FIG. 14-7. Densification (with time and temperature as variables) of MgO plus 10% liquid. (From Cannon, J. H., and J. White, "Some Rate Processes in Ceramics," *Trans. Brit. Ceram. Soc.*, **39**, 377, 1956.)

time and temperature. Longer times at lower temperatures are equivalent to shorter times at higher temperatures. The amount of shrinkage also varies with the initial porosity. This is demonstrated in Fig. 14-8 for a toroid-shaped magnet which has been pressed from one end only (Section 13-3). The resulting unequal bulk density (and porosity) permitted an unequal shrinkage.

Unequal shrinkage and/or warpage may also develop during firing if: (1) the distribution of temperature within the product is not equalized, (2) the time the material is kept at sintering temperature varies markedly from point to point (e.g., thick sections versus thin sections), (3) there is particle segregation during forming which causes variations in particle size or in composition, (4) there is anisotropic shrinkage from preferred particle orientation, (5) flow occurs because of stresses caused by gravity during the high-temperature period of firing, or (6) there is friction between the kiln base and the shrinking ware. The proper attention to forming, drying, and firing procedures can eliminate many of the above unequal dimensional changes. In addition, it is sometimes possible to introduce advance compensation for the anticipated nonuniformities.

Occasionally there is volume expansion during firing. One of two situations may contribute to this: (1) there is a phase change from a more dense to a less dense modification, e.g., quartz → cristobalite in silica compositions, or (2) a gas evolves after initial vitrification. The latter produces bloating, and may even

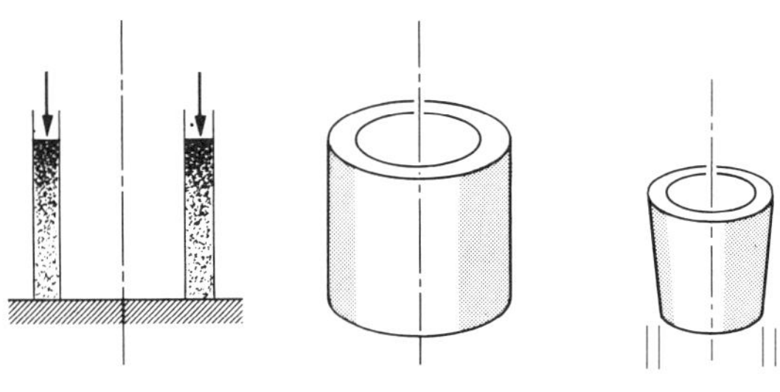

FIG. 14-8. Dimensional changes in a cylindrical magnet. Unequal shrinkage can result from forming procedures if the packing density is not controlled (Fig. 13-12). (a) single-end pressing; (b) stripped and dried; (c) fired.

be intentionally incorporated into a process if a light, insulating product is desired. An example of the bloating mechanism may be cited for a body with a low sintering temperature, which contains some $CaSO_4$. A glassy liquid may start to form as low as 1000°C (1830°F). Calcium sulfate (calcined gypsum) dissociates to CaO and SO_3 at temperatures in excess of 1400°C (2550°F), depending on the partial pressure of SO_3. When the evolving SO_3 is entrapped within a vitreous phase, it forms an internal pressure that can viscously deform the glass around the gas bubble.

Firing procedures. There are many types of furnaces or kilns used for the firing or sintering of ceramic products. The two main categories are: (1) periodic and (2) continuous. The periodic kiln requires the loading and sealing of the kiln prior to heating; afterward, it is cooled and eventually unloaded. This cycle may require a relatively few hours (for small electrical-circuit components) to more than a week (for large construction materials such as brick and tile).

In continuous firing, the formed and dried ceramic products move through the kiln. As in the periodic kiln, the range in sizes of products from electrical-circuit components to construction products involves a proportionate range in sizes and schedules of the kiln itself. The continuous kiln is generally more elaborate and expensive than the periodic kiln, but at the same time it permits certain operating economies if the production schedule warrants. For example, the air used for combustion is often preheated by being first used for cooling fired products. Further, the products of combustion may be used to preheat the incoming ware. This counter-current flow of gases and products increases the efficiency of fuel consumption (Fig. 14–9). Another economy is possible in the tunnel type of continuous kiln if the same cars are used for both drying and firing. This makes it unnecessary to handle the product between its forming and its eventual removal as a finished fired ceramic.

The control of the firing operation is highly automated these days. Control involves a scheduled heating rate, followed by a specified "hold" time, and then cooling. In addition it is frequently necessary to maintain controls over the oxidation-reduction aspects of the kiln atmosphere. The heating rate of dried ceramics is limited, since a steep thermal gradient within the product causes cracks and hence should be avoided. Also, the constituents of some ceramics undergo major volume changes as a result of heating (e.g., the low-to-high transformation of quartz, Section 3–6).

The time required for a specific product to remain at firing temperature varies with the temperature. This was illustrated in Fig. 14–7, where the higher the temperature the more rapid the sintering, as evidenced by densification. In effect, the firing operation integrates time and temperature in such a way that the same results may be obtained by longer times at lower temperatures. For this reason, kiln operators use pyrometric cones, which have been standardized so that the integrated "heat work" may be readily apparent. These pyrometric cones include a series of 60 to 70 compositions with progressively higher softening (glass forming) temperatures. When shaped as a cone and inclined at a

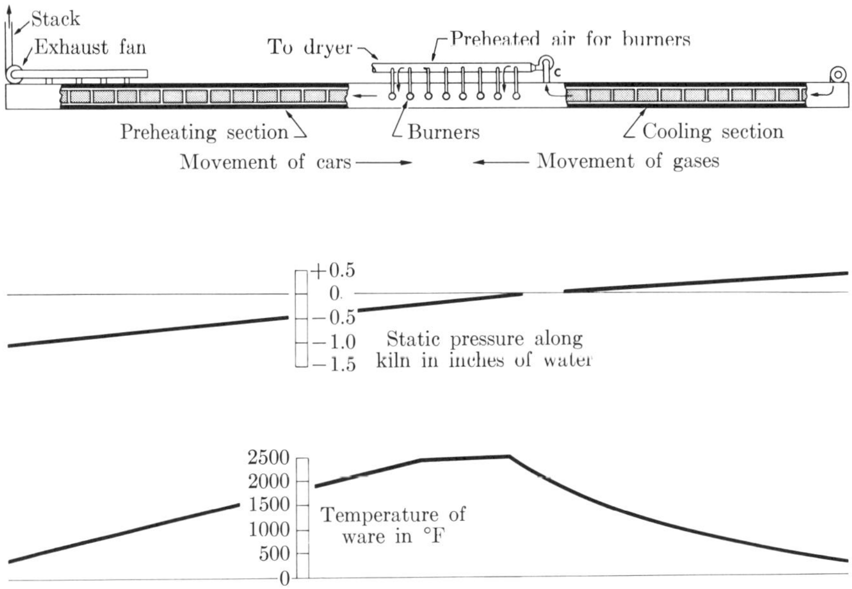

FIG. 14-9. Tunnel kiln. Fuel economy is achieved by having the ware on the cars move counter-currently to the gases. (From Norton, F. H., *Elements of Ceramics*. Reading, Mass.: Addison-Wesley, 1952, page 140.)

slight angle from the vertical, these standards deform slowly by viscous flow of the glass. The "endpoint" of their deformation is a specific function of both time and temperature. The most important use of pyrometric cones is to control the firing of silicate ceramics which also develop a viscous glass.

The oxidation level of the firing operation is usually controlled by means of the fuel-to-air ratio. This control measure means that there can be variations in oxygen pressures from 10^{-4} to 10^{-1} atm. When atmospheres outside this range are required, it is generally more economical to use muffles to enclose the ceramic product in a preselected atmosphere. Muffles are also used to isolate the products of the combustion process from the ware, e.g., in firing for glazing. This is particularly desirable if the fuel used contains deleterious impurities such as sulfur.

It is sometimes necessary to modify the oxidation level of the atmosphere during cooling. Such is the case for magnetic ceramics which require specific Fe^{3+}/Fe^{2+} ratios. The equilibrium for the reaction

$$2Fe^{2+} + \tfrac{1}{2}O_2 \rightleftarrows 2Fe^{3+} + O^{2-} \qquad (14\text{-}6)$$

varies with temperature. Because it is an exothermic reaction as written, the partial pressure of oxygen must be decreased during cooling or else the Fe^{3+}/Fe^{2+} ratio will increase.

EXAMPLE 14–3. In magnetite (Fe_3O_4) the thermodynamic value $\Delta H°$ is $-56{,}000$ cal/mole for the reaction expressed in (14–6). What oxygen partial pressure would be required to maintain the same $Fe^{3+}/Fe^{2+}/O^{2-}$ ratios at 1200°C which were established at 1400°C with a $p_{O_2} = 0.01$ atm?

Answer:

$$\ln K = -\frac{\Delta F°}{RT}, \tag{14–7}$$

$$\log\left[\frac{(Fe^{3+})^2(O^{2-})}{(Fe^{2+})^2(p_{O_2})^{1/2}}\right] = \frac{-\Delta H°}{4.575T} + \frac{\Delta S°}{4.575},$$

$$0.5 \log p_{O_2} - \frac{\Delta H°}{4.575T} = \frac{-\Delta S°}{4.575} + \log\frac{(Fe^{3+})^2(O^{2-})}{(Fe^{2+})^2},$$

$$(0.5)(-2.0) + 7.3 = \text{const} = 6.3 \qquad \text{at } 1673°\text{K},$$

$$\log p_{O_2} = 2(6.3 - 8.3) = -4 \qquad \text{at } 1473°\text{K},$$

$$p_{O_2} = 0.0001 \text{ atm}.$$

Properties after firing. The goal of firing or sintering is to develop suitable microstructures. Let us now take a look at the relationship between firing procedures and the properties of the end products. Of course, the following remarks are generalizations, and therefore cannot always hold true in every contingency.

The strength of a material is inversely related to both porosity and grain size; i.e., greater strength is possible if there is both low porosity (Fig. 8–5) and small grain size (Fig. 8–6). Since longer firing reduces the porosity but increases the grain size, it is desirable to limit the porosity of the material as much as possible through control of the sizes of the particles (Section 13–3). In this way porosity is minimized without excessive firing. Another technique which holds future promise is the use of inhibitors which retard grain growth without interfering with pore elimination.

The electromagnetic properties of ceramics are strongly affected by the way the materials have been fired. This effect is reflected in the dielectric constant as a space charge (Section 10–4) which comes from the presence of heterogeneities in the phase distributions. Pores reduce the dielectric constant significantly as their number and size increase. This effect is illustrated in Fig. 14–10 for TiO_2 ceramics.

The longer the firing, the greater the dielectric strength, because additional firing yields a denser product; hence there are fewer internal paths available for surface breakdown. Where dielectric strength is a paramount consideration, special procedures must be used to reduce porosity to essentially zero (or glazes must be added).

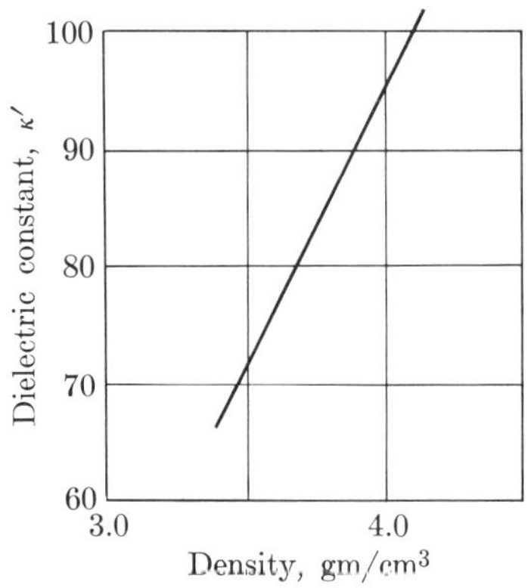

FIG. 14–10. Dielectric constant versus fired density of TiO_2 ceramics. (Compare with Fig. 14–11.) Porosity interferes with domain alignment. (After Economos, G., "Effect of Microstructure on the Electrical and Magnetic Properties of Ceramics," *Ceramic Fabrication Processes*. New York: John Wiley & Sons, 1958, page 204.)

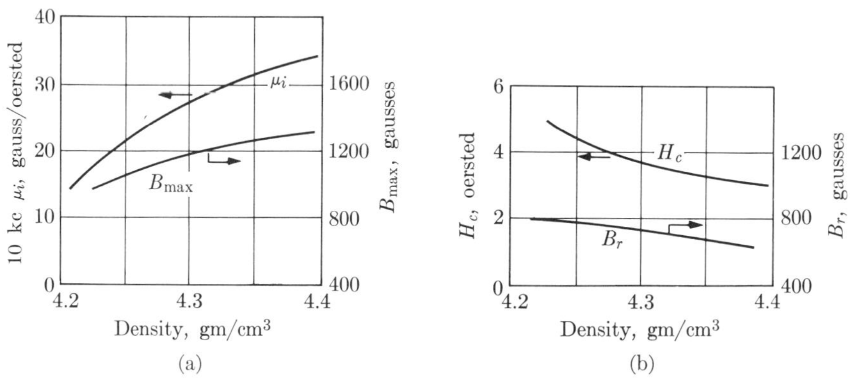

FIG. 14–11. Magnetic properties versus fired density of magnesioferrite ceramics: (a) initial permeability μ_i and maximum magnetization $B_{\max}$; (b) coercive force H_c and remanent magnetization B_r. The latter properties decrease because there are fewer pore walls to prevent domain wall movement. (After Economos, G., "Magnetic Ceramics: III, Effects of Fabrication Techniques," *Journ. Amer. Ceram. Soc.*, **38**, 335–340, 1955.)

In magnetic ceramics two relationships are developed as a consequence of firing history. (1) The initial magnetic permeability μ_i and the maximum magnetization $B_{\max}$ (Fig. 10–27) increase whenever porosity is reduced, because there is greater interaction between adjacent grains (Fig. 14–11a). (2) An increase in grain size reduces the remanent magnetization B_r and the coercive force H_c, because the larger the grains the easier it is for domain walls to move (Fig. 14–11b). Thus, in the manufacture of most soft magnetic ceramic products, the goal is great density and large grain size. In contrast, a permanent magnet requires high values of $B_{\max}$, B_r, and H_c. This means exercising careful control in times and temperatures of firing to assure densification *without* concurrent grain growth.

14-4 Annealing and tempering

After ceramic products have been fired, thermal stresses (discussed in Section 9-5) must either be removed or appropriately controlled. Annealing is used to remove stresses; tempering or "thermal conditioning" is used to control stresses by introducing surface compressions as required.

Annealing. At the annealing-point temperature T_a, where the viscosity is $10^{13.0}$ poises, a major part of any pre-existing thermal stress can be removed quickly. Lillie* gives the equation

$$t = \frac{e^{-0.07(T-T_a)}}{3.28 \times 10^{-6}\sigma} \tag{14-8}$$

for the time t in seconds required to lower the residual stress from any very high value to the desired value of σ (psi) at temperature T (°C). Thus, at the annealing-point temperature, any stress can be lowered within 15 minutes to less than 375 psi, if "coarse" annealing is adequate to the purpose. "Fine" annealing to 50 psi can be accomplished in less than two hours. The higher the temperature the shorter the annealing time.

The time-consuming part of the annealing process is the cooling after annealing, because care must be taken not to reintroduce stresses. Thus, slow cooling is required, particularly if thick sections are involved. The cooling rate R in °C/sec has been estimated by Kingery† to be:

$$R = \frac{\sigma_t}{\alpha_l 10^9 d^2}. \tag{14-9}$$

The maximum allowable tension stress is σ_t; α_l is the linear thermal expansion (0 − 300°C); d is the whole thickness when both surfaces are allowed to cool at the same rate. This semiempirical relationship is derived from our knowledge of thermal diffusivity, of the modulus of elasticity, and of Poisson's ratio, as well as our ability to predict the amount of thermal contraction.

The limitation of the cooling rate is most severe in the temperature range between the annealing-point temperature T_a and the strain-point temperatures T_s. Below T_s, residual stresses do not develop because the strain is elastic. Here the only hazard is the possibility that the elastic stresses may exceed the fracture strength.

Example 14-4. Estimate and compare the high-temperature time required for annealing glass 1.0 in. thick at (a) 600 and (b) 630°C. Strain-point temperature = 580°C;

---

* Lillie, H. R., "Basic Problems in Glass Annealing," *Glass Industry*, **31**, 355-58, 382 (1950); *Cer. Abs.*, **34**, 23*f* (1951).

† Kingery, W. D., *Introduction to Ceramics*. New York: John Wiley & Sons (1960), p. 644.

annealing-point temperature = 600°C; maximum allowable stress = 500 psi; α_l = 10^{-5} in/in/°C.

Answer: (a) At 600°C,

$t = 1/(3.28 \times 10^{-6})(500)$ = 10 min;

$R = (500)/(10^{-5})(10^{9})(1)^2 = 0.05°\text{C/sec}$ = 7 min between 600 and 580°C;

17 min total.

(b) At 630°C,

$t = e^{-2.1}/(328 \times 10^{-6})(500)$ = 1 min;

$R = 0.05°\text{C/sec}$ = 17 min between 630 and 580°C;

18 min total.

Tempering. Glass is tempered by means of air or oil quenches from temperatures above the annealing point. Equation (9–13a) can be redefined so that the surface stress σ is related to the differential between the surface temperature T and the mean temperature T_m,

$$\sigma = \frac{E\alpha_l(T_m - T)}{1 - \mu}, \tag{14–10}$$

where E is the modulus of elasticity and μ is Poisson's ratio.

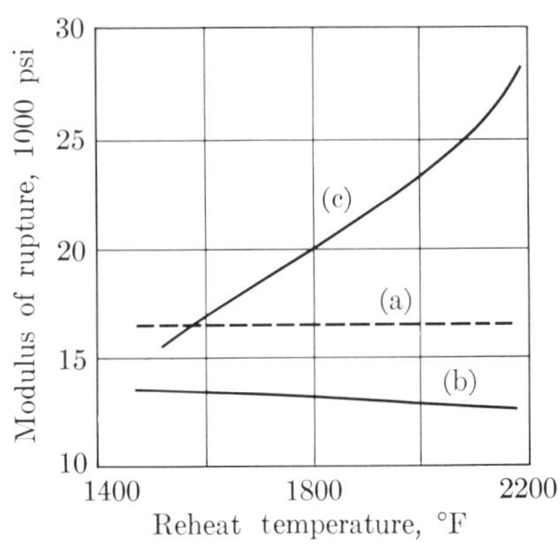

FIG. 14–12. Thermal conditioning of steatite ceramics: (a) modulus of rupture as received; (b) cooled within the kiln; (c) air-blast quench. Quenches from temperatures above 2200°F produced cracks. (After Smoke, E. J., and J. H. Koenig, *Thermal Properties of Ceramics.* New Brunswick, N. J.: Rutgers University Press, 1958, page 46.)

The effect of tempering on the physical properties of glass is important because the compressive surface stresses must be overcome before the glass is subjected to fracturing stresses. The same principle applies to crystalline ceramics, where tempering is known as "thermal conditioning." Data on thermal conditioning from Smoke and Koenig are presented in Fig. 14–12. Air-blast-quenched steatite (i.e., magnesium silicate) ceramics have higher moduli of rupture than do annealed samples. However, there is a limit to the usefulness of this technique, because quenching from excessively high temperatures produces cracking.

14-5 Crystallization

In the field of ceramic materials, two aspects of crystallization have recently been attracting attention. The first is controlled devitrification of glass, which yields a polycrystalline ceramic product. The second involves the production of single crystals of ceramic compounds, which are being used more and more in technical applications, particularly in the electronics industry.

Devitrification. Controlled devitrification of glass presents interesting potentialities. Chief among these is the possibility of using the versatile glass-forming operations to develop intricate shapes of nonporous ceramic products, then crystallizing (devitrifying) them so that when the products are used they will not undergo viscous deformation. In general, we can say that the properties of ceramic products made by this method would be superior to those of many of the products fabricated by present ceramic procedures, because porosity would be absent.

The successful control of the devitrification of glass centers around the control of nucleation within the glass. We pointed out in Section 4-5 that nucleation occurs very slowly at temperatures immediately below the liquid-to-crystal equilibrium point (i.e., the true melting temperature) because there is very little driving force ($F_{gl} - F_{cryst}$, where F is the volume free energy) to provide the energy for the new surface. If, however, a nucleating surface is provided, crystal growth can continue more readily because the surface-to-volume ratio decreases. In normal glasses, there is only a small statistical probability that a stable nucleus will develop. However, the introduction of artificial nuclei may serve the same purpose. These artificial nuclei need not be of the same composition as the bulk solution. The chief requirement is a low-energy interface between the nucleus and the crystal, which we may achieve by selecting nuclei similar in structure and interatomic spacing to the crystallizing phase. Better-known examples of artificial nucleation include cloud seeding by silver iodide crystals, and austenite nucleation in stainless steel ingots by fine periclase (MgO) particles.

Nucleation of glass may be achieved by dispersing colloidal metal within the glass. Under oxidizing conditions, metals such as copper, gold and silver may be dissolved in glass as cations. Each of these metals is readily reduced to the metallic state, provided that some other atom may be concurrently oxidized. For example,

$$2Cu^+ + Sb^{3+} \rightarrow 2Cu^0 + Sb^{5+} \tag{14-11a}$$

or

$$Ag^+ + Mn^{2+} + \text{photon} \rightarrow Ag^0 + Mn^{3+}. \tag{14-11b}$$

Both thermal and radiation procedures may be used to activate the oxidation-reduction reaction. The metal ions (and/or atoms) can readily diffuse to many local sites and develop colloidal particles. These particles, in turn, serve as nuclei for the growth of crystalline ceramic phases. In practice, nucleation is followed by low-temperature heat treatment to permit growth of the colloidal metal particles, and then a higher-temperature treatment to crystallize the glass. Stookey* has indicated that Al_2O_3-SiO_2-Li_2O-K_2O compositions are typical of the glasses used, and that Li_2SiO_5 and quartz are common crystalline phases which develop. The crystal size is usually less than one micron.

Since the above procedures have significant potentialities for ceramic fabrication, we can be sure that engineers will give them considerable attention, and hence that much information which is currently proprietary will become more generally available.

Single crystals. In order to prepare single crystals, we must isolate a single nucleus, which will then undergo continuous growth. Three procedures have been developed: (1) flame and zone fusion, (2) hydrothermal solutions, and (3) fused melts.

The procedure for flame fusion is illustrated in Fig. 14–13. If this process is to be usable, the crystal must solidify congruently so that the only alteration is the melting of the powder in the flame and then its crystallization on the boule. If an initial seed crystal (nucleus) is introduced and care is taken to avoid subsequent nucleation (e.g., by nonmelted powder), it is possible to grow exceptionally large single crystals. Furthermore, we may position the seed so that it produces boules with prescribed orientation; oriented bicrystals may be formed from duplex seeds; and compositions may be chosen with specific solid solution additions; for example, Cr^{3+} may be substituted for Al^{3+} in corundum.

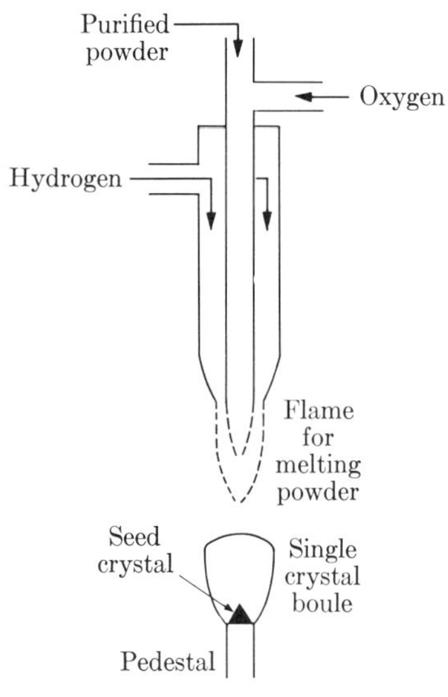

FIG. 14–13. Schematic drawing of flame fusion. The powder is melted by the flame. The droplets crystallize as they accumulate onto the seeded boule.

A rod of polycrystalline material is melted locally in zone fusion. As the heated zone is moved along the rod, crystallization proceeds from the initially established nucleus or inserted seed crystal, thus producing large single crystals.

Many useful single crystals either do not solidify congruently or have polymorphic phase changes which preclude direct-fusion techniques. Prime among these is quartz (SiO_2) which must be formed below 573°C (1065°F) in order to avoid dimensional changes when there is a phase change (Section 3–6). In the Na_2O-H_2O-SiO_2 system (or $NaOH$-H_2O-SiO_2) quartz crystallizes directly from

* Stookey, S. D., "Nucleation," *Ceramic Fabrication Processes,* New York: John Wiley & Sons (1958), pp. 189–195.

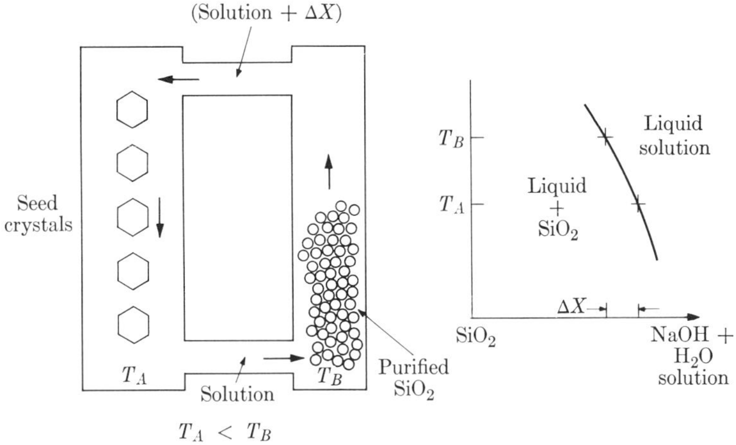

FIG. 14–14. Schematic drawing of the hydrothermal process for the formation of quartz. The temperature T_B on the solution side is a few degrees higher than the temperature T_A on the precipitation side. Therefore, ΔX of SiO_2 is precipitated from the solution as it circulates from the warmer to the cooler sides. High pressure is required.

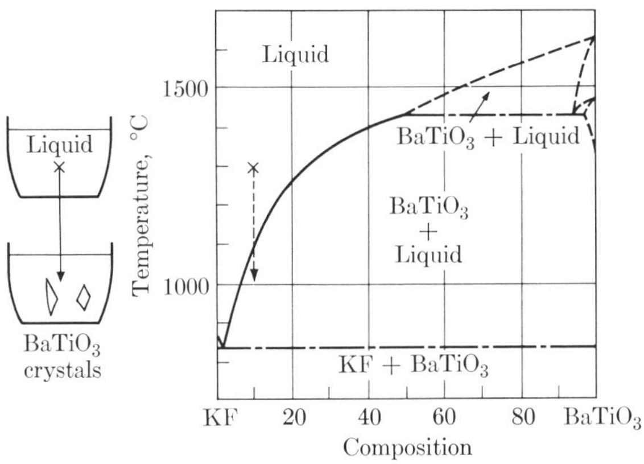

FIG. 14–15. Growth from fused melts. (Adapted from DeVries, R. C., "Observations on Growth of $BaTiO_3$ Crystals from KF Solutions," *Journ. Amer. Ceram. Soc.*, **42,** 547–558, 1959.)

the high-temperature (and high-pressure) liquid. The term "hydrothermal" has been used because the reactions are analogous to those encountered in hydrothermal geological crystallization (Section 13–2). As shown in Fig. 14–14, the whole process depends on a change in chemical activity of the SiO_2. The silica is dissolved in the alkaline aqueous solution at one temperature and precipitated from solution at a slightly lower temperature onto seed crystals. Close temperature control is required. The temperature differential, $T_B - T_A$, should be as great as possible to accelerate the reaction; however, excessive supercooling will cause spontaneous nucleation, which we must avoid if we wish to obtain single crystals.

Crystallization from fused melts is similar in principle to the hydrothermal process. However, the use of fused melts rather than water as the solute removes the necessity for high pressures. An example, given by DeVries, of phase relationships which permit this type of crystal growth is shown in Fig. 14–15. Again a change in solubility is utilized to produce controlled super-saturation. Slow cooling is required in order for seed crystals to grow and in order to prevent spontaneous nucleation.

References for Further Reading

14–1. KINGERY, W. D., editor, *Ceramic Fabrication Processes*. New York: John Wiley & Sons, and Cambridge, Mass.: The Technology Press, 1958. The following chapters present information on the thermal processing of ceramics. Chapter 14, "Warping and Its Control," by F. H. Norton. Chapter 15, "Recrystallization and Sintering in Ceramics," by J. E. Burke. Chapter 16, "Sintering in the Presence of a Liquid Phase," by W. D. Kingery. Chapter 17, "The Hot Pressing of Ceramics," by P. Murray, D. T. Livey, and J. Williams. Chapter 19, "Nucleation," by S. D. Stookey. Chapter 21, "Effect of Microstructure on the Electrical and Magnetic Properties of Ceramics," by G. Economos. Chapter 22, "Effect of Microstructure on the Mechanical Properties of Ceramic Materials," by R. L. Coble.

14–2. KINGERY, W. D., *Introduction to Ceramics*. New York: John Wiley & Sons, 1960. Chapter 10 presents nucleating crystal growth and solid-state reactions. Chapter 12 presents grain growth, sintering, and vitrification. Advanced undergraduate, upperclass, and graduate level.

14–3. NORTON, F. H., *Elements of Ceramics*. Reading, Mass.: Addison-Wesley, 1952. Chapters 12–16 present the thermal processing of ceramics at an introductory level.

14–4. NORTON, F. H., *Refractories*, third edition. New York: McGraw-Hill, 1949. Although specialized, the principles presented here are applicable to the manufacture of tonnage ceramics of various kinds.

14–5. STOOKEY, S. D., and R. D. MAURER, "Catalyzed Crystallization of Glass—Theory and Practice," *Progress in Ceramic Science*, Vol. 2 (J. E. Burke, editor). New York: Pergamon Press, 1962, pp. 77–101. This is the best summary to date of the principles involved in the newer devitrification processes.

Problems

14–1. A ceramic magnet in the shape of a toroid weighs 14.2 gm dry, 14.4 gm saturated with water, and 11.3 gm suspended in water. (a) What is its porosity? (b) Bulk density? (c) Apparent density?

14–2. A clay has the following pertinent characteristics: plastic weight, 86 gm; suspended plastic weight*, 46 gm; dry weight, 62 gm; saturated weight after drying*, 77 gm; suspended weight after drying*, 42.5 gm. Calculate (a) the linear drying shrinkage (based on original plastic dimensions), and (b) the water of plasticity. [*Suspended in, and saturated with, kerosene. Specific gravity = 0.82.]

14–3. The clay of problem 14–2 is fired with the following results: fired weight, 56 gm, length before firing, 2.1 cm, length after firing, 1.87 cm, saturated weight (with water), 62 gm. (a) What is the apparent porosity? (b) the apparent specific gravity?

14–4. The materials which were mixed in Example 14–2 (page 266) are equilibrated at 1400°C (2550°F). (a) What phase(s) are present? (b) What is the composition of each? (c) How much of each?

14–5. An Al_2O_3 insulator has the following laboratory data:

	Length	Porosity
After 2 hr at temperature	1.97 in.	—
After 4 hr at temperature	1.93 in.	4.2

(a) What was the porosity at the end of 2 hr? (Assume no phase changes in the intervening time.) (b) Estimate the time required to attain 2.0 percent porosity. (c) What assumptions are required for your calculations?

14–6. (Refer to Example 14–3, page 273.) Given that the oxygen pressure and the oxygen ion concentration remain the same as the temperature is reduced from 1400°C to 1200°C, how much will the Fe^{3+}/Fe^{2+} ratio change? (For computation purposes, consider the Fe^{3+}/Fe^{2+} ratio to be R at 1400°C.)

14–7. (Refer to Example 14–4, page 275.) Determine the optimum annealing temperature for minimum annealing and cooling time for this glass.

14–8. Most annealed commercial glasses have moduli of elasticity of about 10^7 psi, strengths of about 10^4 psi, and Poisson's ratios of about 0.2. Specify a maximum allowable thermal expansion for glass to be used in tumblers.

APPENDIXES

Appendix A

SELECTED CONSTANTS

N	Avogadro's number	$= (0.602\ldots)(10^{24})$ molecules/mole
e	electron charge	$= (1.6\ldots)(10^{-19})$ coulomb/electron $= (4.803\ldots)(10^{-10})$ esu
h	Planck's constant	$= (6.62\ldots)(10^{-27})$ erg·sec
k	Boltzmann's constant	$= (1.38\ldots)(10^{-16})$ erg/°K, or $= (1.38\ldots)(10^{-23})$ joule/°K
R	gas constant	= 1.987 cal/(°K·mole)
ev	electron volt	$= (1.602\ldots)(10^{-12})$ erg $= (0.38\ldots)(10^{-19})$ cal
A	angstrom unit	$= 10^{-8}$ cm
c	velocity of light	$= (2.998)(10^{10})$ cm/sec

Appendix B

GLOSSARY OF TERMS AS APPLIED TO CERAMIC MATERIALS*

ABSORPTION (moisture). Weight of water absorbed, divided by the dry weight of a ceramic material.

ACCEPTOR LEVELS. Energy levels formed within the energy gap by a deficiency of electrons.

ACID RESISTANCE (enamels). Resistance of a porcelain enamel to acid attack. ASTM tests have been established.

ACTIVATOR (luminescent). Ions which are photon emitters.

ADSORPTION, CHEMICAL. Surface adherence; accompanied by the formation of primary bonds.

ADSORPTION, PHYSICAL. Adsorption through secondary bonds.

AGGLOMERATED. Bonded aggregate.

AGGREGATED. Packed particles.

ALLOTROPIC. *See* POLYMORPHISM.

AMPHOTERIC ELEMENT. An element which may behave as a metal (loses electrons) or as a nonmetal (accepts or shares electrons) depending on its environment.

ANELASTICITY. Time-dependent strain in the elastic range.

---

* For a more complete glossary, see E. C. Van Schoick, *Ceramic Glossary*. Columbus, O.: American Ceramic Society, 1963.

ANHEDRAL. Lacking planar surfaces.

ANNEALING (glass). The process of removing residual stresses by controlled cooling below the annealing point.

ANNEALING POINT. Temperature at which the viscosity of glass is 10^{13} poises. Formerly defined as $10^{13.4}$ poises. (The internal stresses are substantially relieved in 15 minutes at this temperature.)

ANTIFERROELECTRIC. Spontaneous electrical polarization with equal numbers of dipoles in opposite directions.

ANTIFERROMAGNETIC. Spontaneous magnetic orientation of atoms with equal magnetic moments aligned in opposite directions.

ARCH BRICK. A brick with both large faces equally inclined toward a long, narrow side (for use in arch construction).

BAFFLE WALL. Deflection wall to redirect flame or gas flow within furnace.

BALL CLAY. A highly plastic sedimentary clay with light-colored firing characteristics (usually high in organic matter).

BALL MILL. A rotating cylinder containing balls or pebbles for grinding a charge.

BARN. Unit of cross-section probability equivalent to 10^{-24} cm^2.

BASE-EXCHANGE CAPACITY. The number of millequivalents of NaOH required to neutralize the exchangeable H^+ ions adsorbed onto 100 gm of colloidal material.

BASIC REFRACTORIES. Refractories with a high MgO and/or CaO content. Many basic refractories also contain Cr_2O_3.

BAUXITE. An Al_2O_3 ore with gibbsite as the major constituent phase.

BENTONITE. A "rock" with montmorillonite as the major constituent phase.

BINARY DIAGRAMS. Phase diagrams of two-component systems.

BIREFRINGENCE. Variation of wavelengths with optical polarization.

BISQUE. Unglazed, once-fired whiteware or pottery.

BLUNGER. Equipment to thoroughly blend a slip.

BODY. Mixture of ceramic raw materials with properties suitable for forming drying, and firing.

BOHR MAGNETON. The net magnetic moment arising from electron "spins."

BONDS, CHEMICAL. *See* COVALENT BONDS, IONIC BONDS, METALLIC BONDS, PRIMARY BONDS, SECONDARY BONDS, VAN DER WAALS BONDS.

BONE CHINA. *See* CHINA, BONE.

BOULE. A ball-shaped single crystal prepared by flame fusion.

BOUNDARIES, LARGE-ANGLE. Boundaries between adjacent grains.

BOUNDARIES, SMALL-ANGLE. Two-dimensional defects consisting of aligned dislocations.

BRAVAIS LATTICES. *See* Section 3–3.

BURGERS VECTOR. Dislocation displacement. The Burgers vector is parallel to a screw-dislocation line and perpendicular to an edge-dislocation line.

BURNING. The term "firing" is preferred.

CALCINATION. Thermal dissociation with gas evolution.

CARRIER (charge). Electrons, electron holes, or positive and negative ions which are mobile under a potential gradient.

CASTABLE REFRACTORY. A refractory composition containing a hydraulic cement.

CEMENTS, HYDRAULIC. A cement which sets with a hydration reaction. See Section 13–6.

CEMENTS, POLYMERIC (inorganic). A cement which sets with an additive polymerization reaction. See Section 13–6.

CEMENTS, REACTION. Cements which set with a condensation polymerization (hydrolysis) reaction.

CERAMIC (*adj.*). Pertaining to ceramic(s).

CERAMIC(s) (*noun*). Product containing ceramic materials (*q.v.*). (CERAMIC GLOSSARY. Any of a class of inorganic, nonmetallic products which are subjected to a high temperature during manufacture or use.)

CERAMIC MATERIALS. Materials which contain compounds of metallic and nonmetallic elements. Examples: MgO, $BaTiO_3$, SiO_2, SiC, clays, spinels, mullite, glasses, etc.

CERMET. Product containing an intimate mixture of metallic and ceramic phases.

CHECKERS. Heat regenerators for a furnace.

CHINA. Vitreous ceramic whiteware (normally used for nontechnical purposes). It is produced by bisque-firing at a moderate temperature, followed by glazing at lower temperatures.

CHINA, BONE. A highly translucent china made with a large proportion of bone ash (calcium phosphate).

CHINA, HOTEL. A semitranslucent body with greater thickness, higher strength, and harder glazes than domestic china. Absorption must be less than 0.3%.

CLAY, RESIDUAL. A clay formed *in situ* by geological reactions.

CLAY, SEDIMENTARY. A clay which has been transported from its point of origin before deposition.

CLAY MINERALS. Kaolinite is the most common clay mineral. Others include attapulgite, dickite, halloysite, hectorite, illite, montmorillonite, and nontronite. *See* Appendix C.

CLAYS. Fine-grained mineral aggregates. (1) Clay minerals, (2) finely crystalline quartz, and (3) finely crystalline calcite are the most common constituent phases of clay.

COERCIVE FORCE. Reversed field required to neutralize magnetization or electrical polarization.

CONE. *See* PYROMETRIC CONES.

COLLOID. A finely divided solid (commonly <1 micron). With a colloid, the surface properties become significant.

COLOR CENTERS. Color absorption within a crystal as a result of a point defect (vacancy, interstitialcy, or substitution) which produces an electronic transition.

COMPATIBILITY TRIANGLE. Subsystem of compatible phases in a ternary system.

COMPONENT. A compositional variable in a material's system. (*See* Section 6–1).

CONGRUENT MELTING. Melting without reaction so that the solid and liquid compositions are identical.

CONODE. Isothermal construction line between two equilibrated phases.

COORDINATION NUMBER. The number of nearest-neighbor atoms for any specific atom.

COVALENT BONDS. Interatomic attractions arising from the sharing of valence electrons.

CRAZING. Polygonized cracking, in glazes and enamels, which develops from tensile stresses.

CREEP. Time-dependent strain occurring during stress.

CROSS-SECTION (nuclear). Probability of neutron–atom interaction (expressed as an area).

CROWN OPTICAL GLASS. A low-dispersion, relatively low-index glass. Commonly used in the converging elements of lenses. Any glass with a Nu-value of more than 55.0 (or between 50.0 and 55.0 if the index is more than 1.60).

CRYSTALS. Atomic structures with long-range order. (Euhedral surfaces are not required.)

CULLET. Broken glass from production waste; often recycled through the melting furnace.

CURIE TEMPERATURE. Temperature of spontaneous magnetization or electrical polarization.

DEFECT STRUCTURE. *See* Section 3–5.

DEFLOCCULATION. Dispersion of colloidal particles within a suspension.

DENSITY, APPARENT. Weight per apparent volume.

DENSITY, BULK. Weight per bulk volume.

DENSITY, TRUE. Weight per true volume.

DEVITRIFICATION. Crystallization of a glass.

DIELECTRIC CONSTANT. The ratio of electrical permittivity within a material to the electrical permittivity in a vacuum.

DIELECTRIC STRENGTH. Breakdown potential gradient, $\Delta V/\Delta x$.

DIFFERENTIAL THERMAL ANALYSIS. Quantitative procedure for determining the temperature and extent of endothermic and exothermic reactions.

DIHEDRAL ANGLE. The angle of penetration of one phase or grain between two adjacent grains.

DILATANCY. Behavior whereby there is an *apparent* decrease in the liquid content of a colloidal structure during mechanical agitation (because the volume increases).

DIOCTAHEDRAL. Atoms filling $\frac{2}{3}$ of the 6-fold interstices. (*See* Fig. 3–16.)

DIPOLE MOMENT. Product of the dipole charge and the dipole length.

DISLOCATION. A linear crystal defect. (*See* Section 3–5.)

DISPERSION. Variation in refractive index with wavelength.

DISSIPATION FACTOR. Tan δ, when δ is the loss angle, or $(90 - \theta)$, θ being the phase angle.

DOMAINS. Crystal volumes of like ferromagnetic or ferroelectric alignment.

DONOR LEVELS. Energy levels formed in the energy gap by excess electrons.

DRAWING (glass). Continuous forming of sheet, tube, or fibrous glass from molten glass.

DRY-PROCESS (enamel). Process whereby dry powdered enameling materials are applied to a preheated surface.

DTA. *See* DIFFERENTIAL THERMAL ANALYSIS.

EFFLORESCENCE. A white powder that is the product of leaching from the interior onto the surface of a ceramic product.

ELECTRICAL DIPOLE. Displaced center of positive and negative charges.

ELECTROLUMINESCENCE. Luminescence which is activated by an electrical potential.

ELECTRON HOLES. Electron-deficient sites which provide acceptor energy levels. These are positive charge carriers.

ENAMEL (vitreous or porcelain). Glassy coating for metals.

ENERGY BAND. Energy spectrum of valence electrons in a polyatomic material. Conduction is not significant if the energy band is filled.

ENERGY GAP. Forbidden part of the energy spectrum of valence electrons. If the lower energy band is filled, electrons must be activated across this gap before electronic conduction is realized.

ENGOBE. A slip used as an underglaze coating.

EUHEDRAL. Planar external outlines which reveal the crystal structure.

EUTECTIC. Adjective describing temperature or composition for the intersection of two or more liquidus surfaces (or lines).

EXSOLUTION. Solid precipitation from a solid solution.

FERRIELECTRIC. Unbalanced orientation of electrical dipoles. Intermediate between ferroelectric and antiferroelectric.

FERRIMAGNETISM. Unbalanced orientation of magnetic moments. Intermediate between ferromagnetism and antiferromagnetism.

FERROELECTRIC. Spontaneous electrical polarization with all dipoles in the same direction. The polarity can be reversed by an external electrical field.

FERROMAGNETISM. Spontaneous magnetic orientation of all magnetic moments in the same direction. The orientation can be reversed by an external magnetic field.

FERROSPINEL. Spinel with iron as the chief trivalent ion.

FICTIVE TEMPERATURE. *See* TEMPERATURE, FICTIVE.

FIREBRICK (general). A refractory brick. (Specific) An aluminosilicate brick of fireclay composition.

FIRECLAY. A refractory clay approximating kaolinite in composition.

FIRING. High-temperature treatment which sinters particles into a coherent product with desired properties.

FLINT OPTICAL GLASS. A high-dispersion and usually high-index glass. Commonly used in diverging elements of lenses. Any glass with a Nu-value of less than 50.0 (or between 50.0 and 55.0 if the index is less than 1.60).

FLOCCULATION. Aggregation of colloidal particles within a suspension.

FLUX. A low-melting ceramic admixture.

FRENKEL DEFECT. An atomic displacement resulting in a vacancy and an interstitialcy.

FRIT. Powdered glass employed as a constituent of a glaze or vitreous enamel.

FUSED QUARTZ. *See* SILICA GLASS.

FUSED SILICA. *See* SILICA GLASS.

FUSION TEST. *See* PYROMETRIC CONE EQUIVALENT.

GLASS. A product of fusion which has cooled to a rigid condition without crystallizing. By general definition, glass is inorganic. The most common (but not exclusive) glass-forming compound is SiO_2, which may dissolve a considerable quantity of other oxides within its structure. The most typical structural characteristic of glass is the presence of short-range atomic order, and the absence of long-range order.

GLASS. *Also see* CROWN OPTICAL GLASS, ENAMEL, FLINT OPTICAL GLASS, FRIT, GLAZE, PLATE GLASS, PYREX®, SILICA GLASS, TEMPERED GLASS.

GLASS FORMERS. *See* NETWORK FORMERS.

GLAZE. Vitreous (glassy) coating for a ceramic body.

GLOST FIRING. Kiln firing of bisque ware after glazing.

GRADE (abrasives). Index of friability of bonded abrasive products.

GRAINS. Individual crystals within a microstructure. This term may also be used in processing to indicate a particle or grog grain.

GREEN WARE. Formed but as yet unfired products.

GROG. Coarse refractory particles, commonly polycrystalline.

GROUND COAT. The initial coat of a porcelain enamel.

HIGH-HEAT DUTY (refractories). Fireclay brick with a PCE of at least 32½. There are additional ASTM requirements.

HOTEL CHINA. *See* CHINA, HOTEL.

HUMIDITY DRYING. Process of heating in a moisture-saturated atmosphere and then drying after the product is hot and the water viscosity is lower.

IMMISCIBILITY. Separation into two phases (specifically two liquid phases).

INDEX OF REFRACTION. Ratio of light velocity in a vacuum to light velocity within a material.

INTERMEDIATE HEAT DUTY (refractories). Fireclay brick with a PCE of at least 29. There are additional ASTM requirements.

INTERMEDIATES. Oxides whose structural behavior in glass falls between the network formers and network modifiers.

INTERSTICES. Openings between atoms or particles.

INTERSTITIALCY. Crystal imperfection involving an atom in an interstitial position.

INVARIANT POINT. Temperature with the maximum number of phases as defined by phase rule. *See* VARIANCE.

INVERSION. Polymorphic transformation.

ION EXCHANGE. Replacement of one ion by another without altering the basic structure.

IONIC BONDS. Interatomic bonds arising from coulombic attractions.

ISOSTATIC PRESSING. Pressing with hydrostatic compaction.

JIGGERING. Forming by a rotating mold and a contoured tool to provide the two surfaces.

KAOLIN. A soft, fine-grained rock with kaolinite as the major constituent phase.

KEY BRICK. A brick with both narrow sides equally inclined toward the small end.

KILN. A furnace for the production of ceramic products. (Preferred prounounciation: kĭl.)

KILN, CONTINUOUS. A kiln which has zones for heating, high-temperature holding, and cooling, through which the product passes.

KILN, PERIODIC. A kiln which is filled, heated, held at temperature, cooled and unloaded, using a batch operation.

LASER. Acronym for Light Amplification by Stimulated Emission of Radiation.

LEACHING. Process of selective solution.

LEHR. A furnace or oven for annealing glass. (Pronunciation: leer.)

LIMESTONE. Rock composed predominantly of calcite.

LIQUIDUS. The locus of temperatures above which there is only liquid. Also a liquid solubility limit with the second phase being a solid.

LOSS FACTOR (capacitive). Product of dielectric constant κ' and the dissipation factor, tan δ, or κ' tan δ.

LOSS FACTOR (inductive). The ratio of the dissipation factor, tan δ, to the permeability μ, or tan δ/μ.

LOW-HEAT DUTY (refractories). Fireclay brick with a PCE of at least 19. There are additional ASTM requirements.

LUMINESCENCE. Emission of light from activated electrons.

MADELUNG CONSTANT. The ratio of energy of an ion in a three-dimensional solid to the energy between two single ions.

MAGNETON. *See* BOHR MAGNETON.

MAT (*or* MATTE). A low-gloss, nonreflective surface (eggshell).

MATTE. Sulfide liquid with a high melting temperature.

METALLIC BONDS. Interatomic attractions utilizing "free" electrons.

METALLIC MATERIALS. Materials which contain elements that readily release their electrons (as contrasted to ceramic materials and nonmetallic materials). Examples: Al, Mg, brass, steel, $CuAl_2$, $MoSi_2$, etc.

MICROSTRUCTURE. A structure with heterogeneities that can be seen through a microscope.
MILLER INDICES. *See* Section 3–3.
MODERATING RATIO. Measure of efficiency of energy reduction without neutron loss:

$$M = \frac{\sigma_s \delta}{\sigma_a},$$

where σ_s and σ_a are the scattering and absorption cross sections respectively, and δ is the fractional energy loss per collision.
MODULUS OF RUPTURE. Calculated maximum stress (outer fiber) in test-beam loading. Applies only with nonductile materials.
MOHS SCALE. Arbitrary hardness scale: talc = 1, diamond = 10.
MONOLITHIC. A wall without joints; *e.g.*, concrete.
MUFFLE. Furnace wall which isolates the products from the combustion atmosphere.
n-TYPE. *See* SEMICONDUCTOR, *n*-TYPE.
NETWORK FORMERS. Oxides which contribute to the polymerization of glass, such as SiO_2, B_2O_3, GeO_2, P_2O_5, As_2O_5.
NETWORK MODIFIERS. Oxides which contribute to the depolymerization of glass, such as BaO, CaO, K_2O, MgO, Na_2O, PbO.
NINE-INCH EQUIVALENT (refractories). 100 cubic inches of wall (which is equivalent to a 9 in. $\times$ $4\frac{7}{16}$ in. $\times$ $2\frac{1}{2}$ in. standard refractory brick).
NONMETALLIC MATERIALS. Materials that contain elements which readily accept or share additional electrons (as contrasted to metallic materials and ceramic materials). Examples: O_2, CH_4, CH_3COOH, oils, plastics, wood, etc.
NU-VALUE. Reciprocal dispersive power:

$$\nu = \frac{n_D - 1}{n_F - n_C}.$$

OPACIFIER. A phase which decreases the transparency of a glass, and thereby increases the diffuse reflectance.
ORDER, LONG-RANGE. Repetition of coordination over many atomic distances. The resulting structure is crystalline.
ORDER, SHORT-RANGE. Identical first-neighbor coordination of atoms. Typical of glassy structures.
p-TYPE. *See* SEMICONDUCTOR, *p*-TYPE.
PACKING FACTOR. Ratio of true volume to bulk volume. Also (1.0 − porosity).
PASTE. Plasticized colloidal solid. It has a low but finite yield point.
PCE. *See* PYROMETRIC CONE EQUIVALENT.
PERMEABILITY (magnetic). Ratio to magnetic flux density to magnetic field strength (B/H).
PERIODIC KILN. *See* KILN, PERIODIC.
PETROGRAPHY. Analytical procedure utilizing transmitted polarized light.
PHASE. A structurally homogeneous part of a materials system.
PHASE RULE. A relationship between the number of observed phases, the variance, and the number of environmental factors in a materials system. (*See* Section 6–1).
PHONON. Quantized atomic vibrations. The main mechanism of thermal conductivity in ceramic materials.
PHOSPHORESCENCE. Delayed luminescence.

PIEZOELECTRICITY. Electrical polarity varying with mechanical pressure.

PLASTER (of Paris). Partially dehydrated gypsum.

PLATE GLASS. Sheet glass which has been ground and polished to give two parallel surfaces.

POLARIZATION (dielectric). Displacement of the centers of positive and negative charges.

POLYMERIZATION. Process of growing large molecules (or structures) from atoms or small molecules.

POLYMORPHISM. More than one crystal structure for a specific composition: *e.g.*, quartz, tridymite and cristobalite for SiO_2.

PORCELAIN. A vitreous whiteware produced by bisque-firing at a low temperature followed by glazing at a relatively high temperature.

PORCELAIN ENAMEL. A vitreous coating for metals.

POROSITY, APPARENT. Porosity of open pores only.

POROSITY, TRUE. Porosity of open and closed pores.

PORTLAND CEMENT. The most common hydraulic cement of construction. The phases most commonly present before hydration are Ca_2SiO_4, Ca_3SiO_5, $Ca_3Al_2O_6$, and Ca_2AlFeO_5.

POTTERY. Nontechnical hollow or flatware containing clay, and generally having significant porosity.

PRIMARY BONDS. Covalent, ionic, and metallic bonds. These have energies of the order of magnitude of 100 kcal/mole.

PRIMARY CLAY. *See* CLAY, RESIDUAL.

PYREX. A brand of borosilicate glass.

PYROELECTRICITY. Electrical polarity resulting from temperature variations.

PYROLYTIC GRAPHITE. Graphite formed by pyrolysis of a carbonaceous gas.

PYROMETRIC CONE EQUIVALENT. The number of that standard pyrometric cone whose tip would touch the supporting plaque simultaneously with a cone of the material being investigated. ASTM.

PYROMETRIC CONES. Small, slender three-sided pyramids of a ceramic material for use in determining the time–temperature effect of heating. A standardized series includes between 60 and 70 compositions with progressively higher softening temperatures.

QUALITY FACTOR. Reciprocal of the dissipation factor or $1/\tan \delta$, where δ is the loss angle.

QUARTZ, FUSED. *See* SILICA GLASS.

QUARTZITE. A hard rock with quartz as the major constituent phase.

RADIATION DAMAGE. Radiation alteration of properties through modification of the crystalline or electronic structures.

REFRACTION. The redirection of light rays as they change media.

REFRACTORIES. Materials which can withstand high temperatures. They are usually ceramic materials.

REGENERATION. A reversing heat exchanger for preheating combustion air (and gaseous fuels) from waste heat of the exhaust gases.

REMANENCE. Remaining magnetization or electrical polarization following the removal of the magnetic or electrical field.

RESORPTION. Decrease or disappearance of a solid phase during cooling as a result of equilibrium changes.

RHEOLOGY. Study of flow and flow properties.

RHEOPEXIC. Converse of thixotropic. *Also see* DILATANCY.

RUBY. Corundum (Al_2C_3) containing Cr^{3+} ions in solid solution.

SAFETY GLASS. Most commonly a "sandwich" of plastic between two sheets of glass, i.e., laminated safety glass. Also tempered glass and wire safety glass.

SAGGER. A refractory box for protecting the ceramic product from flames during firing.

SAGGING (glass). Process of shaping sheet glass by gravity subsidence until it conforms to a mold.

SAND. Particles ranging from approximately 0.1 to 5 mm in diameter. (Normally silica.)

SCHOTTKY DEFECT. An atomic vacancy.

SECONDARY BONDS. Van der Waals bonds. These have energies of less than 10 kcal/mole.

SECONDARY CLAY. *See* CLAY, SEDIMENTARY.

SEED (glass). A very small gas inclusion.

SEMICONDUCTOR, *n*-TYPE. Charge carriers are electrons (negative).

SEMICONDUCTOR, *p*-TYPE. Charge carriers are electron holes (positive).

SEMICONDUCTORS. Materials with resistivities intermediate between metals ($\rho < 10^{-1}$ ohm/cm) and insulators ($\rho > 10^3$ ohm/cm).

SENSITIZER (LUMINESCENCE). Ions added to a luminescent material to increase the efficiency of electron activation.

SHALE. A soft, fine-grained rock of partially indurated clay.

SILICA, FUSED. *See* SILICA GLASS.

SILICA GLASS. Noncrystalline SiO_2. It has short-range order but no long-range order. Sometimes inappropriately called "quartz."

SILICONE. Polymerized silicon compounds, with organic side radicals.

SILOXANE. A silicone with alternating silicons and oxygens.

SINTERING. To agglomerate particles by high-temperature treatment.

SKEW BRICK. A brick with one narrow side inclined so that the two large faces are unequal in size.

SLAG. Oxide liquids (exclusive of the commercial glasses) with a high melting temperature.

SLAKE. To crumble in water.

SLIP (liquid). A liquid suspension of colloidal ceramic materials.

SLIP (plastic). To glide along crystal planes.

SLIP CASTING. Forming a ceramic product from a suspension (slip) by dewatering it in a porous mold until a semirigid article is produced.

SLURRY. *See* SLIP (liquid).

SOFTENING POINT. Temperature at which the viscosity of glass is approximately $10^{7.6}$ poises. (The actual viscosity at the softening point varies slightly with glass density.)

SOLIDUS. The locus of temperatures below which there is only solid. Also, solid solubility limit, with the second phase being a liquid.

SOLVUS. Solid solubility limit, with the second phase being a solid.

SPALLING. Cracking from stresses induced by volume changes.

SPECIFIC GRAVITY (BULK, APPARENT, AND TRUE). *See* DENSITY (BULK, APPARENT, AND TRUE).

SPINEL. A ceramic compound with the basic composition of $R^{2+}R_2^{3+}O_4$. The divalent ions R^{2+} are commonly Mg^{2+}, Fe^{2+}, Mn^{2+}, Ni^{2+}, Zn^{2+}; the trivalent ions, R^{3+}, are commonly Fe^{3+}, Al^{3+}, Cr^{3+}. Occasionally $R^{1+}R^{3+}$ combinations may substitute for $2R^{2+}$; $MgAl_2O_4$ is the prototype spinel.

Spinel, inverse. Spinel with the divalent cations in 6-fold (octahedral) sites, and the trivalent cations in 6-fold and 4-fold (tetrahedral) sites.

Spinel, normal. Spinel with all divalent cations in 4-fold (tetrahedral) sites, and all trivalent cations in 6-fold (octahedral) sites.

Squareness ratio. Ratio of magnetization at $(-H/2)$ to the magnetization at (H), where H is the field strength of the completed hysteresis loop.

Stacking fault. A missing plane in a stacking sequence. (*See* Section 3–5.)

Strain point. Temperature at which the viscosity of glass is $10^{14.5}$ poises. Formerly defined at $10^{14.6}$ poises. (The internal stresses are substantially relieved in 4 hours at this temperature.)

Structure (abrasives). Porosity of bonded abrasive products.

Superduty (refractories). Fireclay refractories with a PCE of at least 33. There are additional ASTM requirements.

Temper (glass). To introduce compressive surface stresses into the glass by rapid cooling from above the annealing point.

Temper (moisture). To mix the appropriate quantity of water into a body until it has proper forming plasticity.

Temperature, fictive. The transition temperature between a supercooled liquid and a glass. (*See* Section 4–4.)

Tempered glass. Glass which has been rapidly cooled from above the annealing point to induce residual compressive stresses in the surface.

Ternary diagrams. Phase diagrams of 3-component systems.

Tetrahedron, SiO_4. A silicon atom surrounded by four oxygen atoms. Each oxygen must either receive an extra electron or be shared by two tetrahedra.

Thermal growth. *See* Thermal ratcheting.

Thermal ratcheting. Irreversible volume expansion accompanying thermal cycling.

Thermistor. Material or device with a large temperature dependence of resistivity.

Thixotropic. Behavior whereby there is a decrease in the apparent viscosity of a colloidal structure during mechanical agitation. Transformation from gel to sol by mechanical agitation.

Tie line. *See* Conode.

Tile. A thin-walled ceramic structural product.

Transducer. Device for converting input energy of one form to output energy of another form.

Transformation, displacive. A change in crystal symmetry as a result of changes in bond length or bond angles (as contrasted to reconstructive transformations). The short-range order is unchanged; the long-range order is changed.

Transformation, glass. The transition from a supercooled liquid to a true glass. (*See* Section 4–4.)

Transformation, reconstructive. Polymorphic phase change which requires bond breaking (as contrasted to displacive transformation).

Transition metals. Elements with partially filled subvalence shells. (*See* Table 2–1.)

Trioctahedral. Atoms filling all of the 6-fold interstices. (*See* Fig. 3–16.)

Twin. A stacking reversal within a crystal structure.

Vacancy. Crystal imperfection involving a missing atom.

Van der Waals bonds. Interatomic attractions arising from electrical dipoles, either natural or induced.

Variance (phase rule). The number of unassigned variables in a materials system. (*See* Section 6–1.)

VARISTOR. Material or device with nonlinear current/voltage response.
VITREOUS. Glasslike.
VITRIFICATION. The reduction of porosity through the formation of a glassy bond in a ceramic product.
VITRIFICATION, INCIPIENT. Initial formation of glass within a ceramic body.
VITRIFIED. Describing a nonporous, glass-bonded ceramic material.
VOLUME, APPARENT. True volume + closed-pore volume.
VOLUME, BULK. True volume + open-pore volume + closed-pore volume.
VYCOR. Trade name for 96% silica glasses.
WATER, INTERLAYER OR INTERPARTICLE. Water among the particles which is responsible for the major volume changes during drying shrinkage.
WATER, LATTICE. Molecular water at specific lattice sites.
WATER, PORE. Free water in the pores of unfired ceramics (not adsorbed).
WEDGE BRICK. A brick with both large faces equally inclined toward the small end.
WHITEWARE. Ceramic products that develop a white-fired color. (Usually applied to domestic products or electrical porcelain, but sometimes includes other products with similar production methods.)
WETTING. Liquid contact angle of less than 90°.

Appendix C

COMMON CERAMIC PHASES

Name	General Composition	Remarks
Akermanite	$Ca_2MgSi_2O_7$	$Si_2O_7^{6-}$ is an ion (*See* Fig. 3–15b.)
Alabandite	MnS	MgO structure
Albite	$NaAlSi_3O_8$	A plagioclase feldspar
Anatase	TiO_2	Polymorph of rutile
Andalusite	Al_2SiO_5	Metastable
Anhydrite	$CaSO_4$	——
Anorthite	$CaAl_2Si_2O_8$	Forms a solid solution with albite
Apatite	$Ca_5P_3O_{12}F$	F^- may be replaced by $(OH)^-$; Mohs hardness is 5
Asbestos	——	Fibrous silicate with a double silicate chain
Attapulgite	$Mg_5Si_8O_{20}[OH]_6 \cdot 4H_2O$	A clay mineral with a chain rather than a sheet structure; $2Al^{3+}$ may replace $3Mg^{2+}$
(Austenite)	fcc Fe	Intermediate temp. form; may dissolve carbon interstitially
Baddeleyite	ZrO_2	Must be transformed to the cubic form and stabilized (for example, with CaO) before use
Bauxite	——	Primary source of Al_2O_3; gibbsite is the major phase
Bentonite	——	A "rock" containing montmorillonite as the major phase
Berlinite	$AlPO_4$	Same structure as quartz (*See* Fig. 13–18.)
Biotite	$[KAl(Mg, Fe)_3Si_3O_{10}(OH)_2]_2$	Black mica; multilayered silicate (*See* Table 3–5.)
Blende	ZnS	Synonym for sphalerite
Borazon (trade name)	BN	β-SiC structure; hardness of diamond (*See* Fig. 3–2.)

COMMON CERAMIC PHASES *(Continued)*

Name	General Composition	Remarks
Borax	$Na_2B_4O_7 \cdot 10H_2O$	——
Brookite	TiO_2	Polymorph of rutile
Brownmillerite	Ca_2AlFeO_5	Constituent of cement clinker
Brucite	$Mg(OH)_2$	Layered structure (*See* Fig. 3–16a.)
Bunsenite	NiO	MgO structure
Calcite	$CaCO_3$	Chief mineral of limestone; Mohs hardness is 3 (*See* Fig. 3–14.)
(Cementite)	Fe_3C	——
Chlorite	(*See* Table 3–5)	Micaceous hybrid of hydroxide and silicate structures
Chromite	$FeCr_2O_4$	A spinel with $R^{2+} = Fe^{2+}$ and $R^{3+} = Cr^{3+}$
Clay	——	*See* Appendix B
Coesite	SiO_2	High-pressure form of silica
Cordierite	$Mg_2Al_4Si_5O_{18}$	——
Corundum	Al_2O_3	Hcp oxygens with $\frac{2}{3}$ of 6-fold sites filled; Mohs hardness is 9
Cristobalite	SiO_2	High-temperature form of silica (*See* Fig. 3–11a.)
Cuprite	Cu_2O	The substitution of $(Cu^{2+} + \square)$ may occur for $2Cu^+$ (*See* Fig. 3–12b.)
Devitrite	$Na_2Ca_3Si_6O_{16}$	Most common devitrification product of soda-lime-silica glasses
Diamond	C	High-pressure form of carbon; Mohs hardness is 10
Dickite	$[Al_2Si_2O_5(OH)_4]_4$	Isomorphous with kaolinite, but with a different stacking arrangement
Diopside	$CaMg(SiO_2)_2$	——
Dolomite	$CaMg(CO_3)_2$	——
Enstatite	$MgSiO_3$	Contains an $(SiO_3)_x^{2-}$ chain (*See* Fig. 3–15c.)
Fayalite	Fe_2SiO_4	An Fe^{2+} olivine

(continued)

COMMON CERAMIC PHASES *(Continued)*

Name	General Composition	Remarks
Feldspar	$R^{+}AlSi_3O_8$	$R^{+}Si^{4+}$ may be replaced by $R^{2+}Al^{3+}$
(Ferrite)	bcc Fe	Ambient polymorph of iron
Fluorite	CaF_2	(*See* Fig. 3–10.) Mohs hardness is 4
Fluorspar	CaF_2	Same as fluorite
Forsterite	Mg_2SiO_4	An Mg^{2+} olivine
Garnet	$R_3^{2+}R_2^{3+}Si_3O_{12}$	Most magnetic garnets have the Si^{4+} replaced with Fe^{3+}, becoming $R_3^{3+}Fe_5^{3+}O_{12}$
Gehlenite	$Ca_2Al_2SiO_7$	Forms a solid solution with akermanite
Gibbsite	$Al(OH)_3$	Two-dimensional structure (*See* Fig. 3–16b.)
Graphite	C	Stable form of carbon; two-dimensional structure
Gypsum	$CaSO_2(OH)_4$ (*or* $CaSO_4 \cdot 2H_2O$)	Naturally occurring, and also the hydrated form of plaster of Paris; Mohs hardness is 2
Halite	NaCl	Prototype for MgO
Halloysite	$[Al_2Si_2O_5(OH)_4 \cdot 2H_2O]_2$	Polymorph of kaolinite, which contains lattice water between the sheets; (as a consequence the sheets become cylindrical)
Hectorite	*See* montmorillonite	Montmorillonite, with Al^{3+} replaced by $Mg^{2+}(+R^{+})$ (*See* Table 3–5.)
Hematite	Fe_2O_3	Corundum structure
Hercynite	$FeAl_2O_4$	A spinel with $R^{2+} = Fe^{2+}$, and $R^{3+} = Al^{3+}$
Illite	Complex	Clay mineral with multiple silicate layers
Ilmenite	$FeTiO_3$	——
Kaolin	——	A "rock" containing kaolinite as the major phase

COMMON CERAMIC PHASES *(Continued)*

Name	General Composition	Remarks
Kaolinite	$[Al_2Si_2O_5(OH)_4]_2$	The most common clay mineral; two-dimensional silicate structure (*See* Table 3–5 and Fig. 3–17.)
Keatite	SiO_2	High-pressure form of silica
Kyanite	Al_2SiO_5	Metastable
Larnite	Ca_2SiO_4	High-temperature form; not an olivine
LeChatelierite	SiO_2	A glass
Leucite	$KAlSi_2O_6$	——
Lime	CaO	Hydrates very readily
Magnesite	$MgCO_3$	——
Magnetite	Fe_3O_4	A magnetic spinel (*See* Section 10–6.)
Manganosite	MnO	MgO structure
(Martensite)	Fe	Metastable; transition phase between austenite and ferrite; highly distorted
Merwinite	$Ca_3MgSi_2O_8$	——
Mica	$[R^+Al_3Si_3O_{10}(OH)_2]_2$	3 R^{2+} may replace $2Al^{3+}$; multilayered silicate (*See* Table 3–5.)
Microcline	$KAlSi_3O_8$	A potash feldspar polymorph
Monticellite	$MgCaSiO_4$	——
Montmorillonite	$[(Al,(MgR^+))_2Si_4O_{10}(OH_2)]_2$	Has high base-exchange properties (*See* Table 3–5.)
Mullite	$Al_6Si_2O_{13}$	Al/Si ratio may be varied; only stable aluminosilicate
Muscovite	$[KAl_3Si_3O_{10}(OH)_2]_2$	White mica; multilayered silicate (*See* Table 3–5.)
Nephelite	R^+AlSiO_4	——
Nontronite	See montmorillonite	Montmorillonite with Al^{3+} replaced by Fe^{3+}
Olivine	$R_2^{2+}SiO_4$	A family of orthosilicates; the radius of R^{2+} must be about 0.8 A

(continued)

COMMON CERAMIC PHASES *(Continued)*

Name	General Composition	Remarks
Orthoclase	$KAlSi_3O_8$	A potash feldspar polymorph; Mohs hardness is 6
Periclase	MgO	Same structure as NaCl (*See* Fig. 3–9b.)
Perovskite	$CaTiO_3$	The same structure as $BaTiO_3$ (*See* Fig. 3–13.)
Phlogopite	$[KAlMg_3Si_3O_{10}(OH, F)_2]_2$	A mica; multilayered structure; most common artificial mica (*See* Table 3–5.)
Plagioclase	$[(NaSi), (CaAl)]AlSi_2O_8$	A family of feldspars; albite and anorthite are end members
Plaster of Paris	$Ca_2S_2O_7(OH)_2$ (*or* $CaSO_4 \cdot \frac{1}{2}H_2O$)	Hydrates to gypsum
Pyrite	FeS_2	Fool's gold
Pyrophyllite	$[Al_2Si_4O_{10}(OH)_2]_2$	Al^{3+} analog to talc (*See* Table 3–5.)
Pyrrhotite	$Fe_{<1}S$	Hcp structure of S^{2-} with Fe^{2+} (and a few $Fe^{3+} + \square$) in the 6-fold sites
Quartz	SiO_2	Ambient form; most common of *all* solid phases; Mohs hardness is 7
Quartzite	——	A rock containing quartz as the major phase
Ruby	——	Corundum with some substitution of Cr^{3+} for Al^{3+} (red)
Rutile	TiO_2	Most common form of TiO_2 (*See* Fig. 3–12a.)
Sand	——	*See* Appendix B
Sapphire	——	Corundum with cobalt in solid solution (blue)
Serpentine	$[Mg_3Si_2O_5(OH)_4]_2$	Mg^{2+} analog to kaolinite (*See* Table 3–5.)
Sillimanite	Al_2SiO_5	Metastable
Shale	——	*See* Appendix B
Sphalerite	ZnS	Sometimes called zincblende; has significant covalency (*See* Fig. 3–9c.)

COMMON CERAMIC PHASES *(Continued)*

Name	General Composition	Remarks
Spinel	$R^{2+}R_2^{3+}O_4$	A family of minerals with the structure of $MgAl_2O_4$ (*See* Section 10–6.)
Steatite	——	Not a phase; materials made with large amounts of talc
Talc	$[Mg_3Si_4O_{10}(OH)_2]_2$	Triple-layered silicate; Mohs hardness is 1 (*See* Table 3–5.)
Tobermorite	*See* Section 13–6	Usually amorphous; chief product of portland cement hydration
Topaz	$Al_2(F,OH)_2SiO_4$	Mohs hardness is 8
Tridymite	SiO_2	Intermediate-temperature form of silica (*See* Fig. 3–11b.)
Villiaumite	NaF	NaCl structure
Wollastonite	$CaSiO_3$	——
Wurtzite	ZnS	*See* Fig. 3–9d.
Wüstite	FeO	MgO structure; it has defects in it if oxidized; $2Fe^{3+} + \square$ replace $3Fe^{2+}$
Zincite	ZnO	Same structure as wurtzite
Zircon	$ZrSiO_4$	——

Appendix D

TABLE OF ELEMENTS AND OXIDES

(Based on C^{12} = 12.000)

Element	Symbol	Atomic number	Atomic weight	Valence, most common	Approximate Ionic radius, A (Coordination number = 6)	Oxide (solid)			
						Formula	Formula weight	Density, gm/cm^3	Melting point, °C
Hydrogen	H	1	1.008	+	—	H_2O	18.016	0.92	0.0
Helium	He	2	4.003	Inert	—	—	—	—	—
Lithium	Li	3	6.94	+	0.78	Li_2O	29.88	2.013	Subl.*
Beryllium	Be	4	9.01	2+	0.34	BeO	25.01	3.02	2575
Boron	B	5	10.81	3+	0.25	B_2O_3	69.62	1.84	577
Carbon	C	6	12.000	4+	~0.2	CO_2	44.00	1.53	Subl.
Nitrogen	N	7	14.007	3−	—	—	—	—	—
Oxygen	O	8	15.999	2−	1.32	O_2	32.00	—	—
Fluorine	F	9	19.00	—	1.33	—	—	—	—
Neon	Ne	10	20.18	Inert	—	—	—	—	—
Sodium	Na	11	22.99	+	0.98	Na_2O	61.98	2.27	Subl.
Magnesium	Mg	12	24.31	2+	0.78	MgO	40.31	3.65	2800
Aluminum	Al	13	26.98	3+	0.57	Al_2O_3	101.94	4.0	2050
Silicon	Si	14	28.09	4+	0.39	SiO_2	60.09	2.65	1710
Phosphorus	P	15	30.97	5+	0.2–0.4	P_2O_5	141.94	2.39	Subl.
Sulfur	S	16	32.06	2−	1.74	SO_3	80.06	—	17
Chlorine	Cl	17	35.45	—	1.81	—	—	—	—
Argon	Ar	18	39.95	Inert	—	—	—	—	—
Potassium	K	19	39.10	+	1.33	K_2O	94.19	2.32	—
Calcium	Ca	20	40.08	2+	1.06	CaO	56.08	2.62	2572

* Sublimates.

† Decomposes.

Scandium	Sc	21	44.96	3+	0.83	Sc_2O_3	137.92	3.86	—
Titanium	Ti	22	47.90	4+	0.64	TiO_2	79.90	4.26	1640
Vanadium	V	23	50.94	3+	0.65	V_2O_3	149.88	4.87	1970
Chromium	Cr	24	52.00	3+	0.64	Cr_2O_3	152.00	5.21	1900
Manganese	Mn	25	54.94	2+	0.91	MnO	70.94	5.43	1840
Iron	Fe	26	55.85	2+	0.83	FeO	71.85	5.7	1420
				3+	0.67	Fe_2O_3	159.70	5.2	D†
Cobalt	Co	27	58.93	2+	0.82	CoO	74.93	5.68	D
Nickel	Ni	28	58.71	2+	0.78	NiO	74.71	7.45	2090
Copper	Cu	29	63.54	+	0.96	Cu_2O	143.14	6.0	1235
Zinc	Zn	30	65.37	2+	0.83	ZnO	81.37	5.61	>1800
Gallium	Ga	31	69.72	3+	0.62	Ga_2O_3	187.44	6.44	1900
Germanium	Ge	32	72.59	4+	0.44	GeO_2	104.59	6.3	1086
Arsenic	As	33	74.92	3+	0.69	As_2O_3	197.84	3.86	Subl.
Selenium	Se	34	78.96	2−	1.91	SeO_3	126.96	3.6	—
Bromine	Br	35	79.92	—	1.96	—	—	—	—
Krypton	Kr	36	83.80	Inert	—	—	—	—	—
Rubidium	Rb	37	85.47	+	1.49	Rb_2O	186.95	3.72	D
Strontium	Sr	38	87.62	2+	1.27	SrO	103.62	4.7	2430
Yttrium	Y	39	88.91	3+	1.06	Y_2O_3	225.82	4.84	2410
Zirconium	Zr	40	91.22	4+	0.87	ZrO_2	123.22	5.49	2700
Niobium	Nb	41	92.91	5+	0.69	Nb_2O_5	265.8	4.6	1520
Molybdenum	Mo	42	95.94	4+	0.68	MoO_2	127.94	6.44	Subl.
Technicium	Tc	43	99	?	?	—	—	—	—
Ruthenium	Ru	44	101.1	4+	0.65	RuO_2	133.1	7.0	—
Rhodium	Rh	45	102.9	3+	0.68	RhO	118.9	—	—
Palladium	Pd	46	106.4	2+	0.9	PdO	122.4	8.31	D
Silver	Ag	47	107.9	+	1.13	Ag_2O	231.8	7.14	D
Cadmium	Cd	48	112.4	2+	1.03	CdO	128.4	8.3	D

(continued)

Appendix D

TABLE OF ELEMENTS AND OXIDES (Continued)

(Based on $C^{12} = 12.000$)

Element	Symbol	Atomic number	Atomic weight	Valence, most common	Approximate Ionic radius, A (Coordination number = 6)	Oxide (solid)			
						Formula	Formula weight	Density, gm/cm^3	Melting point, °C
Indium	In	49	114.8	3+	0.92	In_2O_3	277.6	7.18	—
Tin	Sn	50	118.7	4+	0.74	SnO_2	150.7	7.0	1127
Antimony	Sb	51	121.8	3+	0.90	Sb_2O_3	291.5	5.2	652
Tellurium	Te	52	127.6	2—	2.11	TeO_3	175.6	5.08	D
Iodine	I	53	126.9	—	2.20	—	—	—	—
Xenon	Xe	54	131.3	Inert or 4+	?	?	—	—	—
Cesium	Cs	55	132.9	+	1.65	Cs_2O	281.8	4.36	D
Barium	Ba	56	137.3	2+	2.17	BaO	153.3	5.72	1923
Lanthanum	La	57	138.9	3+	1.22	La_2O_3	325.8	6.51	2315
Cerium	Ce	58	140.1	3+	1.18	Ce_2O_3	328.3	6.9	1690
Praeseodymium	Pr	59	140.9	3+	1.16	Pr_2O_3	329.8	6.88	—
Neodymium	Nd	60	144.3	3+	1.15	Nd_2O_3	336.5	7.24	1900
Promethium	Pm	61	147	3+	?	?	—	—	—
Samarium	Sm	62	150.4	3+	1.13	Sm_2O_3	348.8	7.43	—
Europium	Eu	63	152.0	3+	1.13	Eu_2O_3	352.0	7.42	—
Gadolinium	Gd	64	157.3	3+	1.11	Gd_2O_3	362.6	7.4	—
Terbium	Tb	65	158.9	3+	1.09	Tb_2O_3	365.8	—	—
Dysprosium	Dy	66	162.5	3+	1.07	Dy_2O_3	373.0	7.81	—
Holmium	Ho	67	164.9	3+	1.05	?	—	—	—
Erbium	Er	68	167.3	3+	1.04	Er_2O_3	368.6	8.64	—

Thulium	Tm	69	168.9	3+	1.04	Tm_2O_3	386.6	—	—
Ytterbium	Yb	70	173.0	3+	1.02	Yb_2O_3	394.0	9.17	—
Lutecium	Lu	71	175.0	3+	0.99	—	—	—	—
Hafnium	Hf	72	178.5	4+	0.84	HfO_2	210.5	9.7	2812
Tantalum	Ta	73	180.9	5+	0.68	Ta_2O_5	441.8	8.74	D
Tungsten	W	74	183.9	4+	0.68	WO_3	231.9	7.16	2130
Rhenium	Re	75	186.2	4+	?	ReO_4	250.22	8.4	D
Osmium	Os	76	190.2	4+	0.67	OsO_2	222.2	7.91	—
Iridium	Ir	77	192.2	4+	0.66	IrO_2	224.2	3.15	—
Platinum	Pt	78	195.1	2+	0.8	PtO	211.1	14.9	555
Gold	Au	79	197.0	+	1.37	Au_2O	410.4	3.6	D
Mercury	Hg	80	200.6	2+	1.12	Hg_2O	417.2	9.8	D
Thallium	Tl	81	204.4	3+	1.05	Tl_2O_3	456.8	10.2	759
Lead	Pb	82	207.2	2+	1.32	PbO	223.2	9.5	888
Bismuth	Bi	83	209.0	3+	?	Bi_2O_3	466.0	8.5	860
Polonium	Po	84	210	?	?	?	—	—	—
Astatine	At	85	210	?	?	?	—	—	—
Radon	Rn	86	222	Inert	—	—	—	—	—
Francium	Fr	87	223	+	?	?	—	—	—
Radium	Ra	88	226	2+	1.4	?	—	—	—
Actinium	Ac	89	227	3+	?	?	—	—	—
Thorium	Th	90	232	4+	1.10	ThO_2	264	9.69	3050
Protactinium	Pa	91	231	?	?	?	—	—	—
Uranium	U	92	238.1	4+	1.05	UO_2	270.1	10.9	2176

Appendix E

PROPERTIES OF SELECTED COMMERCIAL CERAMIC MATERIALS

(Taken from numerous sources)

Material		Density, gm/cm^3	Thermal conductivity, $\left(\frac{cal \cdot cm}{°C \cdot cm^2 \cdot sec}\right)^*$, at 68°F	Thermal expansion, $\left(\frac{in.}{in. \cdot °F}\right)^\dagger$, at 68°F	Electrical resistivity, ohm·cm‡, at 68°F	Modulus of elasticity, psi, at 68°F
Al_2O_3	T	3.8	0.07	5×10^{-6}	10^{12}	53×10^6
Brick						
Building	B	2.3	0.0014	5×10^{-6}	—	—
Fireclay	B	2.1	0.002	2.5×10^{-6}	10^8	15×10^6
Paving	B	2.5	—	2×10^{-6}	—	—
Silica	B	1.75	0.002	—	10^8	—
Concrete	B	2.4	0.0025	7×10^{-6}	—	—
Glass						
Plate	T	2.5	0.0018	5×10^{-6}	10^{14}	9×10^6
Borosilicate	T	2.4	0.0025	2×10^{-6}	10^{16}	8×10^6
Silica	T	2.2	0.003	0.3×10^{-6}	10^{20}	10×10^6
Vycor	T	2.2	0.003	0.4×10^{-6}	10^{18}	10×10^6
Wool	T	0.05	0.0006	—	—	—
Graphite (ave.)	T	2.2	0.1	3×10^{-6}	10^{-5}	—
MgO	T	3.7	0.1	5×10^{-6}	10^{10}	30×10^6
SiC	T	3.2	0.2	2.5×10^{-6}	10^0	50×10^6
SiO_2 (quartz)	T	2.65	0.03	7×10^{-6}	—	—
TiC	T	4.3	0.07	4×10^{-6}	10^{-4}	45×10^6

* Multiply by 0.806 to get (Btu·in.)/(°F·ft²·sec)
† Multiply by 1.8 to get (cm)/(cm·°C)
‡ Multiply by 0.393 to get (ohm·in.)

B = Bulk volume properties
T = True volume properties

Appendix F

SELECTED PHASE DIAGRAMS

These phase diagrams are selected (1) to include those most frequently encountered in considering ceramic materials, (2) to illustrate specific reactions, and (3) to provide additional diagrams for study problems.

Phase diagrams of other systems are available in *Phase Diagrams for Ceramists*, 1964 edition by E. M. Levin, C. R. Robbins, and H. F. McMurdie. Columbus, O.: American Ceramic Society. Also see Reference 6–3 for recent modifications of selected ternary silicate diagrams.

Coordinates are expressed in *weight percent* and *Celsius* (centigrade) scale unless otherwise stated.

————————	Established boundaries
----------------	Estimated boundaries
▬ – ▬ –	Three-phase invariant temperatures

Single-Component Systems

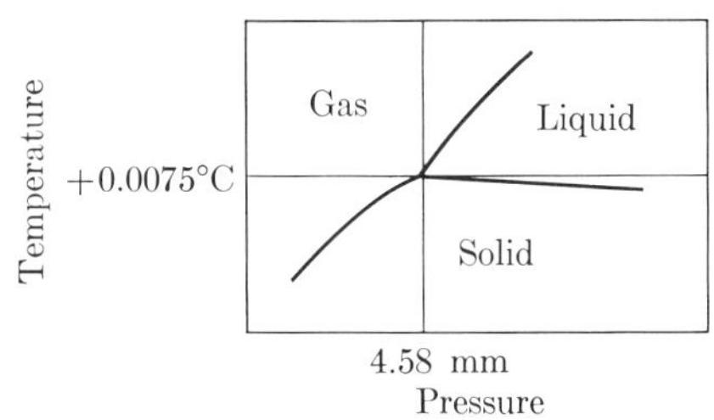

Fig. F–1. The H_2O system.

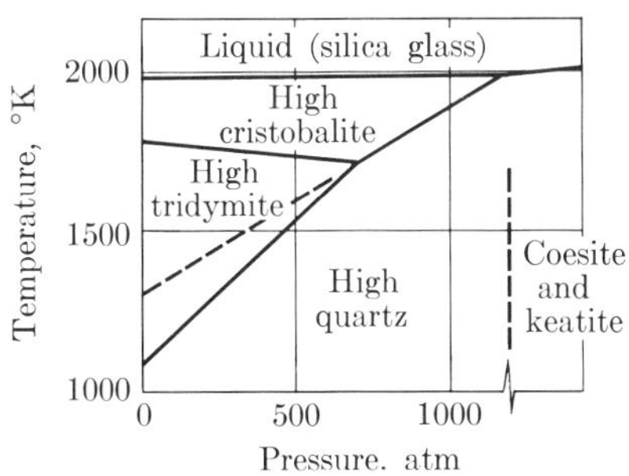

Fig. F–2. The SiO_2 system.

Binary Systems

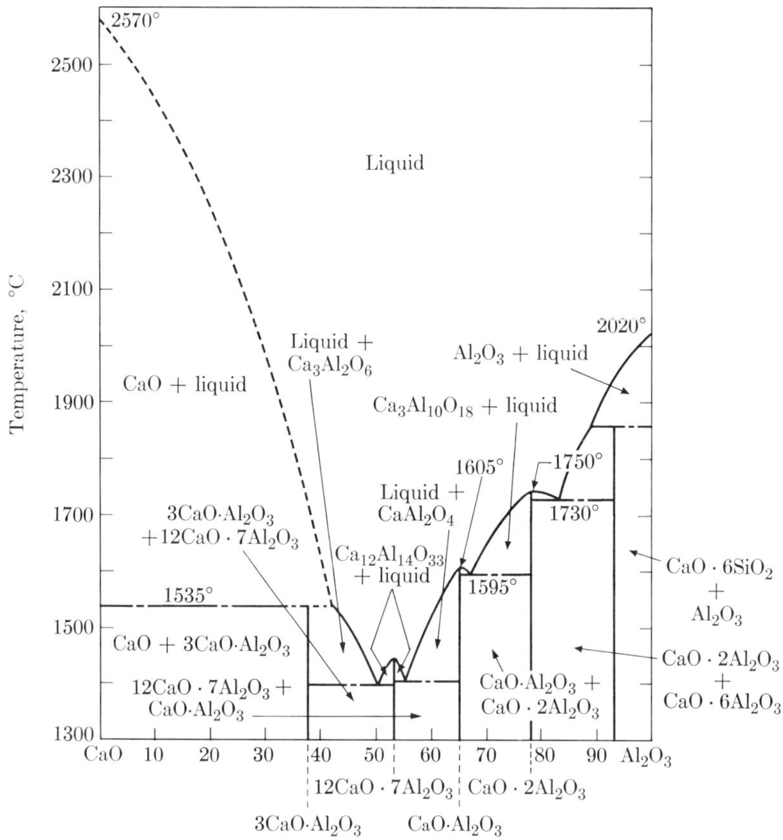

FIG. F–3. **Al_2O_3-CaO.** Adapted from G. A. Rankin and F. E. Wright, *Amer. Journ. of Sci.*, 4th series, **39,** 11 (1915); and E. F. Osborn and A. Muan, Reference 6–3.

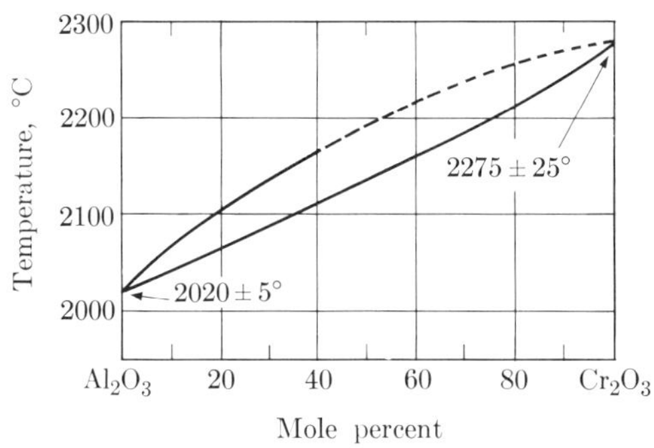

FIG. F–4. **Al_2O_3-Cr_2O_3.** Adapted from E. N. Bunting, *Bureau of Standards, Journ. of Res.*, **6** (6), 948 (1931); and E. F. Osborn and A. Muan, Reference 6–3.

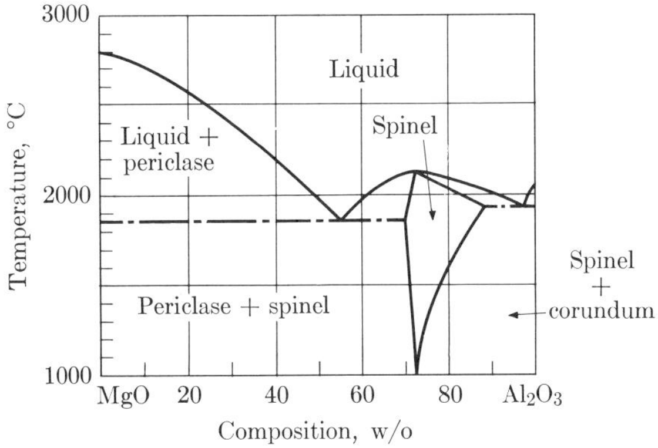

FIG. F–5. **Al_2O_3-MgO.** Adapted from G. A. Rankin and H. E. Merwin, *Journ. Amer. Chem. Soc.*, **38,** 571 (1916); and D. M. Roy and E. F. Osborn, *Journ. Amer. Ceram. Soc.*, **36,** 149 (1953).

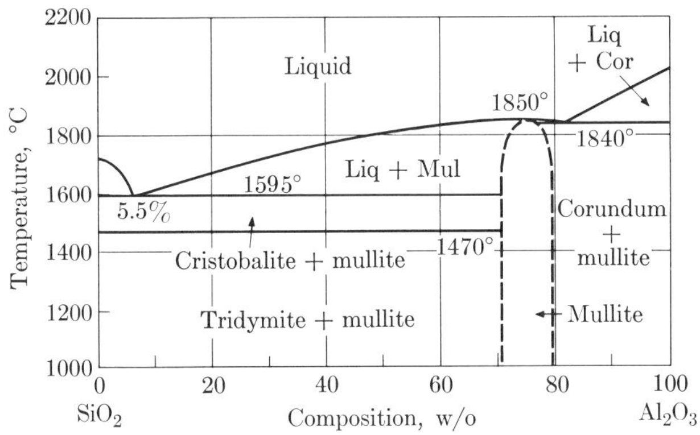

FIG. F–6. **Al_2O_3-SiO_2.** Adapted from S. Aramaki and R. Roy, *Journ. Amer. Ceram. Soc.*, **42,** 644 (1959).

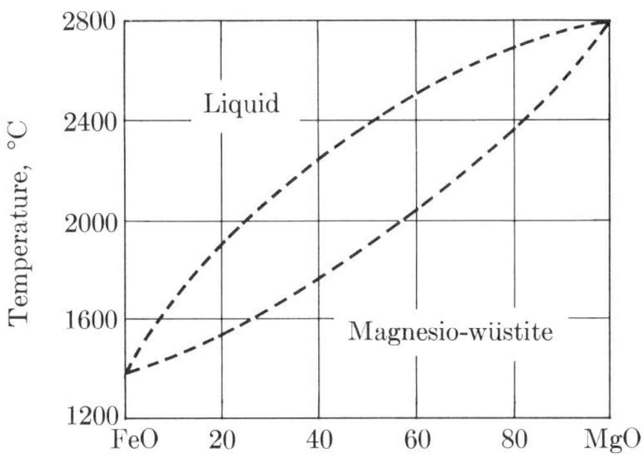

FIG. F–7. **FeO-MgO.** Adapted from N. L. Bowen and J. F. Schairer, *Amer. Journ. Sci.*, 5th series, **29,** 153 (1935); and E. F. Osborn and A. Muan, Reference 6–3.

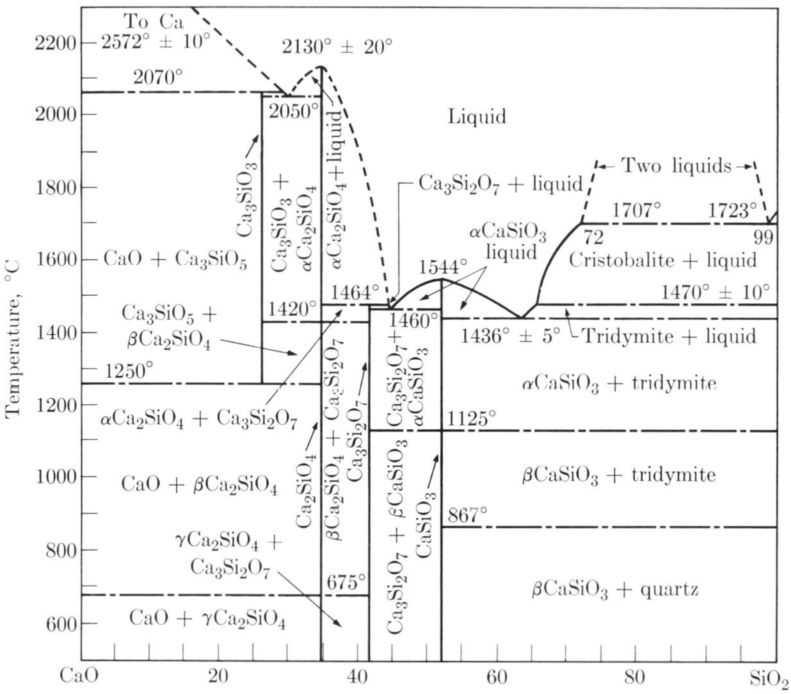

FIG. F–8. **CaO-SiO_2.** Adapted from E. M. Levin, H. F. McMurdie, and F. P. Hall, *Phase Diagrams for Ceramists*, Columbus, O.: American Ceramic Society, 1956, p. 48; and E. F. Osborn and A. Muan, Reference 6–3.

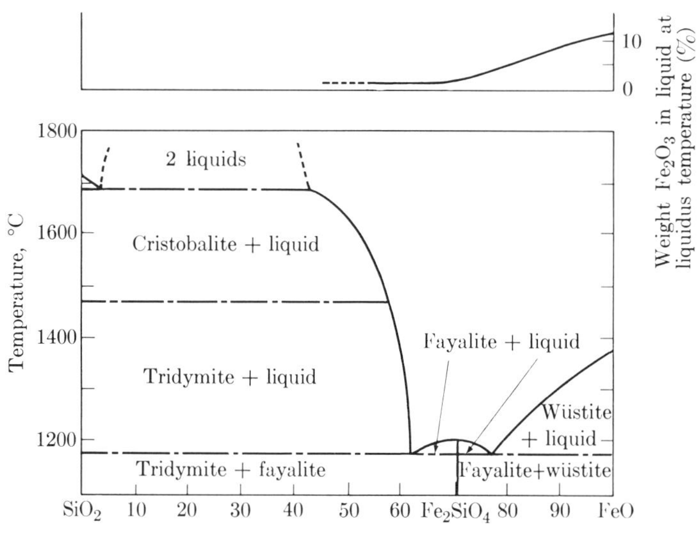

FIG. F–9. **FeO-SiO_2.** Adapted from N. L. Bowen and J. F. Schairer, *Amer. Journ. Sci.*, 5th series, **24,** 200 (1932); and E. F. Osborn and A. Muan, Reference 6–3.

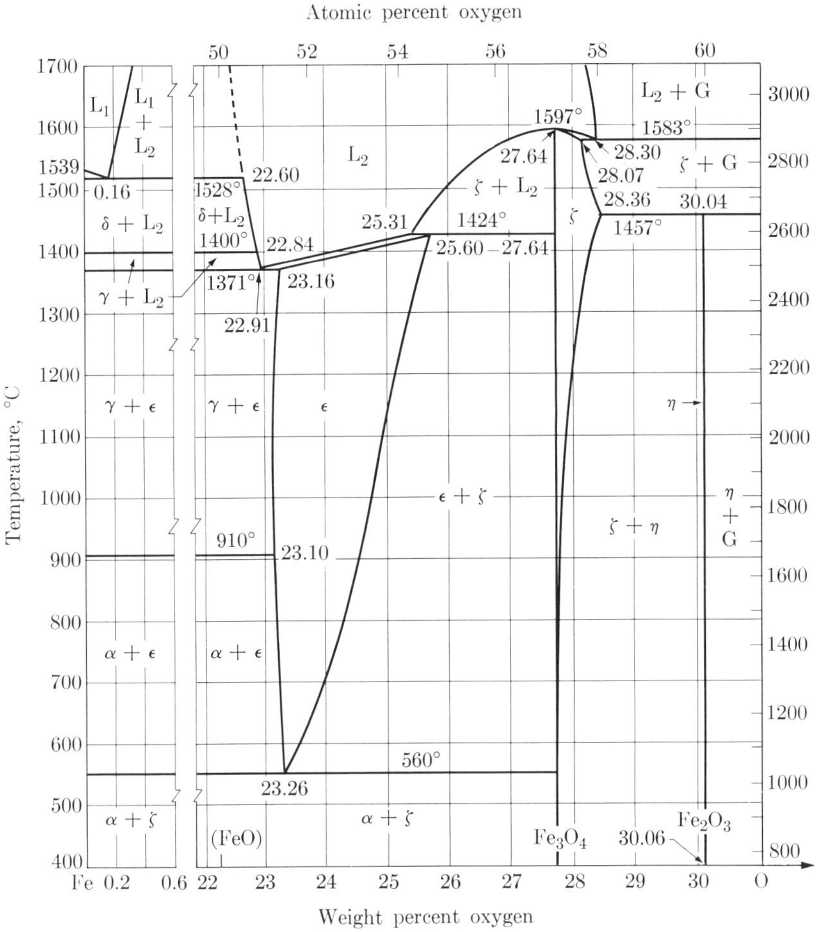

Fig. F–10. **Fe-O.** L. S. Darken and R. W. Gurrey, "Iron-Oxygen," *Metals Handbook,* Cleveland, O.: American Society of Metals, 1948, p. 1212.

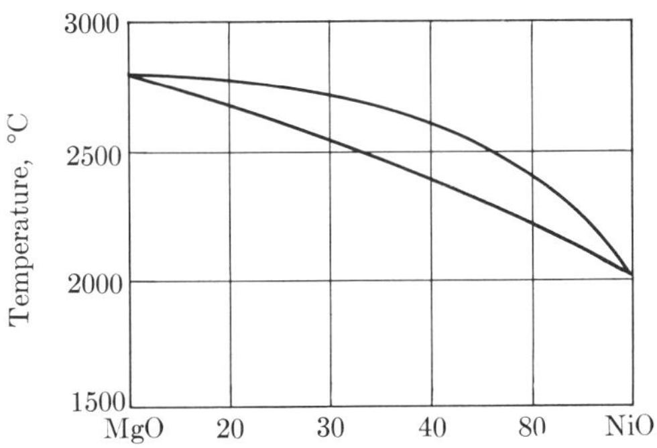

Fig. F–11. **MgO-NiO.** Adapted from H. Wartenburg and E. Prophet, *Z. Anorg. u. Allgem. Chem.,* **208,** 379 (1932).

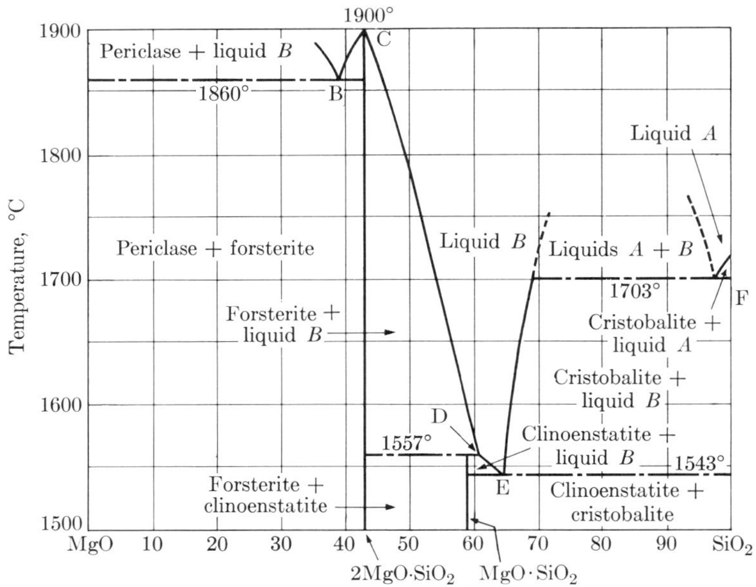

FIG. F–12. **MgO-SiO$_2$.** Adapted from N. L. Bowen and O. Anderson, *Amer. Journ. Sci.*, 4th series, **37,** 488 (1914); and E. F. Osborn and A. Muan, Reference 6–3.

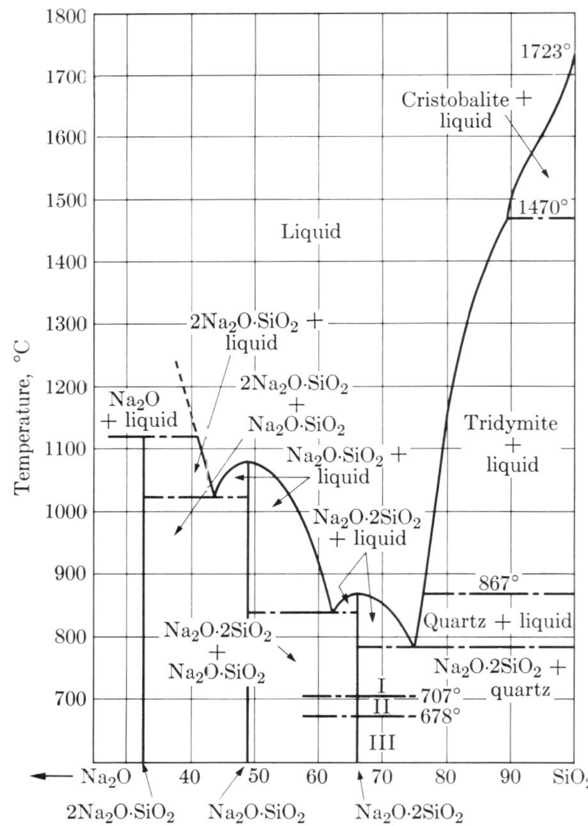

FIG. F–13. **Na$_2$O-SiO$_2$.** Adapted from F. C. Kracek, *Journ. Phys. Chem.*, **34,** 1588 (1930), and *Journ. Amer. Chem. Soc.*, **61,** 2869 (1939).

Ternary Systems

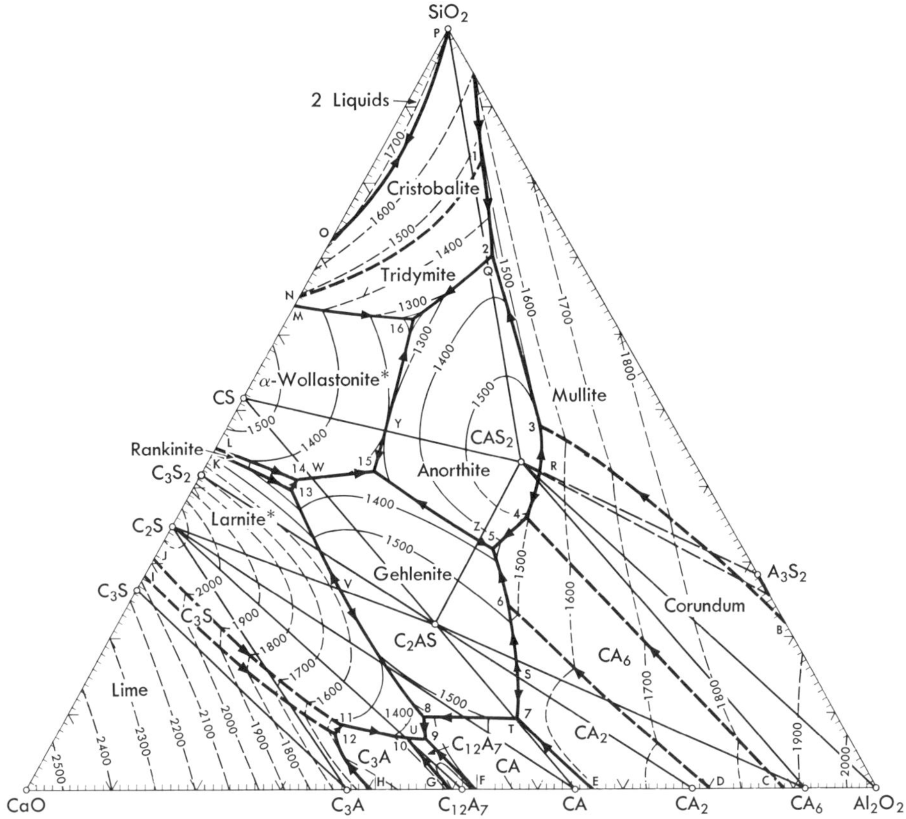

Compound	Melting temp., °C	Phase name	Composition		
			Al_2O_3	CaO	SiO_2
A-Al_2O_3	~2020	Corundum	100	—	—
C-CaO	~2570	Lime	—	100	—
S-SiO_2	1723	Cristobalite	—	—	100
S-SiO_2		Tridymite	—	—	100
S-SiO_2		Quartz*	—	—	100
CS	1544	Wollastonite*	—	48	52
A_3S_2	~1850	Mullite	72	—	28
CAS_2	1553	Anorthite	37	20	43
C_2AS	1593	Gehlenite	37	41	22
C_3S_2	*D*†	Rankinite	—	58	42
C_2S	~2130	Larnite*	—	65	35
C_3S	*D*†	—	—	74	26
C_3A	*D*†	—	38	62	—
$C_{12}A_7$	1455	—	51	49	—
CA	~1605	—	64.5	35.5	—
CA_2	~1750	—	78.5	21.5	—
CA_6	*D*†	—	91.5	8.5	—

*Name of 25°C polymorph
†*D* = decomposes

Invariant Points									
Ternary					Binary				
	Temp., °C	Al_2O_3	CaO	SiO_2		Temp., °C	Al_2O_3	CaO	SiO_2
1	1470	12	5	83	A	~1590	6	—	94
2	1345	20	10	70	B	~1840	78	—	22
3	1512	37	15	48	C	~1850	89	11	—
4	1495	41	23	36	D	~1730	80	20	—
5	1380	39	29	32	E	~1595	66.5	33.5	—
6	1475	45	31	24	F	1400	53	47	—
7	1500	53	37.5	9.5	G	1395	50	50	—
8	1380	42	48.5	9.5	H	1535	41	59	—
9	1335	43.5	49.5	7	I	~2070	—	72	28
10	1335	41.5	52	6.5	J	~2050	—	70	30
11	1455	33	58.5	8.5	K	1464	—	55.5	44.5
12	1470	33	59.5	7.5	L	1460	—	55	45
13	~1315	12	49	39	M	1436	—	36	64
14	~1310	12	47	41	N	1470	—	35	65
15	1266	20	38	42	O	1707	—	27	73
16	1170	15	23	62	P	1707	—	1	99
					Q	1368	19.5	11.5	70
					R	1547	39.5	18.5	42
					S	1552	50	35	15
					T	1512	52.5	38	9.5
					U	1335	41	52.5	6.5
					V	1545	24	50	26
					W	1318	13	46	41
					Y	1307	37	30	33
					Z	1385	19	34	47

FIG. F–14. **Al_2O_3-CaO-SiO_2.** Adapted from E. F. Osborn and A. Muan, *Phase Equilibrium Diagrams of Oxide Systems*, Columbus, O.: American Ceramic Society, 1960, Plate I.

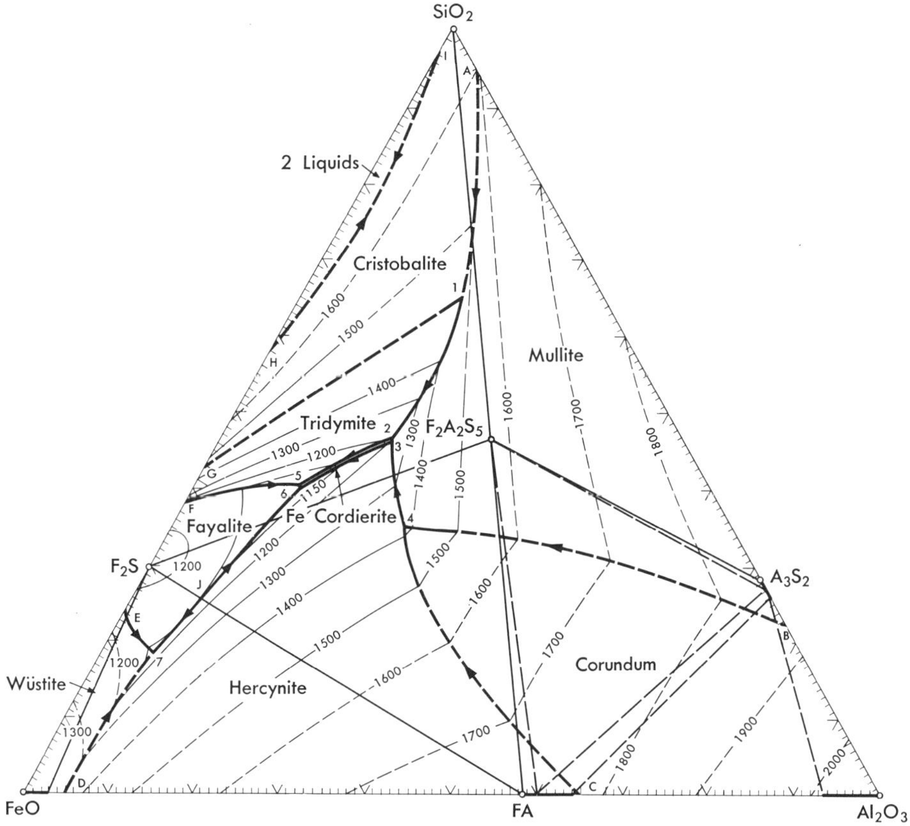

Compound	Melting temp., °C	Phase name	Composition		
			Al_2O_3	FeO	SiO_2
A-Al_2O_3	~2020	Corundum	100	—	—
F-FeO	1369	Wüstite	—	100	—
S-SiO_2	1723	Cristobalite	—	—	100
S-SiO_2		Tridymite	—	—	100
S-SiO_2		Quartz*	—	—	100
A_3S_2	~1850	Mullite	72	—	28
FA	~1780	Hercynite	58.5	41.5	—
F_2S	1205	Fayalite	—	71.5	29.5
$F_2A_2S_5$	*D*†	Fe-Cordierite	31.5	22.5	46

* Name of 25°C polymorph
† *D* = decomposes

Invariant Points										
Ternary					Binary					
	Temp., °C	Al_2O_3	FeO	SiO_2		Temp., °C	Al_2O_3	FeO	SiO_2
1	1470	19	16	65	A	~1590	6	—	94
2	1210	20	33.5	46.5	B	~1840	78	—	22
3	1205	20	34	46	C	~1750	65	35	—
4	1380	27	38	35	D	1329	5	95	—
5	1083	12	47.5	40.5	E	1177	—	76	24
6	1088	12.5	48	39.5	F	1178	—	62	38
7	1148	6	75.5	18.5	G	1470	—	57.5	42.5
					H	1698	—	42	58
					I	1698	—	3.5	96.5
					J	1150	8	66.5	25.5

FIG. F–15. **Al_2O_3-FeO-SiO_2.** Adapted from E. F. Osborn and A. Muan, *Phase Equilibrium Diagrams of Oxide Systems*, Columbus, O.: American Ceramic Society, 1960, Plate 9.

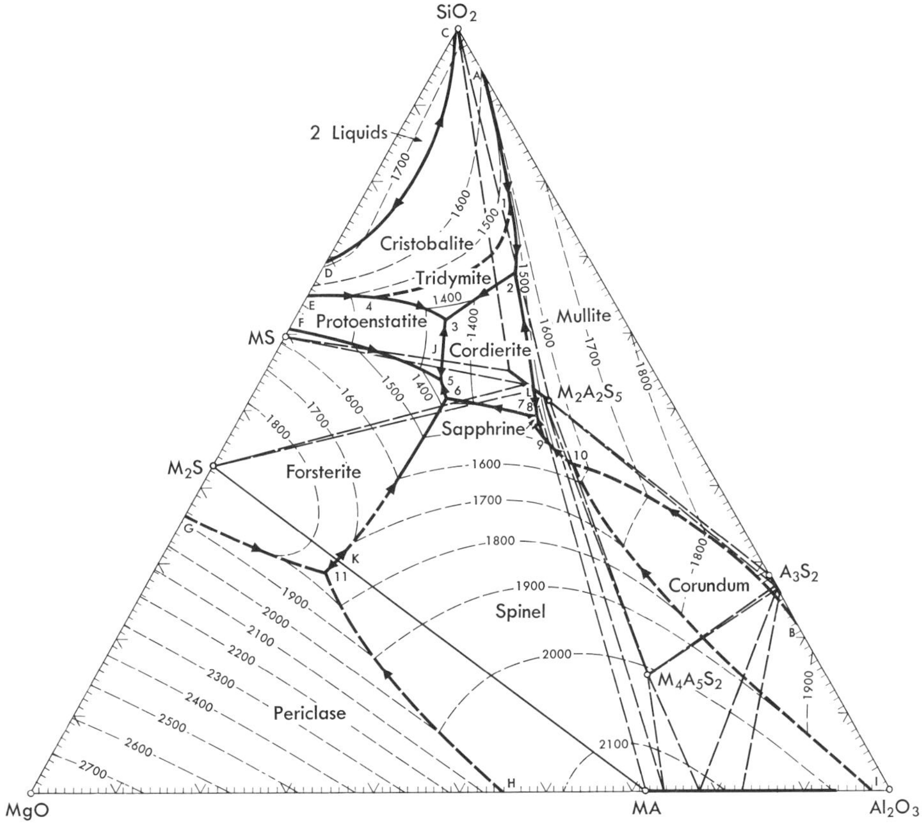

Compound	Melting temp., °C	Phase name	Composition Al_2O_3	MgO	SiO_2
A-Al_2O_3	~2020	Corundum	100	—	—
M-MgO	~2800	Periclase	—	100	—
S-SiO_2	1723	Cristobalite	—	—	100
S-SiO_2		Tridymite	—	—	100
S-SiO_2		Quartz*	—	—	100
MS	D†	Enstatite*	—	40	60
M_2S	~1900	Forsterite	—	57	43
A_3S	~1850	Mullite	72	—	28
MA	~2135	Spinel	71.5	28.5	—
$M_2A_2S_5$	D†	Cordierite	35	14	51
$M_4A_5S_2$	D†	Sapphrine	65	20	15

*Name of 25°C polymorph
†D = decomposes

Invariant Points									
Ternary					Binary				
	Temp., °C	Al_2O_3	MgO	SiO_2		Temp., °C	Al_2O_3	MgO	SiO_2
1	1470	18	5.5	76.5	A	~1590	6	—	94
2	1440	22.5	9.5	68	B	~1840	78	—	22
3	1335	17.5	20.5	62	C	1703	—	1	99
4	1470	8.5	26.5	65	D	1703	—	31	69
5	1365	21	25	54	E	1543	—	35	65
6	1370	23	25.5	51.5	F	1557	—	39	61
7	1453	33.5	17.5	49	G	~1860	—	64	36
8	1460	34.5	16.5	49	H	~1850	55	45	—
9	1482	37	17	46	I	~1925	98	2	—
10	1578	42	15	43	J	1367	20	24	56
11	~1710	20	51	29	K	~1710	21	49	30
					L	1465	32	15	53

FIG. F–16. **Al_2O_3-MgO-SiO_2.** Adapted from E. F. Osborn and A. Muan, *Phase Equilibrium Diagrams of Oxide Systems*, Columbus, O.: American Ceramic Society, 1960, Plate 3.

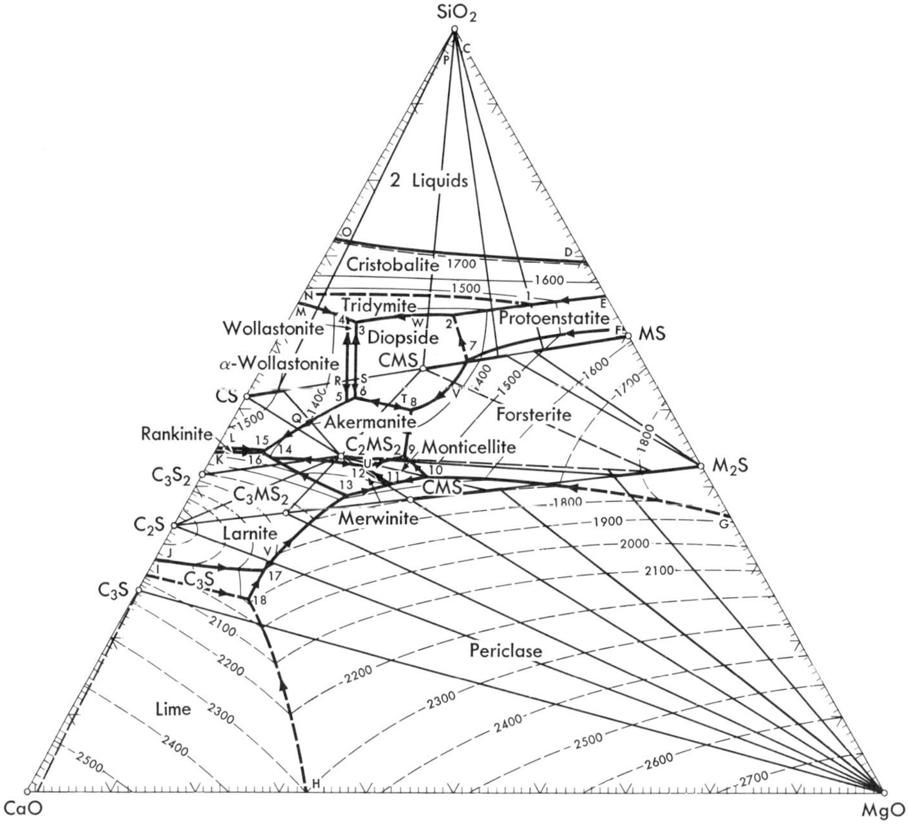

Compound	Melting temp., °C	Phase name	Composition CaO	MgO	SiO_2
C-CaO	~2570	Lime	100	—	—
M-MgO	~2800	Periclase	—	100	—
S-SiO_2	1723	Cristobalite	—	—	100
S-SiO_2		Tridymite	—	—	100
S-SiO_2		Quartz*	—	—	100
CS	1544	Wollastonite*	48	—	52
C_3S_2	*D*†	Rankinite	58	—	42
C_2S	~2130	Larnite*	65	—	35
C_3S	*D*†	—	74	—	26
MS	*D*†	Enstatite*	—	40	60
M_2S	~1900	Forsterite	—	57	43
CMS_2	1391	Diopside	26	18.5	55.5
C_2MS_2	1454	Akermanite	41	15	44
CMS	*D*†	Monticellite	36	26	38
C_3MS_2	*D*†	Merwinite	51.5	12	36.5

*Name of 25°C polymorph
†*D* = decomposes

Invariant Points									
Ternary					Binary				
	Temp., °C	CaO	MgO	SiO_2		Temp., °C	CaO	MgO	SiO_2
1	1470	9	27	64	C	1703	—	1	99
2	~1375	19	19	62	D	1703	—	31	69
3	1320	30.5	8	61.5	E	1543	—	35	65
4	1336	31.5	7	61.5	F	1557	—	39	61
5	1360	37	12	51	G	~1860	—	64	36
6	1350	35.5	13	51.5	H	~2300	67	33	—
7	1390	20	23.5	56.5	I	~2070	72	—	28
8	1357	30	20	50	J	~2050	70	—	30
9	1430	33.5	22.5	44	K	1464	55.5	—	44.5
10	1502	32.5	26	41.5	L	1460	55	—	45
11	1490	37.5	22.5	40	M	1436	36	—	64
12	1436	39	18.5	42.5	N	1470	35	—	65
13	1575	43	18	39	O	1707	27	—	73
14	1400	49	7	44	P	1707	1	—	99
15	1376	50	5.5	44.5	Q	1400	44.5	8	47.5
16	1379	50	6	44	R	1368	35.5	10.5	54
17	~1790	57.5	13.5	29	S	1358	34.5	11.5	54
18	~1850	61	13.5	25.5	T	1367	32.5	17	50.5
					U	1450	42.5	14.5	43
					V	~1800	56	14	30

FIG. F–17. **CaO-MgO-SiO_2.** Adapted from E. F. Osborn and A. Muan, *Phase Equilibrium Diagrams of Oxide Systems*, Columbus, O.: American Ceramic Society, 1960, Plate 2.

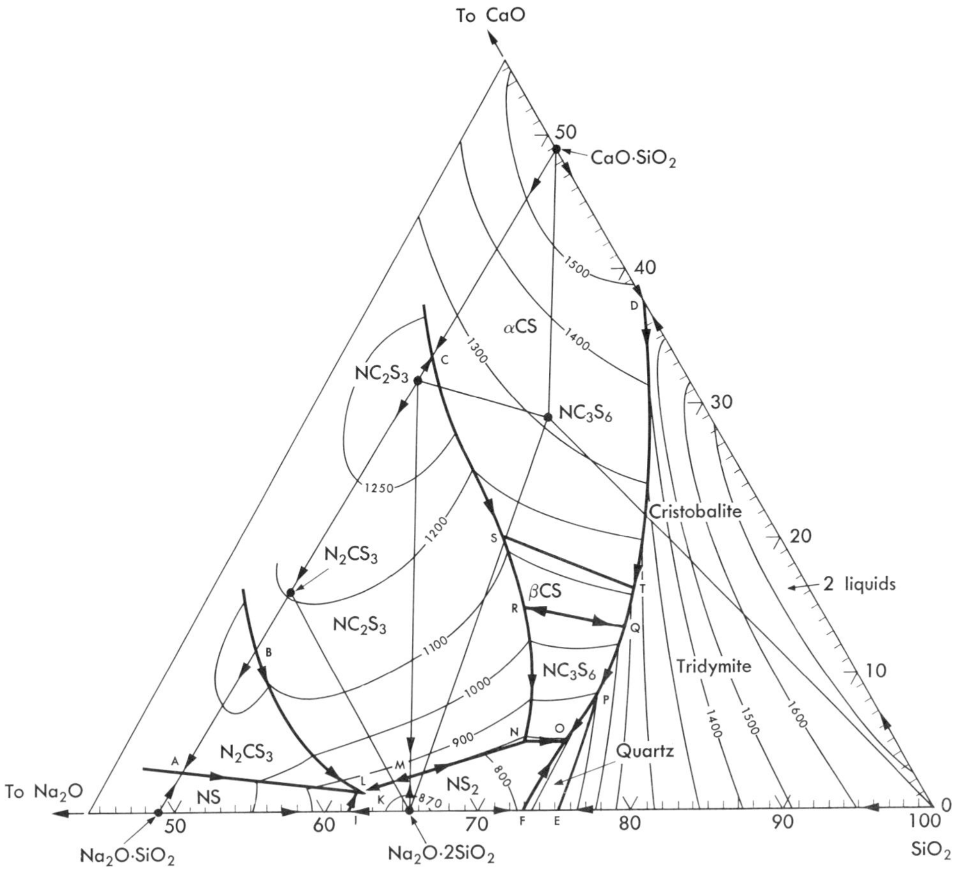

Compound	CaO	Na_2O	SiO_2	Temp.	
SiO_2			100.0	1723	M
$\alpha CaO \cdot SiO_2$	48.3		51.7	1544	M
$\beta CaO \cdot SiO_2$	48.3		51.7	1180	I
$Na_2O \cdot SiO_2$		50.8	49.2	1088	M
$Na_2O \cdot 2SiO_2$		34.1	65.9	874	M
$2Na_2O \cdot CaO \cdot 3SiO_2$	15.6	34.4	50.0	1141	D
$Na_2O \cdot 3CaO \cdot 6SiO_2$	28.5	10.5	61.0	1047	D
$Na_2O \cdot 2CaO \cdot 3SiO_2$	31.6	17.5	50.9	1284	M

M = Melting point
D = Decomposition point
I = Inversion point

Point		Crystal phases	CaO	Na_2O	SiO_2	Temp.
△	A	$NS\text{-}N_2CS_3$	3.0			1060
•	B	$N_2CS_3\text{-}NC_2S_3$	11.5			1141
△	C	$NC_2S_3\text{-}\alpha CS$	33.0			1280
△	D	$\alpha CS\text{-}S$	37.0		63.0	1436
⊙	E	T-quartz		24.3	75.7	870
△	F	$Quartz\text{-}NS_2$		26.4	73.6	790
*	K	$NS_2\text{-}NS\text{-}N_2CS_3$	1.8	37.5	60.7	821
○	L	$N_2CS_3\text{-}NC_2S_3\text{-}NS_2$	2.0	36.6	61.4	827
○	N	$NS_2\text{-}NC_2S_3\text{-}NC_3S_6$	5.2	24.1	70.7	740
*	O	$NC_3S_6\text{-}Q\text{-}NS_2$	5.2	21.3	73.5	725
⊙	P	$Q\text{-}NC_3S_6\text{-}T$	7.0	18.7	74.3	870
○	Q	$T\text{-}\beta CS\text{-}NC_3S_6$	12.9	13.7	73.4	1035
○	R	$NC_3S_6\text{-}NC_2S_3\text{-}\beta CS$	14.5	19.0	66.5	1030
•	S	$\beta CS\text{-}NC_2S_3\text{-}\alpha CS$	19.5	17.7	62.8	1110
•	T	$\alpha CS\text{-}S\text{-}\beta CS$	15.6	11.4	73.0	1110
△	I	$NS\text{-}NS_2$		38.0	62.0	840

△ Binary eutectic
* Ternary eutectic
• Decomposition point
○ Reaction point
⊙ Inversion point
C = CaO N = Na_2O
S = SiO_2 Q = Quartz
T = Tridymite

FIG. F–18. **$CaO\text{-}Na_2O\text{-}SiO_2$.** Adapted from G. W. Morey and N. L. Bowen, *Journ. Glass Tech.*, **9,** 232 (1925).

List of Symbols

Symbol	*Meaning*	*Commonly Used Unit**
a	lattice constant (x-axis)	angstrom
a/o	atom percent	dimensionless
A	angstrom	angstrom
A	area	cm^2
A	Madelung constant	dimensionless
APF	atomic packing factor	dimensionless
α	absorption coefficient	dimensionless
α_l	coefficient of linear expansion	$(°C)^{-1}$
α_v	coefficient of volume expansion	$(°C)^{-1}$
b	lattice constant (y-axis)	angstrom
B	magnetization	gauss
c	lattice constant (z-axis)	angstrom
c	phonon velocity	cm/sec
c	velocity of light	cm/sec
c_p	specific heat capacity	cal/gm·°C
C	capacitance	microfarad
C	component	——
C	concentration	dimensionless
C_p	molar heat capacity (constant pressure)	cal/mole·°C
C_v	molar heat capacity (constant volume)	cal/mole·°C
CLF	capacitive loss factor	dimensionless
CN	coordination number	——
cnfs	critical normal fracture stress	psi
css	critical shear stress	psi
γ	interfacial energy	erg/cm^2
γ	shear strain	dimensionless
d	electron orbital	——
d	thickness	cm
D	diameter	cm
D	diffusion coefficient	cm^2/sec
D	linear dimension	cm
δ	fractional energy loss	dimensionless
δ	loss angle	degree
e	electron charge	coul
E	electric field	volt/cm
E	energy	erg
E	environmental factors	——
E	voltage	volt
E	Young's modulus of elasticity	psi

* Since technical usage often is not limited to a single set of units, the practice in this book is to follow the more common engineering units in each individual case.

Symbol	*Meaning*	*Commonly Used Unit*
ϵ	axial strain	dimensionless
ϵ	dielectric permittivity	farad/cm
η	viscosity coefficient	dyne·sec/cm^2
f	electron orbital	——
f	frequency	sec^{-1}
f	function	——
F	force	gm·cm/sec^2
F	fraction	dimensionless
F	free energy	cal/mole
F_p	packing coefficient	dimensionless
G	shear modulus	psi
h	Planck's constant	erg·sec
h	reciprocal of x-intercept	dimensionless
h	thermal diffusivity	cm^2/sec
H	enthalpy	cal/mole
H	magnetic field	oersted
H_c	coercive force	oersted
I	electric current	ampere
I	intensity	dimensionless
ILF	inductive loss factor	cm/henry
J	diffusion flux	(cm^2·sec)$^{-1}$
k	Boltzmann's constant	ergs/°K
k	proportionality constant	dimensionless
k	reciprocal of y-intercept	dimensionless
k	thermal conductivity	(cal·cm)/(cm^2·°C·sec)
K	equilibrium constant	dimensionless
K	permeability coefficient	cm^4/sec
κ'	relative dielectric constant	dimensionless
l	reciprocal of z-intercept	dimensionless
l/o	linear percent	dimensionless
L	length	in.
L	liquid	——
λ	phonon mean free path	cm
λ	wavelength of light	cm
M	moderating ratio	dimensionless
μ	charge mobility	cm^2/(volt·sec)
μ	dipole moment	coul·cm
μ	magnetic permeability	gauss/oersted
μ	Poisson's ratio	dimensionless
n	charge carriers	cm^{-3}
n	index of refraction	dimensionless
n	neutron flux	(cm^2·sec)$^{-1}$
N	number	——
N_0	Avogadro's number	(molecular weight)$^{-1}$
ν	frequency of light	sec^{-1}
ν	reciprocal relative dispersion	dimensionless

Symbol	*Meaning*	*Commonly Used Unit*
O	origin	——
p	electron orbital	——
p	partial pressure	atm
P	electronic polarization	coul/cm^2
P	phase	——
P	porosity, or volume pore fraction	dimensionless
P	pressure	dynes/cm^2
PF	packing factor	dimensionless
Q	activation energy	cal/mole
Q	quality factor	dimensionless
r	cation radius	angstrom
r	interatomic distance	angstrom
R	anion radius	angstrom
R	cooling rate	°C/sec
R	gas constant	cal/(mole·°K)
R	reflectance	dimensionless
R	resistance	ohm
ρ	density	gm/cm^3
ρ	resistivity	ohm·cm
s	electron orbital	——
s	volumetric specific heat	cal/(cm^3·°C)
S	entropy	cal/(mole·°K)
S	solid	——
S	strength	psi
SRI	spalling resistance index	(Btu·in.)/(hr·ft^2)
σ	axial stress	psi
σ	conductivity	(ohm·cm)$^{-1}$
σ	cross section	cm^2
σ	standard deviation	——
σ_c	concentrated stress	psi
σ_f	fracture stress	psi
σ_i	ionic conductivity	(ohm·cm)$^{-1}$
σ_n	normal stress	psi
σ_s	shear stress	psi
t	time	sec
T	temperature	°C
τ	relaxation time	sec
τ	shear stress	psi
θ	dihedral angle	degree
θ	phase angle	degree
θ	stress angle	degree
U	bonding energy	cal/mole
v	velocity	cm/sec
v/o	volume percent	dimensionless
V	valence	——
V	variance	——

Symbol	*Meaning*	*Commonly Used Unit*
V	voltage	volt
V	volume	cm^3
w/o	weight percent	dimensionless
W	watts	volt·amp
x	distance	cm
Z	number of charges on an ion	——
□	atom hole	——
( )	Miller indices	——
{ } or (( ))	related Miller indices	——
[ ]	crystal direction	——
⟨ ⟩ or [[ ]]	related crystal directions	——

Answers to Problems

Chapter 2

2–1. (a) $1s^2 2s^2 6p^2$ (b) $1s^2 2s^2 2p^6 3s^2 3p^6 3d^{10} 4s^2 4p^6$ (c) $1s^2 2s^2 2p^6 3s^2 3p^6 3d^3$ (d) $1s^2 2s^2 2p^6 3s^2 3p^6 3d^4$

2–2. (a) 1 (b) 0 (c) 2 (d) 0 (e) 4 (f) 3 (g) 0 (h) 5 (i) 4 (j) 3 (k) 1 (l) 0 (m) 4 (n) 5 (o) 3 (p) 4 (q) 2 (r) 3 (s) 0 (t) 1 (u) 0

2–3. (a) 23 kcal/mole (b) $n = 9$, only coulombic and electronic forces

2–4. (a) 2.1 (b) 1.4 (c) 2.1 (d) 1.1

2–5. 1.6

Chapter 3

3–2. (a) 4.1 A (b) 4.1 A (c) 0.7 A, 0.37 A (d) 4.0 gm/cm^3 (e) 2.8 A, 1.14 A

3–4. 3.65 gm/cm^3

3–5. 3.4 gm/cm^3

3–6. (a) 3.99 A (b) $a_0 = 3.98$ A, $c_0 = 4.02$ A

3–7. (a) First = O^{2-}, second = Ba^{2+}, third = Ti^{4+} (b) $6O^{2-}$, $8Ba^{2+}$, $6Ti^{4+}$

3–8. (a) 1.8 A (radius) (b) 1.34 A (radius)

3–9. 2580 A

3–10. (a) 9.5 w/o (b) 15.8 w/o (c) 84.2 w/o

3–11. (a) 74.6 w/o Fe, 76 w/o, 76.4 w/o, 77 w/o (b) 8.6×10^{21} holes/cm^3, 4.6×10^{21}, 4.1×10^{21}, 2.5×10^{21}

3–12. 164°

Chapter 4

4–1. 45.5%

4–2. 2.2 gm/cm^3

4–3. (a) 1.5 (b) 0.5 (c) 19.3 (d) 2.13

4–4. 2.5

4–5. 60.5 w/o sand, 20.0 w/o soda ash, 19.5 w/o calcite

4–6. Hydrolysis and calcination

Chapter 5

5–1. 90 A

5–2. 15%

5–3. 0.4 cal/gm

5–4. $\gamma_{\text{sul/oxi}} = 700$ ergs/cm^2; $\gamma_{\text{sul/sul}} = 790$ ergs/cm^2; $\gamma_{\text{sul/liq}} = 450$ ergs/cm^2; $\gamma_{\text{oxi/liq}} = 410$ ergs/cm^2

5–5. (a) 540 ergs/cm^2 (b) $\gamma_{\text{gas/sol}} = 610$ ergs/cm^2 (c) $\gamma_{\text{gas/sol}} \leq 540$ ergs/cm^2 (d) $\gamma_{\text{liq/sol}} = 510$ ergs/cm^2 (e) Cannot change energies independently (f) Rigid solid

Chapter 6

6–1. Liq, 40 w/o, Mul, 60 w/o

6–2. 75.5 w/o MgO

6–3. (a) 6.5 w/o (b) 6.0 w/o (c) 4.0 w/o (d) 4.0 w/o (e) 100 w/o (f) 67 w/o (g) 100 w/o (h) 67 w/o

6–4. (a) Anorthite (b) Tridymite (c) Wollastonite

6–5. (a) CS_6, Geh, Ano (c) Cor, Mul, Ano (e) Lim, C_3A, C_3S, Lar, $C_{12}A_7$ (g) 2Liq, Cri, Tri, Wol, Ano

			w/o CaO	w/o Al_2O_3	w/o SiO_2
6–6. (a)	1600°C:	Liq	25.5	54.0	20.5
		CA_6	8.5	91.5	—
	1500°C:	Liq	29.5	45.0	25.5
		CA_6	8.5	91.5	—
	1400°C:	Liq	30.0	40.0	30.0
		CA_6	8.5	91.5	—
		Geh	41.0	37.0	22.0
(c)	1800°C:	Liq	7.0	59.0	34.0
		Cor	—	100.0	—
	1700°C:	Liq	10.0	48.0	42.0
		Cor	—	100.0	—
		Mul	—	75.0	25.0
	1600°C:	Liq	13.5	40.5	46.0
		Cor	—	100.0	—
		Mul	—	75.0	25.0
(e)	1600°C:	Liq	62.0	33.0	5.0
	1500°C:	Liq	60.5	34.0	5.5
		Lim	100.0	—	—
	1400°C:	Liq	53.0	40.0	7.0
		C_3A	62.0	38.0	—
		Lar	65.0	—	35.0
(g)	1700°C:	Liq	22.5	1.5	76.0
		Cri	—	—	100.0
	1600°C:	Liq	27.0	2.0	71.0
		Cri	—	—	100.0
	1500°C:	Liq	30.0	2.5	67.5
		Cri	—	—	100.0
	1400°C:	Liq	33.0	4.0	63.0
		Tri	—	—	100.0
		Wol	48.0	—	52.0
	1300°C:	Liq	28.0	9.5	62.5
		Tri	—	—	100.0
		Wol	48.0	—	52.0
	1200°C:	Liq	24.0	14.0	62.0
		Tri	—	—	100.0
		Wol	48.0	—	52.0

6–7. (a) 1600°C: Liq, 96 w/o, CA_6, 4 w/o
1500°C: Liq, 79 w/o, CA_6, 21 w/o
1400°C: Liq, 53 w/o, CA_6, 30 w/o, Geh, 17 w/o

(c) 1800°C: Liq, 73 w/o, Cor, 27 w/o
1700°C: Liq, 47 w/o, Cor, 32 w/o, Mul, 21 w/o
1600°C: Liq, 37 w/o, Cor, 32 w/o, Mul, 31 w/o

(e) 1600°C: Liq, 100 w/o
1500°C: Liq, 96 w/o, Lim, 4 w/o
1400°C: Liq, 7 w/o, C_3A, 80 w/o, Lar, 13 w/o

(g) 1700°C: Liq, 63 w/o, Cri, 37 w/o
1600°C: Liq, 52 w/o, Cri, 48 w/o
1500°C: Liq, 46 w/o, Cri, 54 w/o
1400°C: Liq, 28 w/o, Tri, 63 w/o, Wol, 9 w/o
1300°C: Liq, 10 w/o, Tri, 66 w/o, Wol, 24 w/o
1200°C: Liq, 7 w/o, Tri, 67 w/o, Wol, 26 w/o

6–8. (b) Spi, For, Cord (d) Mul, Cord, Sap (f) Cri, Tri, Pro, Cord

6–9. (b)

		w/o MgO	w/o Al_2O_3	w/o SiO_2
1700°C:	Liq	42.0	24.0	34.0
	Spi	28.0	72.0	—
1600°C:	Liq	36.5	22.5	41.0
	Spi	28.0	72.0	—
	For	57.0	—	43.0
1500°C:	Liq	31.0	23.0	46.0
	Spi	28.0	72.0	—
	For	57.0	—	43.0
1400°C:	Liq	26.0	23.0	51.0
	Spi	28.0	72.0	—
	For	57.0	—	43.0

(d)

1600°C:	Liq	13.0	41.0	46.0
	Mul	—	72.0	28.0
1500°C:	Liq	15.0	35.0	50.0
	Mul	—	72.0	28.0

(f)

1600°C:	Liq	19.0	10.0	71.0
	Cri	—	—	100.0
1500°C:	Liq	22.0	11.0	67.0
	Cri	—	—	100.0
1400°C:	Liq	22.0	15.0	63.0
	Tri	—	—	100.0

	w/o		w/o CaO	w/o Al_2O_3	w/o SiO_2
6–10. (a)	Liq:	9	7.5	17.0	75.5
	Mul:	68	—	72.0	28.0
	Tri:	23	—	—	100.0
(c)	Liq:	36	11.0	24.5	64.5
	Mul:	35	—	72.0	28.0
	Ano:	29	20.0	37.0	43.0
(e)	Liq:	60	28.5	38.0	33.5
	Ano:	40	20.0	37.0	43.0
(g)	C_3S:	14	74.0	—	26.0
	C_3A:	46	62.0	38.0	—
	Lar:	40	65.0	—	35.0

6–11.

(a) Phase		(c) w/o	(b) w/o FeO	w/o Al_2O_3	w/o SiO_2
1600°C:	Liq	27	13	24	63
	Mul	73	—	72	28
1300°C:	Liq	13	28	20.5	51.5
	Mul	77	—	72	28
	Tri	10	—	—	100
1200°C:	Mul	75	—	72	28
	Tri	10	—	—	100
	Fe-Cord	15	22.5	31.5	46
1100°C			(Same as 1200°C)		

Note: Intersect the $FeSiO_3$-kyanite line and the 75% mullite loci.

Chapter 7

7–1. $D_{36}/D_4 = 1.6$ ($D_{64}/D_4 = 2$)

7–3. MgO, 28 w/o; Al_2O_3, 16 w/o; SiO_2, 56 w/o (a) For, Cord, Pro (b) Cor, Pro, Liq, Spi

7–4. (a) FeO, 18 w/o; Al_2O_3, 55 w/o; SiO_2, 27 w/o (b) $FeO/Al_2O_3 = 35/65$ (c) 50 w/o (d) FeO, 26 w/o; Al_2O_3, 41 w/o; SiO_2, 33 w/o

7–5. See Fig. 7–14

7–6. FeO fluxes SiO_2; matrix contains liquid and fayalite; quartz transforms most readily to cristobalite at surfaces

7–7. (a) 7100 A (b) 710 A (c) 71 A

7–8. 150 A

Chapter 8

8–1. 77 v/o

8–2. 10,000 psi

8–3. (a) 4 bonds/a_0^2 on (100), 2.3 bonds/a_0^2 on (111), easier cleavage on (111) (b) low-index planes such as {110}, {111}, {123}

8–4. (a) 4900 lb (b) 206,000 psi
8–5. $[10\bar{1}](101)$ slip
8–6. (a) $F_{[001]} = 2A\sigma_s$, $F_{[101]} = 4A\sigma_s$ (b) $F_{[111]} = \infty A\sigma_s = \infty$

CHAPTER 9

9–2. Zincite, $17 \times 10^{-6}{}^\circ C^{-1}$; quartz, $37 \times 10^{-6}{}^\circ C^{-1}$; calcite, $13 \times 10^{-6}{}^\circ C^{-1}$; zircon, $14 \times 10^{-6}{}^\circ C^{-1}$; rutile, $22 \times 10^{-6}{}^\circ C^{-1}$
9–3. (a) 18.73 cal/mole at 25°C, 25.68 cal/mole at 227°C, 29.64 cal/mole at 727°C, 31.65 cal/mole at 1223°C (b) 9.05 cal/mole at 25°C, 10.46 cal/mole at 227°C, 11.77 cal/mole at 727°C, 12.73 cal/mole at 1223°C
9–4. (a) l^2/t
(b) Al_2O_3, 0.103 cm^2/sec @ 20°, 0.028 cm^2/sec @ 500°C,
MgO, 0.12 cm^2/sec @ 20°, 0.033 cm^2/sec @ 500°C,
SiO_2 glass, 0.008 cm^2/sec @ 20°, 0.012 cm^2/sec @ 500°C
9–5. 0.34 l/o at 200°C; 0.94 l/o at 300°C; 1.03 l/o at 400°C; 1.1 l/o at 500°C; 1.3 l/o at 600°C
9–6. (a) 142,000 psi (b) 100,000 psi (c) 160,000 psi (d) none
9–8. 0.002 (cal·cm)/(cm^2·°C·sec)
9–9. (a) 7.2 days (b) 10.35 days (c) 13.5 days
9–10. σ_{en} = 8400 psi compression; σ_{st} = 2100 psi tension
9–11. 1200°C; 1550°C

CHAPTER 10

10–1. 30×10^{-6} ohm-cm
10–2. 0.0032 $(\text{ohm-cm})^{-1}$
10–3. (a) $10^{-11.0}$ cm^2/sec, $10^{-8.6}$ cm^2/sec (b) 43,500 cal/mole
10–4. (a) 3.4 (b) 0.068 (c) 50 (d) 88.9°
10–5. 4.3
10–7. 32 Bohr magnetons/unit cell
10–8. 20 Bohr magnetons/unit cell

CHAPTER 11

11–3. (a) ΔR = 20% (b) ΔR = 7%
11–4. 0.06 mm
11–5. 60%
11–6. (a) 81 sec (b) 0.03 (as compared to 15 sec)

CHAPTER 13

13–1. (a) SiO_2, 60 (b) Al_2O_3, 32 (c) K_2O, 2 (d) Na_2O, 1 (e) CaO, 5
13–2. (a) Al_2O_3, 28.2; CaO, 4.2; K_2O, 1.5; Na_2O, 0.5; SiO_2, 53.3; CO_2, 3.2; H_2O, 9.1
(b) Al_2O_3, 32.1; CaO, 4.7; K_2O, 1.7; Na_2O, 0.6; SiO_2, 60.8
13–3. (a) Borax, 30; cryolite, 20; feldspar, 20; flint, 35; soda ash, 6; ZnO, 6
13–4. (a) 8.0% (b) 2.54 gm/cm^3 (c) 2.33 gm/cm^3 (d) 345 cm^3 (e) 375 cm^3
13–5. (b) 903 gm (c) 528 gm
13–6. (a) 0.63 mm (b) 0.063 mm

13–7. (a) 0 w/o = 1.85 gm/cm^3
5 w/o = 1.95 gm/cm^3
10 w/o = 2.06 gm/cm^3
15 w/o = 2.13 gm/cm^3
20 w/o = 2.00 gm/cm^3
25 w/o = 1.87 gm/cm^3

(b) 0 v/o = 1.85 gm/cm^3
10 v/o = 1.95 gm/cm^3
20 v/o = 2.05 gm/cm^3
30 v/o = 2.15 gm/cm^3
40 v/o = 1.99 gm/cm^3
50 v/o = 1.82 gm/cm^3

Chapter 14

14–1. (a) 0.065 (b) 4.6 gm/cm^3 (c) 4.9 gm/cm^3
14–2. (a) 0.045 cm/cm (b) 0.39 gm/gm
14–3. (a) 0.20 (b) 2.34 gm/cm^3
14–4. Mullite, 16 w/o; Al_2O_3, 72; SiO_2, 28
Anorthite, 75 w/o; Al_2O_3, 37; CaO, 20; SiO_2, 43
Liquid, 9 w/o; Al_2O_3, 24; CaO, 11; SiO_2, 65
14–5. (a) 10.1% (b) 6.6 hr
14–6. $R_{1200} = 3.16\ R_{1400}$
14–7. 610°C
14–8. 8×10^{-6}/°C

INDEX

* Letters indicate appendix references.